Tumor Immunology and Cancer Therapy

IMMUNOLOGY SERIES

Editor-in-chief

NOEL R. ROSE

Professor and Chairman
Department of Immunology and Infectious Diseases
The Johns Hopkins University
School of Hygiene and Public Health
Baltimore, Maryland

European Editor

ZDENEK TRNKA

Basel Institute for Immunology
Basel, Switzerland

1. Mechanisms in Allergy: Reagin-Mediated Hypersensitivity, *edited by Lawrence Goodfriend, Alec Sehon and Robert P. Orange*
2. Immunopathology: Methods and Techniques, *edited by Theodore P. Zacharia and Sidney S. Breese, Jr.*
3. Immunity and Cancer in Man: An Introduction, *edited by Arnold E. Reif*
4. *Bordetella pertussis*: Immunological and Other Biological Activities, *J. J. Munoz and R. K. Bergman*
5. The Lymphocyte: Structure and Function (in two parts), *edited by John J. Marchalonis*
6. Immunology of Receptors, *edited by B. Cinader*
7. Immediate Hypersensitivity: Modern Concepts and Development, *edited by Michael K. Bach*
8. Theoretical Immunology, *edited by George I. Bell, Alan S. Perelson, and George H. Pimbley, Jr.*
9. Immunodiagnosis of Cancer (in two parts), *edited by Ronald B. Herberman and K. Robert McIntire*
10. Immunologically Mediated Renal Diseases: Criteria for Diagnosis and Treatment, *edited by Robert T. McCluskey and Guiseppe A. Andres*
11. Clinical Immunotherapy, *edited by Albert F. LoBuglio*
12. Mechanisms of Immunity to Virus-Induced Tumors, *edited by John W. Blasecki*
13. Manual of Macrophage Methodology: Collection, Characterization, and Function, *edited by Herbert B. Herscowitz, Howard T. Holden, Joseph A. Bellanti, and Abdul Ghaffar*
14. Supressor Cells in Human Disease, *edited by James S. Goodwin*

15. Immunological Aspects of Aging, *edited by Diego Segre and Lester Smith*
16. Cellular and Molecular Mechanisms of Immunologic Tolerance, *edited by Tomáš Hraba and Milan Hašek*
17. Immune Regulation: Evolution and Biological Significance, *edited by Laurens N. Ruben and M. Eric Gershwin*
18. Tumor Immunity in Prognosis: The Role of Mononuclear Cell Infiltration, *edited by Stephen Haskill*
19. Immunopharmacology and the Regulation of Leukocyte Function, *edited by David R. Webb*
20. Pathogenesis and Immunology of Treponemal Infection, *edited by Ronald F. Schell and Daniel M. Musher*
21. Macrophage-Mediated Antibody-Dependent Cellular Cytotoxicity, *edited by Hillel S. Koren*
22. Molecular Immunology: A Textbook, *edited by M. Zouhair Atassi, Carel J. van Oss, and Darryl R. Absolom*
23. Monoclonal Antibodies and Cancer, *edited by George L. Wright, Jr.*
24. Stress, Immunity, and Aging, *edited by Edwin L. Cooper*
25. Immune Modulation Agents and Their Mechanisms, *edited by Richard L. Fenichel and Michael A. Chirigos*
26. Mononuclear Phagocyte Biology, *edited by Alvin Volkman*
27. The Lactoperoxidase System: Chemistry and Biological Significance, *edited by Kenneth M. Pruitt and Jorma O. Tenovuo*
28. Introduction to Medical Immunology, *edited by Gabriel Virella, Jean-Michel Goust, H. Hugh Fudenberg, and Christian C. Patrick*
29. Handbook of Food Allergies, *edited by James C. Breneman*
30. Human Hybridomas: Diagnostic and Therapeutic Applications, *edited by Anthony J. Strelkauskas*
31. Aging and the Immune Response: Cellular and Humoral Aspects, *edited by Edmond A. Goidl*
32. Complications of Organ Transplantation, *edited by Luis H. Toledo-Pereyra*
33. Monoclonal Antibody Production Techniques and Applications, *edited by Lawrence B. Schook*
34. Fundamentals of Receptor Molecular Biology, *Donald F. H. Wallach*
35. Recombinant Lymphokines and Their Receptors, *edited by Steven Gillis*
36. Immunology of the Male Reproductive Organs, *edited by Pierluigi E. Bigazzi*
37. The Lymphocyte: Structure and Function, *edited by John J. Marchalonis*
38. Differentiation Antigens in Lymphohemopoietic Tissues, *edited by Masayuki Miyasaka and Zdenek Trnka*
39. Cancer Diagnosis In Vitro Using Monoclonal Antibodies, *edited by Herbert Z. Kupchick*
40. Biological Response Modifiers and Cancer Therapy, *edited by J. W. Chiao*

41. Cellular Oncogene Activation, *edited by George Klein*
42. Interferon and Nonviral Pathogens, *edited by Gerald I. Byrne and Jenifer Turco*
43. Human Immunogenetics: Basic Principles and Clinical Relevance, *edited by Stephen D. Litwin, with David W. Scott, Lorraine Flaherty, Ralph A. Reisfeld, and Donald M. Marcus*
44. AIDS: Pathogenesis and Treatment, *edited by Jay A. Levy*
45. Cell Surface Antigen Thy-1: Immunology, Neurology, and Therapeutic Applications, *edited by Arnold E. Reif and Michael Schlesinger*
46. Immune Mechanisms in Cutaneous Disease, *edited by David A. Norris*
47. Immunology of Fungal Diseases, *edited by Edouard Kurstak*
48. Adoptive Cellular Immunotherapy of Cancer, *edited by Henry C. Stevenson*
49. Colony-Stimulating Factors: Molecular and Cellular Biology, *edited by T. Michael Dexter, John M. Garland, and Nydia G. Testa*
50. Introduction to Medical Immunology: Second Edition, *edited by Gabriel Virella, Jean-Michel Goust, and H. Hugh Fudenberg*
51. Tumor Suppressor Genes, *edited by George Klein*
52. Organ-Specific Autoimmunity, *edited by Pierluigi E. Bigazzi, Georg Wick, and Konrad Wicher*
53. Immunodiagnosis of Cancer: Second Edition, *edited by Ronald B. Herberman and Donald W. Mercer*
54. Systemic Autoimmunity, *edited by Pierluigi E. Bigazzi and Morris Reichlin*
55. Molecular Immunobiology of Self-Reactivity, *edited by Constantin A. Bona and Azad K. Kaushik*
56. Tumor Necrosis Factors: Structure, Function, and Mechanism of Action, *edited by Bharat B. Aggarwal and Jan Vilček*
57. Granulocyte Responses to Cytokines: Basic and Clinical Research, *edited by Ronald G. Coffey*
58. Introduction to Medical Immunology: Third Edition, *edited by Gabriel Virella*
59. Monoclonal Antibodies and Peptide Therapy in Autoimmune Diseases, *edited by Jean-François Bach*
60. Macrophage–Pathogen Interactions, *edited by Bruce S. Zwilling and Toby K. Eisenstein*
61. Tumor Immunology and Cancer Therapy, *edited by Ronald H. Goldfarb and Theresa L. Whiteside*

ADDITIONAL VOLUMES IN PREPARATION

Tumor Immunology and Cancer Therapy

edited by

Ronald H. Goldfarb
Theresa L. Whiteside

Pittsburgh Cancer Institute
and University of Pittsburgh School of Medicine
Pittsburgh, Pennsylvania

CRC Press
Taylor & Francis Group
Boca Raton London New York

CRC Press is an imprint of the
Taylor & Francis Group, an **informa** business

CRC Press
Taylor & Francis Group
6000 Broken Sound Parkway NW, Suite 300
Boca Raton, FL 33487-2742

First issued in paperback 2019

ISBN-13: 978-0-8247-9179-7 (hbk)
ISBN-13: 978-0-367-40218-1 (pbk)

Library of Congress Cataloging-in-Publication Data

Tumor immunology and cancer therapy / edited by Ronald H. Goldfarb,
Theresa L. Whiteside.
p. cm. — (Immunology series ; 61)
Includes bibliographical references and index.
ISBN 0-8247-9179-7 (alk. paper)
1. Tumors—Immunological aspects—Congresses. 2. Cancer—
Immunotherapy—Congresses. I. Goldfarb, Ronald H. II. Whiteside,
Theresa L. III. Series: Immunology series ; v. 61.
[DNLM: 1. Neoplasms—immunology. 2. Neoplasms—therapy. W1
IM53K v.61 1994 / QZ 266 T925 1994]
QR188.6.T864 1994
616.99'406—dc20
DNLM/DLC
for Library of Congress 93-39772
CIP

Visit the Taylor & Francis Web site at
http://www.taylorandfrancis.com

and the CRC Press Web site at
http://www.crcpress.com

Series Introduction

The concepts of tumor immunology and immunotherapy of cancer date from the early days of the twentieth century. After developing the first drug for treatment of syphilis, Paul Ehrlich was supposedly summoned to appear before the German Kaiser, Wilhelm II, a well-known cancer-phobe. After congratulating Ehrlich on his triumph, the Kaiser is reputed to have said, "Now, Herr Professor, you must find a cure for cancer!" "But," Ehrlich protested, "that will be very difficult." At that point, the Kaiser abruptly turned his back on the famous scientist, indicating that he was dismissed.

After this encounter, Ehrlich returned to his laboratory and initiated a series of experiments on cancer immunology. Although he succeeded in establishing a line of tumor cells, he never was able to develop an immunologic approach to cancer therapy. He did, however, start a pathway of cancer research reaching to the present day and to the book at hand. Along the way, many discoveries of fundamental importance were made. Our knowledge of transplantation immunity, histocompatibility antigens, natural and induced cell-mediated cytotoxicity and cytokine effector mechanisms are but a few examples of a range of immunological phenomena that are the products of research on tumor immunology.

The present volume returns to the original Ehrlichian theme of harnessing the immune response to arrest the unrestrained growth of tumor cells. It is based on the pioneering research of Ronald Herberman, and encapsulates the work of his many students and colleagues. The dramatic progress made during the past few years had kindled a renewed enthusiasm for the realization of immunologic treatment of human cancer.

Noel R. Rose

Foreword

I am very pleased and honored to have so many of my colleagues and present and former associates joined together in this overview of tumor immunology. This occasion leads me to reflect upon the remarkable changes and advances in tumor immunology over the past 25 years that I have been involved in the field. In the late 1960s and early 1970s, there was an early, emerging interest and optimism about tumor-associated antigens and the ability of such antigens to elicit specific T-cell immunity. However, although such a paradigm was readily demonstrable with a variety of virus and chemically induced tumors in experimental rodents, it was very difficult to obtain analogous positive results with spontaneous experimental tumors. Coupled with problems of lack of consistently positive results and early immunotherapeutic approaches at the clinical level, much skepticism developed about the practical applicability of immunologic approaches for the treatment of human cancer. During the past few years, however, there has been a remarkable turnaround and re-emergence of extensive investigative activity and optimism. This new spirit is well reflected by several of the contributions to this volume, including detailed characterization of human tumor-associated antigens and their relationship to the major histocompatibility complex, sophisticated evaluation of tumor-infiltrating lymphocytes and restricted-usage T-cell receptor genes by T cells obtained from tumor-bearing patients, and promising approaches to tumor vaccines and gene therapy of cancer.

The other impressive change in the field over the last 20 years or so has been the widespread and almost universal acceptance of the existence of natural cell-mediated immunity against cancer, with natural killer (NK) cells representing the particular central focus of attention. NK cells have been well documented to be a consistently demonstrable subpopulation of effector cells, with very interesting immunobiologic characteristics. They also have been shown to play a substantial role in resistance against the development of tumor metastases and, as discussed in some of the chapters here, may be of practical usefulness for immunotherapy.

The contributions to this volume not only provide a good picture of the recent progress in tumor immunology, they also provide a glimpse of the exciting prospects for future developments in the field.

Ronald B. Herberman

Preface

An international symposium entitled "Tumor Immunology: Basic Mechanisms and Prospects for Therapy" was held May, 1991, in Pittsburgh at the Westin William Penn Hotel to honor Dr. Ronald B. Herberman, Director of the Pittsburgh Cancer Institute, on his 50th birthday. We collectively conceived and organized this gathering, along with Dr. Jules Heisler, to bring together an international roster of friends and colleagues who are basic scientists and clinicians to discuss the latest developments in tumor immunology. The attendees included many former students and colleagues of Dr. Herberman who have had active research careers in the field of tumor immunology. The occasion attracted researchers from all parts of the United States as well as visitors from many nations, including Denmark, Finland, France, Germany, Israel, Italy, Japan, Korea, Sweden, The Netherlands, and the United Kingdom. Presentations at the Symposium focused on the latest research on tumor immunology in six sessions: tumor-associated antigens and the role of antibodies and cellular immunity in cancer, the characteristics and basic mechanisms of antitumor effector cells, animal models in tumor biology, the tumor microenvironment and immune effector cells, immunomodulation of antitumor mechanisms and the immunotherapy of cancer.

This volume emphasizes future strategies for the use of immunologic approaches in cancer treatment and is therefore a fitting commemoration in honor of Dr. Herberman's accomplishments. Since this volume is in his honor, a few words concerning Dr. Herberman are in order. Dr. Herberman is nationally and internationally recognized as having played a seminal role in contributing to and championing tumor immunology as an important and promising direction for cancer research, diagnosis and intervention. Dr. Herberman has played a critical role for many years as a world-class opinion leader in the field of tumor immunology and in the identification of novel strategies for the immunologic treatment of cancer. He is particularly well recognized for his laboratory-based approaches to cancer, which address the gaps between findings in the laboratory and their application in the clinical treatment of cancer. Many of his colleagues have recognized Dr. Herberman's persistence, scientific vision, synthetic formulation, clarity of thought and capacity to find solutions. Moreover, Dr. Herberman's work ethic, management skills and organizational capabilities are well known to all of Dr. Herberman's professional colleagues. These skills have clearly contributed to his premier role in tumor immunology for which he has been recognized with many awards, including a Lifetime Science Award from the Institute for Advanced Studies in Immunology and Aging for his pioneering work in cellular immunology, the Solomon A.

Berson Medical Alumni Achievement Award in Clinical Science from his alma mater, New York University, and the Commonwealth of Pennsylvania Governor's Award for Excellence in the Sciences for his pioneering work in cancer immunology and biological therapeutics in cancer. As Thomas Detre, M.D., President of the University of Pittsburgh Medical Center, has noted, "It is clear that Dr. Ronald B. Herberman has not only made major discoveries and other substantial contributions to the field of tumor immunology but, equally important, has actively fostered the application of this information to novel approaches to cancer therapy."

The symposium was attended by many old friends, colleagues and associates of Dr. Herberman, including several whom he had not seen for many years. Dr. Herberman noted that: "To see a number of people who worked with me when they were very young and to see how much they've done since they left my lab has been most impressive to me." It is well known that Dr. Ronald B. Herberman has a long list of young and established scientists he has launched or interacted with in long-standing and on-going collaborative efforts. The authors who contributed to the chapters in this volume are well known and accomplished scientists, many of whom have had the pleasure of working closely with Dr. Herberman over the years. We believe that the high quality and thoroughness of the chapters in this volume are fitting tribute to Dr. Herberman. We thank the authors for their contributions to this volume, which is a state-of-the-art compendium of contemporary basic and clinically applicable and innovative approaches to tumor immunology. The editors are confident that this timely and exciting volume will be appreciated by students, researchers and clinicians in the area of tumor immunology and cancer biology and that it will stimulate new research ideas, discussion and directions in the important field of tumor immunology and immunotherapy.

We wish to acknowledge the expert editorial assistance of Ms. Barbara Klewien and Ms. Jeanne Nepa of the Pittsburgh Cancer Institute. We deeply appreciate their dedication, proficiency and professionalism in attending to multiple details and stylistic issues that were prerequisite to the completion of this volume in a timely fashion and in its current form.

Ronald H. Goldfarb
Theresa L. Whiteside

Contents

Contributors

Clement O. Akogyeram, Ph.D. Departments of Pharmacology, Biochemistry and Cancer Center, Howard University College of Medicine, Washington, D.C.

Yair Argon, Ph.D. Duke Comprehensive Cancer Center, Duke University Medical Center, Durham, North Carolina

Emira Ayroldi, M.D., Ph.D. Institute of Pharmacology, University of Perugia, Perugia, Italy

Timothy T. Back Biological Carcinogenesis and Development Program, Program Resources, Inc./DynCorp, Frederick, Maryland

Per Basse, M.D., Ph.D. Pittsburgh Cancer Institute, and University of Pittsburgh School of Medicine, Pittsburgh, Pennsylvania

Robert C. Bast, Jr., M.D. Duke Comprehensive Cancer Center, Duke University Medical Center, Durham, North Carolina

Mirsada Begovic, M.D., Ph.D. Pittsburgh Cancer Institute, and Department of Pathology, University of Pittsburgh School of Medicine, Pittsburgh, Pennsylvania

Arie Belldegrun, M.D. Immunotherapy Laboratory, Division of Urology, Department of Surgery, UCLA School of Medicine, Los Angeles, California

Andrew Berchuck, M.D. Duke Comprehensive Cancer Center, Duke University Medical Center, Durham, North Carolina

Malaya Bhattacharya-Chatterjee, Ph.D. University of Kentucky, Lexington, Kentucky

D. Kay Blanchard, Ph.D. H. Lee Moffitt Cancer Center and Research Institute, University of South Florida College of Medicine, Tampa, Florida

Cinda M. Boyer, Ph.D. Duke Comprehensive Cancer Center, Duke University Medical Center, Durham, North Carolina

Lorenza Cannarile, Ph.D. Institute of Pharmacology, University of Perugia, Perugia, Italy

Federica Cavallo, Ph.D. Institute of Microbiology, University of Turin, Turin, Italy

Jennie Crews, M.D. Duke Comprehensive Cancer Center, Duke University Medical Center, Durham, North Carolina

Francesca D'Adamio, Ph.D. Institute of Pharmacology, University of Perugia, Perugia, Italy

Luciano D'Adamio, Ph.D. Institute of Pharmacology, University of Perugia, Perugia, Italy

Giovanna Damia, M.D. Laboratory of Experimental Immunology, Biological Response Modifiers Program, National Cancer Institute-Frederick Cancer Research and Development Center, Frederick, Maryland

Albert B. DeLeo, Ph.D. Pittsburgh Cancer Institute, and Department of Pathology, University of Pittsburgh School of Medicine, Pittsburgh, Pennsylvania

Domenico Delfino, M.D., Ph.D. Institute of Pharmacology, University of Perugia, Perugia, Italy

Julie Y. Djeu, Ph.D. H. Lee Moffitt Cancer Center and Research Institute, University of South Florida College of Medicine, Tampa, Florida

Marc S. Ernstoff, M.D. Department of Medicine, Dartmouth-Hitchcock Medical Center, Hanover, New Hampshire

Roberto Fagnani, Ph.D. Hybritech, Inc., San Diego, California

Connie R. Faltynek, Ph.D. Sterling Research Group, Sterling Drug, Inc., Malvern, Pennsylvania

Lee Ann Feeney Applied Immune Sciences, Inc., Santa Clara, California

Kenneth A. Foon, M.D. University of Kentucky, Lexington, Kentucky

Guido Forni, M.D. CNR-Immunogenetics and Histocompatibility Center, Torino, Italy

Luigi Frati, M.D. Department of Experimental Medicine, University of Rome "La Sapienza", Rome, Italy

Mirella Giovarelli, Ph.D. Institute of Microbiology, University of Turin, Turin, Italy

Angela Gismondi, Ph.D. Department of Experimental Medicine, University of Rome "La Sapienza," Rome, Italy

Ronald H. Goldfarb, Ph.D. Pittsburgh Cancer Institute, and Department of Pathology, University of Pittsburgh School of Medicine, Pittsburgh, Pennsylvania

Elieser Gorelik, M.D., Ph.D. Pittsburgh Cancer Institute, and Department of Pathology, University of Pittsburgh School of Medicine, Pittsburgh, Pennsylvania

Theresa A. Gregorio Laboratory of Experimental Immunology, Biological Response Modifiers Program, National Cancer Institute-Frederick Cancer Research and Development Center, Frederick, Maryland

M. Eilene Gruys Laboratory of Experimental Immunology, Biological Response Modifiers Program, National Cancer Institute-Frederick Cancer Research and Development Center, Frederick, Maryland

Alberto Gulino, M.D. Department of Experimental Medicine, University of L'Aquila, L'Aquila, Italy

Tuula Helander, Ph.D. Department of Pathology, University of Helsinki, Helsinki, Finland

Ronald B. Herberman, M.D. Pittsburgh Cancer Institute and Departments of Medicine and Pathology, University of Pittsburgh School of Medicine, Pittsburgh, Pennsylvania

Naoya Inoue, M.D. Department of Late Effect Studies, Radiation Biology Center, Kyoto University, Kyoto, Japan

Juha Jääskeläinen, M.D. Department of Pathology, University of Helsinki, Helsinki, Finland

Yoshitaka Kariya, M.D. Department of Late Effect Studies, Radiation Biology Center; Kyoto University, Kyoto, Japan

Rolf Kiessling, M.D., Ph.D. Department of Immunology, Karolinska Institute, Stockholm, Sweden

John M. Kirkwood, M.D. Pittsburgh Cancer Institute and Department of Medicine, University of Pittsburgh School of Medicine, Pittsburgh, Pennsylvania

Richard P. Kitson, Ph.D. Pittsburgh Cancer Institute and Departments of Neurobiology, Anatomy and Cell Science, Molecular Genetics and Biochemistry and Pathology, University of Pittsburgh School of Medicine, Pittsburgh, Pennsylvania

Kristin L. Komschlies, Ph.D. Biological Carcinogenesis and Development Program, Program Resources, Inc./DynCorp, Frederick, Maryland

Panu Kovanen Department of Pathology, University of Helsinki, Helsinki, Finland

Yaron Lidor, M.D. Duke Comprehensive Cancer Center, Duke University Medical Center, Durham, North Carolina

T. Juhani Linna, M.D., Ph.D.[1] Applied Immune Sciences, Inc., Santa Clara, California

Eva Lotzová, Ph.D. Section of Natural Immunity, Department of General Surgery, The University of Texas, M. D. Anderson Cancer Center, Houston, Texas

Lisa Maier, M.D. Duke Comprehensive Cancer Center, Duke University Medical Center, Durham, North Carolina

Fabrizio Mainiero, M.D. Department of Experimental Medicine, University of Rome "La Sapienza," Rome, Italy

Anatoly Malygin, M.D. Department of Pathology, University of Helsinki, Helsinki, Finland

Anna Mäenpää, M.D. Department of Pathology, University of Helsinki, Helsinki, Finland

Graziella Migliorati Institute of Pharmacology, University of Perugia, Perugia, Italy

Enrico Mihich, M.D. Grace Cancer Drug Center, Roswell Park Cancer Institute, New York State Department of Health, Buffalo, New York

---

[1]Present address: Syntex Research, Division of Syntex (U.S.A.) Inc., Palo Alto, California

Michele Milella, M.D. Department of Experimental Medicine, University of Rome "La Sapienza," Rome, Italy

Dewey J. Moody, Ph.D. Applied Immune Sciences, Inc., Santa Clara, California

Stefania Morrone, Ph.D. Department of Experimental Medicine, University of Rome "La Sapienza," Rome, Italy

Norihiko Okamoto, M.D. Department of Late Effect Studies, Radiation Biology Center; Kyoto University, Kyoto, Japan

Thomas B. Okarma, M.D., Ph.D. Applied Immune Sciences, Inc., Santa Clara, California

Robert K. Oldham, M.D. Biological Therapy Institute, Franklin, Tennessee

Gabriella Palmieri, Ph.D. Institute of Biomedical Technologies, University of Rome "La Sapienza," Rome, Italy

Giorgio Parmiani, M.D. Division of Experimental Oncology, Istituto Nazionale Tumori, Milan, Italy

Mario Piccoli, M.D. Department of Experimental Medicine, University of Rome "La Sapienza," Rome, Italy

Craig Reynolds, Ph.D. Biological Resources Branch, Biological Response Modifiers Program, National Cancer Institute, Bethesda, Maryland

Allen R. Rhoads, Ph.D. Departments of Pharmacology, Biochemistry, and Cancer Center, Howard University College of Medicine, Washington, D.C.

Carlo Riccardi, M.D., Ph.D. Institute of Pharmacology, University of Perugia, Perugia, Italy

Angela Santoni, Ph.D. Laboratory of Pathophysiology, Regina Elena Cancer Institute, and Department of Experimental Medicine, University of Rome "La Sapienza," Rome, Italy

Giuseppe Scala, M.D. Department of Biochemisty and Medical Biotechnology, University of Naples, Naples, Italy

Katsuji Sugie, M.D. Department of Late Effect Studies, Radiation Biology Center, Kyoto University, Kyoto, Japan

Tuomo Timonen, M.D. Department of Pathology, University of Helsinki, Helsinki, Finland

Cho Lea Tso, M.S. Immunotherapy Laboratory, Division of Urology, Department of Surgery, UCLA School of Medicine, Los Angeles, California

Atsushi Uchida, M.D., Ph.D. Department of Late Effect Studies, Radiation Biology Center; Kyoto University, Kyoto, Japan

Alesandra Vacca, Ph.D. Department of Experimental Medicine, University of L'Aquila, L'Aquila, Italy

Brent Vose, ICI Pharmaceuticals, Cheshire, United Kingdom

Nikola L. Vujanovic, M.D., Ph.D. Pittsburgh Cancer Institute, and Department of Pathology, University of Pittsburgh School of Medicine, Pittsburgh, Pennsylvania

Ken Wasserman, Ph.D. Pittsburgh Cancer Institute, and Departments of Neurobiology, Anatomy and Cell Science, Molecular Genetics and Biochemistry, and Pathology, University of Pittsburgh School of Medicine, Pittsburgh, Pennsylvania

Sheng Wei, M.D. H. Lee Moffitt Cancer Center and Research Institute, University of South Florida College of Medicine, Tampa, Florida

William L. West, Ph.D. Departments of Pharmacology, Biochemistry, and Cancer Center, Howard University College of Medicine, Washington, D.C.

Theresa L. Whiteside, Ph.D. Pittsburgh Cancer Institute, and Department of Pathology, University of Pittsburgh School of Medicine, Pittsburgh, Pennsylvania

Robert H. Wiltrout, Ph.D. Laboratory of Experimental Immunology, Biological Response Modifiers Program, National Cancer Institute-Frederick Cancer Research and Development Center, Frederick, Maryland

Fengji Xu, M.D. Duke Comprehensive Cancer Center, Duke University Medical Center, Durham, North Carolina

Yinhua Yu, M.D. Duke Comprehensive Cancer Center, Duke University Medical Center, Durham, North Carolina

Acknowledgements

The Pittsburgh Cancer Institute gratefully acknowledges the support provided by:

Abbott Laboratories
Ajinomoto Company, Inc.
BASF Bioresearch Corporation
The Cancer Research Corporation
Cellco Advanced Bioreactors, Inc.
The Honorable Richard J. Cessar
Chugai Pharmaceutical Company, Ltd.
Fujisawa Pharmaceutical Company, Ltd.
The Hillman Company
Hybritech, Inc.
Immunex Research and Development Corporation
Institut Scientifique Roussel
Johnson & Johnson
The R.W. Johnson Pharmaceutical Research Institute
Kaken Pharmaceutical Company, Ltd.
Sonia & Aaron Levinson Philanthropic Fund-United
Jewish Federation
Mr. Scott Limbach
Mr. & Mrs. Andrew W. Mathieson
The Reverend Arthur McNulty-Calvary Episcopal Church
Mr. Thomas H. Nimick, Jr.
Pfizer, Inc.
Queue Systems, Inc.
Mrs. William R. Roesch
Mrs. Frances G. Scaife
Mr. A.W. Schenck III
Sandoz Research Institute/Pharmaceutical Corporation
SmithKline Beecham Pharmaceuticals
Specialty Laboratories, Inc.
Mrs. George H. Taber

Thrift Drug Company
Mrs. John C. Unkovic
The Upjohn Company
VWR Scientific
Wellcome Italia, S.p.A.

PART I

Introduction

Tumor-Associated Antigens and the Role of Antibodies and Cellular Immunity in Cancer

Brent Vose
ICI Pharmaceuticals
Cheshire, United Kingdom

The prospect that immunological responses to antigens selectively expressed on human malignancies offers an important therapeutic or diagnostic modality continues to excite considerable interest. It also continues to excite considerable controversy.

The 'immunostimulants' represent a sizeable proportion of the world anti-cancer market in a segment which is dominated by Picibanil and Krestin, two natural products sold primarily in Japan. Sales in Japan are falling and the products have not secured market penetration in the rest of the world. Newer, widely accepted agents of defined mechanism other than the cytokines have not appeared in spite of high expenditure by industrial and academic institutions.

In order to reverse this essentially negative situation the key goal remains the identification and characterization of tumour selective markers on cancer cells which can be targetted to engender therapeutic benefit. In this respect the advent of new technologies will continue to have a major impact.

A key advance has been the exploitation of monoclonal antibodies. Although hundreds of man years have been spent in raising antibodies, remarkably few tumour selective specificities have emerged: a finding which serves to emphasise the complexity of the tumour target as an antigen-presenting entity. Nevertheless, antibodies are reaching the clinic either as useful diagnostic aids in patient monitoring, e.g., prostate specific antigen (PSA) and carcinoembryonic antigen (CEA) or as therapeutic agents to target toxins and radioisotopes. Several problems such as immunogenicity, tumour cell heterogeneity, and access to the tumour remain, but an increasing number of clinical trials, and the introduction of newer methodologies such as antibody dependent enzyme-prodrug therapy, will provide vehicles for their resolution.

In the field of cellular immunity an enormous leap forward has occurred in the understanding of immune mechanisms and cell-cell interactions at the molecular level. Building on a considerable body of evidence of cell-mediated reaction to malignancy derived over the past 20 years, the challenge remains to define the frequency of expression and biochemical nature of tumour-associated antigens if manipulation of the host response against them are to be therapeutically useful.

Tumour immunology approaches 40 years of age. It continues to struggle for widespread credibility. The availability of new molecular tools offers the means by which useful therapeutic entities might be discovered and important cellular functions manipulated.

1

The Immunogenicity of Foreign Monoclonal Antibodies in Human Disease Applications: Problems and Current Approaches

Roberto Fagnani
Hybritech, Inc.
San Diego, California

I INTRODUCTION

In recent years, polyclonal and monoclonal antibodies (mAbs) have been investigated for their potential applications in the diagnosis and treatment of cancer (1-3), as immunosuppressive agents in organ transplantation (4-6), or in the treatment of autoimmune diseases like rheumatoid arthritis (7) or multiple sclerosis (8). A consequence of the administration of foreign mAbs in humans is the induction of immune response by the host, especially if multiple administrations over an extended period of time are required for either adequate diagnosis or to achieve therapeutic efficacy. This human anti-mouse immunoglobulin response can potentially neutralize the biological activity of the antibodies, resulting in diminished or abolished clinical responses, and possibly in adverse allergic reactions or anaphylactic shock. Thus, the immunogenicity of xenogeneic antibodies might limit their clinical potential to humans.

The initial reports describing the immunogenicity of antibodies in humans were concerned primarily with murine mAbs. The resulting immune response was therefore referred to as "Human Anti-Mouse Antibody" or HAMA. This term has gained widespread acceptance among immunologists. In recent years, however, antibodies of various species, including human, have been tested in clinical trials. Therefore, in this review I will use the more general term "Human Anti-Immunoglobulin Response", or human anti-Ig, when referring to the immunogenicity of various types and species of antibodies in humans.

Several reports have indicated the presence of human anti-Ig responses in patients receiving the administration of foreign mAbs directed against various types of cancer (9) or other diseases. In general, human anti-Ig response develops in up to 50-70% of the patients treated with a single administration of murine antibodies. This response usually develops within 2 weeks of administration and is primarily mediated by antibodies of the IgG class. Usually, human anti-Ig tends to increase in frequency and magnitude after repeated administrations of murine immunoglobulins (10). The majority of this response is usually directed against determinants present in the Fc portion of murine IgG, with the remainder directed either toward the variable or the hypervariable portion of the molecule. The latter antibodies may be directed against idiotypic determinants (11).

Since it would not be feasible to discuss the immunogenic properties of every foreign antibody administered to humans so far, I have summarized in Table 1 the most salient characteristics related to the development of human anti-Ig response from selected clinical trials. These include the type of antibodies tested and their antigens, the doses employed, the disease type and the number of patients, and the frequency of the observed human anti-Ig responses. I have also grouped these antibodies into sections comprising whole antibody molecules, fragments, immuno-toxins/drugs, chimeric, humanized and human antibodies.

Since human anti-Ig can negatively affect the potential clinical applications of mAbs, several approaches are being pursued to circumvent or reduce this response. These include (I) fragmentation, (II) polymer modification, and (III) humanization of mAbs. This review will summarize the current status of research and clinical progress in these areas.

II FRAGMENTATION OF MONOCLONAL ANTIBODIES

The immunogenicity of a given protein is dependent, at least in part, on its molecular size, its relative antigenic content and its residence time in the vascular compartment. For these reasons, the enzymatic fragmentation of mAb with pepsin, to generate $F(ab')_2$ fragments (12), or with papain, to generate Fab (13) may theoretically reduce the resulting human anti-Ig response by removing the immunogenic Fc component, decreasing the molecular size, and significantly reduce the plasma residence time. However, no definite information on the immunogenicity of antibody fragments in humans is presently available. Polyclonal sheep anti-digoxin Fab fragments have been extensively used in the treatment of life-threatening digitalis intoxication since 1976. Although the administration of these fragments has shown partial or complete reversal of digitalis toxicity with few reported allergic reactions (14), the level of human anti-sheep immune responses in the sera of these patients has never been reported. Earlier experimental evidence with $F(ab')_2$ or Fab' of murine mAbs suggested a reduction of their immunogenicity (15). However, human anti-Ig responses still occur after repeated administration of these fragments in humans (9,10).

In recent years, advanced manipulation of the immunoglobulin genes have led to the generation of smaller size fragments containing the antigen binding site of antibodies. These include Fv fragments and minimal recognition units. Fv fragments, also termed single chain Fvs, are the smallest fragment of the antibody containing the complete, univalent antigen-binding site. An Fv is usually composed of the variable light chain amino acid sequence (V_L) of the antibody molecule connected to the variable heavy chain sequence (V_H) by a designed peptide linking the carboxyl terminus of the V_L to the amino terminus of the V_H or vice versa (16-18). Several single chain Fvs have recently been constructed in Escherichia coli with varying levels of antibody affinity relative to that of their corresponding $F(ab')_2$ or Fab' fragment (16,17). Three Fvs, derived from mAbs raised against either tumor-associated antigens (18,19) or the fragment D-Dimer of fibrin (20), have recently been tested in experimental animal models. These molecules possessed rapid plasma elimination rates relative to their corresponding Fab' fragments (18-20), and specific tumor-localization properties (18-19). However, no data relative to the immunogenicity of these molecules in experimental animals or in clinical trials have been reported to date.

Minimal Recognition Units (MRUs) are synthetic peptides comprising the complementary-determining regions (CDRs) of the antibody molecules. The CDRs are the amino acid sequences of the hypervariable domains of an antibody and are responsible for antigen recognition. Thus, MRUs are the smallest portion of the antibody that retains the ability to bind specifically to an antigen with properties theoretically similar to those of the intact antibody. Several MRUs have been recently described, composed of either linear amino acid sequences or conformationally constrained, cyclic peptides. Traub et al. (21) reported the synthesis of a peptide encompassing the sequence of the third CDR of the heavy chain of PAC1, a murine IgM antibody specific for the GpIIb/IIIa fibrinogen receptor on activated platelets. This peptide inhibited the binding of either fibrinogen or PAC1 to their platelet receptor. Williams et al. (22,23) described both linear and cyclic peptides containing the sequence of the second CDR of the light chain of the mAb 87.92.6, specific for the cellular receptor of reovirus type 3. These peptides mimic the ability of the antibody to inhibit DNA synthesis and to down regulate the reovirus receptor. These authors have since extended their findings to the synthesis of conformationally restricted organic compounds that mimic the biological activity of peptide mimetics of the mAb 87.92.6 (24). Thus, MRUs, because of their very small molecular weight, should theoretically show little immunogenicity. To date, however, no such molecules have been analyzed for their pharmacokinetic properties, the ability to bind to their targets in vivo, or their immunogenicity in either animal models or clinical trials.

TABLE 1

THE IMMUNOGENICITY OF FOREIGN ANTIBODIES IN HUMAN

Authors/Refs	Antibody	Antigen	Dose (mg)	# of doses	Disease type	[1]Human anti-Ig Response	Frequency (%)
(1) Whole Antibody Molecule							
Blottiere et al. (75)	17-1A	gp41	200	1-15	Colorectal Ca	30/42	71
Courtenay-Luck et al. (76)	HMFG1	milk fat globule	2-15	2	Ovarian Ca	4/8	50
Shawler et al. (77)	T101 T101	CD5 CD5	10-500 10-500	8 1-6	[2]CTCL [3]CLL	5/10 0/6	50 0
Jaffers et al. (78)	OKT3	CD3	1-5 daily	10-20	Kidney graft	14/21	75
Chatenoud et al. (79)	OKT3	CD3	5 daily	14	Kidney graft	17/17	100
Khazaeli et al. (80)	17-1A	gp41	400	1-4	Colorectal Ca	21/25	84
Hafler et al. (81)	antiT4 antiT11	CD4 CD2	0.2/kg/day	5	Multiple sclerosis	5/7	71
Goldman-Leikin et. al. (82)	T101	CD5	10	2	CTCL	5/5	100
Schroff et al. (83)	T101 T101 9.2.2.7	CD5 CD5 gp250	1-100 1-100 1-500	8 8 8	CTCL CLL Melanoma	4/4 0/11 3/9	100 0 33
Murray et al. (84)	ZME018	gp240	1-20	1	Melanoma	7/17	41
Klein et al. (39)	[4]Poly anti-ferritin	Ferritin	–	1	Hepatoma	65/138	47
Horneff et al. (85)	16H5	CD4	0.3mg/kg	7	Rheumatoid Arthritis	5/10	50
(2) Antibody Fragments							
Kalofonos et al. (15)	HMFG1	milk fat globule	0.15-0.25	1	Lung Ca	0/14	0
Muto et al. (86)	OC-125	CA125	10-70	1	Ovarian Ca	16/123	70%
(3) Immuno-toxins/drugs							
Petersen et al. (87)	KS1/4 vinblastine	gp40	60-1000	4-17	Various Carcinomas	32/44	73
Spitler et al. (88)	[5]XOMAZYME-MEL	p240	3.2-300	5	Melanoma	19/20	95

TABLE 1 (Cont.)

Authors/Refs	Antibody	Antigen	Dose (mg)	# of doses	Disease type	Human anti-Ig Response	Frequency (%)
Avner et al. (89)	[6]Combination	[7]Various antigens	42-4957	up to 10	Various tumors	13/21	62
Hertler et al. (90)	T101-Ricin A	CD5	3/m2	8	CLL	1/4	25
(4) Chimeric antibodies							
LoBuglio et al. (49)	17-1A	gp41	10-40	1-3	Colorectal Ca	1/16	16
Khazaeli et al. (52)	B72.3	TAG-72	3.4-6.9	1-2	Colorectal Ca	7/12	58
(5) CDR-Grafted Humanized antibodies							
Hale et al. (60)	campath-1H	campath-1	1-20	43	Non-Hodgkin lymphoma	0/2	0
Mathieson et al. (61)	campath-1H	campath-1	2	27	Systemic vasculitis	0/1	0
(6) Human antibodies							
Ziegler et al. (69)	HA-1A	lipid A	100	1	Septic shock	0/166	0
Drobyski et al. (71)	MSL-109	CMV	0.050 0.5/kg	–	Bone marrow transplant	0/15	0
Haisma et al. (72)	16.88	--	200--1000	2	Colorectal Ca	0/20	0

[1]Number of human anti-Ig responses/total number of evaluable patients.

[2]CTCL = Cutaneous T Cell Lymphoma

[3]CLL = Chronic Lymphocytic Leukemia

[4]Polyclonal anti-ferritin antibodies of various animal species

[5]Murine anti-melanoma MoAb conjugated to Ricin A chain

[6]Cocktail of combination of anti-tumor MoAbs conjugated to Adriamycine

[7]Various tumor-associated antigens

III POLYMER-MODIFICATION OF ANTIBODIES

In recent years, proteins have been chemically modified with the covalent linkage of water-soluble polymers to reduce their immunogenicity and improve their pharmacokinetic properties in vivo. Examples of these polymers include polyethylene glycol (PEG) of 2-10,000 molecular weight or dextrans of 10,000-500,000 molecular weight.

Abuchowski et al (25,26) have shown that the covalent modification of enzymes like adenosine deaminase with PEG yielded partially active conjugates with prolonged plasma half-lives and low immunogenic characteristics in vivo relative to the unmodified protein. Lee and Sehon (27) have shown that the administration of allergens covalently modified with PEG in experimental animals specifically suppressed an ongoing IgE response. These results have since been extended by other groups to include therapeutically useful enzymes like streptokinase (28), and asparaginase (29), or growth factors like Interleukin-2 (30). Recently, Wilkinson et al (31) extended these studies to the covalent modification of antibodies. These authors generated conjugates which possessed tolerogenic properties in mice toward subsequent administrations of the unmodified antibody. However, PEG-modification resulted in the abrogation of antibody immunoreactivity.

Dextrans of various molecular weights have also been used to modify enzymes or hormones (32,33), yielding conjugates possessing increased plasma half-life, lower antigenic reactivity and increased stability in vitro. Similarly, anti-tumor antibodies have been conjugated to dextrans carrying cytotoxic drugs for the targeted delivery of pharmacologically active agents to tumors in vivo (34-36).

We have modified monoclonal or polyclonal antibodies with oxidized dextran of low molecular weight to generate conjugates with negligible immunogenicity in vivo (37). The murine mAb T101, which recognizes a 65,000 molecular weight antigen on human T lymphocytes and polyclonal rabbit anti-human ferritin (RAF) were modified with dextran (38,39). T101 and RAF have been clinically evaluated in the treatment of, respectively, chronic lymphocytic leukemias and hepatomas, and have been shown to possess substantial immunogenic properties (see Table 1). After modification, antibody-dextran conjugates retained optimal immunoreactivity as well as in vivo pharmacokinetics and tumor-localization properties when tested in experimental animals. In addition, multiple administrations in xenogeneic immunocompetent animals did not elicit measurable response to either the antibody (Figures 1,2) or the dextran portion of these conjugates (37). At present, however, the immunogenicity of polymer-modified murine immunoglobulins in humans is awaiting evaluation in clinical trials.

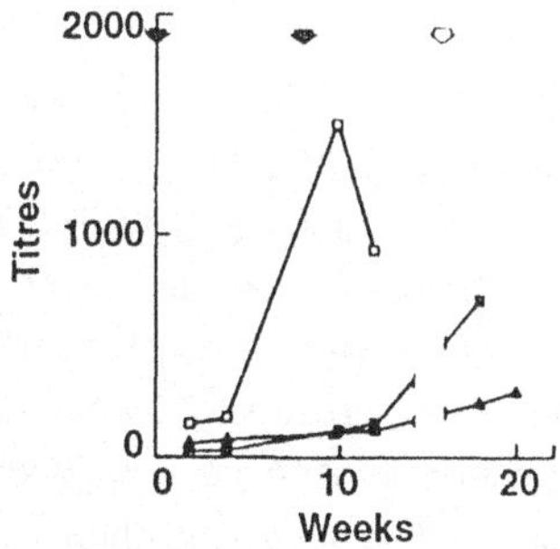

Fig. 1 Immune response to T101-dextran. One hundred μg of T101 (□), or T101 modified with 10 (▲) or 17 moles of dextran/mole of protein (■) were injected intravenously into groups of 7 rabbits each ("solid arrows"). Rabbit antimouse IgG response in serum samples was measured by ELISA as described (37). "Open arrows" indicated challenges with 100 μg of unmodified T101.

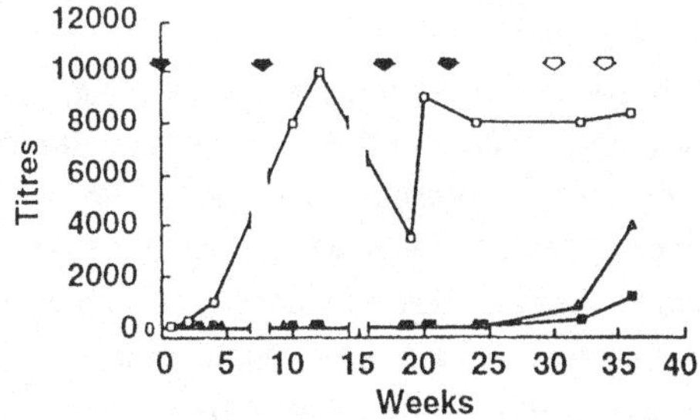

Fig. 2 Immune response to RAF-dextrain. Ten micrograms of unmodified RAF (□), or RAF modified with 8 (▲) or 17 moles of dextran/mole of protein (■), were injected intravenously into groups of 5 mice each ("solid arrows"). The mouse anti-rabbit IgG response in serum samples was measured by ELISA as described (37).

(From Ref. 37, reproduced with permission of the American Association of Cancer Research)

IV HUMANIZATION OF ANTIBODIES

In recent years, the ability to clone the immunoglobulin gene and to express and secrete the resulting genetic constructs in suitable systems has enabled new ways to manipulate the murine immunoglobulin molecule to render it more similar to the human one. This chapter summarizes two such approaches, namely chimerization and CDR - grafting of antibodies. In addition, progress to date with human antibodies will be discussed as well.

A. Chimeric Antibodies

Chimeric mouse/human antibody molecules can be constructed by joining the entire variable domains of the murine mAb to the human Ig constant regions. (Figure 3). This approach theoretically decreases the amount of the antibody molecule which is recognized as foreign by the immune system. Other potential advantages of chimeric antibodies are their prolonged plasma half-lives and their improved effector functions in humans. From a protein engineering stand point, the generation of chimeric antibodies is a relatively straight forward approach, and several murine mAbs have been chimerized so far, with retention of antigen-binding properties and of various effector functions (for review see 40-46). A careful dissection of the relative immunogenicity of chimeric mAbs in an experimental animal model was performed by Bruggemann et al. (47). These authors evaluated the immunogenicity in mice of a human mAb, its human-mouse chimeric derivative, and a fully murine antibody equivalent. The human mAb contained both human C_H domains and V_H frameworks, and represented a model of xenogeneic antibody in their animal system. A chimeric version of this antibody was constructed by joining murine C_H domains to the human V_H frameworks. Ultimately, a syngeneic antibody contained only murine sequences. All these antibodies contained murine light chains. Since these constructs contained different degrees of human residues, they were expected to elicit different degrees of immunogenicity in mice. Indeed, the immunization with xenogeneic antibodies elicited a strong immune response, 80% of which was directed against C_H domains and 10% against V_H frameworks. Chimeric antibodies elicited reduced but considerable antibody response, directed primarily against the V_H frameworks. Only the administration of syngeneic antibodies did not elicit measurable antibody responses. This observation indicated a potential limitation to the chimerization approach: since the entire variable domain of the murine antibody is maintained in the chimeric molecule, it can be perceived as foreign by the immune system. Indeed, variable levels of human anti-mouse immune response have been noted in recently completed clinical trials (Table 1).

One of the chimeric antibodies most extensively studied in humans is chimeric 17-1A, which recognizes a 41 KDa glycoprotein antigen on the surface of gastrointestinal cancer cells (48). This antibody is composed of the variable regions of murine 17-1A and the constant region of a human IgG1 Kappa immunoglobulin. When chimeric 17-1A was administered to colorectal cancer patients, only one patient out of 16 (or 6%) had significant response following the administration of this antibody. In this patient, human anti-Ig response was directed primarily against the murine hypervariable regions of the chimeric antibody (49).

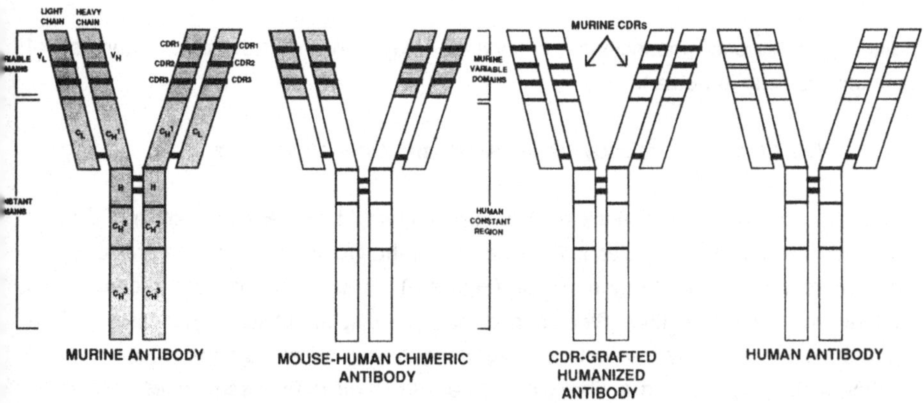

Fig.3 Schematic representation of murine, chimeric, humanized or fully human antibodies. Murine domains are shaded, human domains are plain. A human-chimeric antibody retains the entire V_L and V_H domains from the murine antibody. A humanized antibody retains only the murine CDRs (solid strips). The CDRs of a human antibody are represented as plain strips

Different results were reported when the chimeric B72.3 antibody was tested in clinical trials. This antibody recognizes a high molecular weight mucin (TAG-72) found on adenocarcinoma tumor cells in breast, lung and ovarian cancer (50). Chimeric B72.3 is composed of the variable regions of murine B72.3 joined to the constant regions of human IgG4 heavy and Kappa light chains (51). When chimeric B72.3 was administered to colorectal cancer patients, human anti-Ig response resulted in 7 of 12 patients (58%) (52), a rate similar to that previously reported for the murine B72.3 itself (53). Thus, when administered to the same group of patients, chimeric B72.3 seems more immunogenic than chimeric 17-1A. Because of these findings, Khazaeli et al. (52) suggested that different chimeric monoclonal antibodies may possess different degrees of immunogenicity in humans.

B. CDR Grafting, Antibody "Reshaping", and other approaches

The second approach which is currently pursued toward the generation of "humanized" murine monoclonal antibodies is the insertion of the CDRs of the murine antibody in a human monoclonal antibody molecule (Figure 3). This work has been pioneered by Winter and colleagues (54). Since it involves the transplant of selected smaller regions of the murine antibody molecule, CDR-grafting is theoretically a superior approach to chimerization, yielding an antibody molecule with fewer murine sequences. However, grafting CDR regions defined by sequence variability is often not sufficient to maintain antigen-binding affinity of the humanized antibody. Certain residues of the original murine framework make key contacts with the CDRs that help maintain their conformation. Alteration of these critical contacts by replacement with human frameworks may distort the shape of the CDRs, resulting in reduction or abrogation of antibody affinity. Therefore, to preserve specificity and affinity of a murine antibody, its CDRs, as well as their interaction with one another and with the rest of the variable domains should not be altered. For practical purposes, this represents the need of having to maintain sufficient murine sequences for optimal antibody specificity and affinity while minimizing the overall murine content to decrease the immunogenicity of the humanized antibody. Several laboratories are actively pursuing various methods of humanization of murine antibodies with potential clinical applications. I will summarize here the most recent progress in this field.

Kettleborough et al. (55) recently described a method to "reshape" CDR grafted mAbs 425, which recognizes the epidermal growth factor receptor, to generate humanized antibodies with higher antigen affinity. Using computer modeling, these authors identified framework residues critical for the correct conformation of the CDR loops. Grafting these residues into humanized mAbs resulted in "reshaped" molecules, with avidity for antigen approaching that of the original murine antibody.

Co et al. (56) humanized the murine mAbs Fd79 and Fd138-80 by "reshaping" both the murine variable and constant domains to render them more human-like. These mAbs recognize the gB and gD antigens of the herpes simplex virus. After sequence analysis of the murine antibodies, human V regions with special sequence homology to the murine V regions were chosen from their data-base and used to provide the framework. This provided human V regions that most closely resembled their murine counterparts and were less likely to distort the conformation of the CDRs. This approach enabled the humanization of antibodies possessing high binding specificity and affinity relative to the original antibodies.

Padlan (57) replaced the residues of the exposed framework regions of a murine antibody with residues usually found on the human ones. This resulted in the humanization of only the surface of the murine antibody, with preservation of all internal and contacting residues. Since the antigenicity of a protein is dependent, at least in part, upon the nature of it surface (58), this approach is expected to generate antibodies with reduced immunogenicity by decreasing the probability of interaction with antigen-processing cells (59), while minimally affecting the antibody affinity for the antigen.

To date, two CDR grafted, humanized mAbs have been evaluated in human clinical trials. The mAb Campath-1H (54) recognizes the Campath-1 antigen present on all lymphocytes and monocytes but absent from other blood cells. This antibody is being developed as potential immunosuppressive agent in the treatment of graft-versus-host disease, organ rejection or various lymphoid malignancies. Humanized Campath-1H mAb has been used to treat two patients with Hodgkin's lymphoma (60) and one patient with systemic vasculitis, in the latter case, in combination with a rat mAb to the human CD4 molecule on helper T cells (61). This treatment induced remission of disease, and no antibody response against Campath-1H or the rat mAb was reported (60-61). In either case, however, patients received some form of immunosuppression, and Campath-1H itself is probably immunosuppressive (60).

The other humanized antibody which has been evaluated in vivo is anti-Tac-H, specific for the Interleukin-2 (IL-2) receptor α chain (62) on activated T-cells. Anti-Tac-H is composed by the CDRs of the murine anti-Tac (anti-Tac M), joined to human IgG1 Kappa framework and constant regions (63). This antibody is being developed as a specific immunosuppressive agents in the treatment of allograft rejection, since activated T-cells express the inducible IL-2 receptor in response to antigenic stimulation and participate in the allograft rejection. Although anti-Tac-H has not yet been administered to humans, it has been tested in a preclinical allograft rejection model in primates, where it was compared to its murine counterpart (64). With this model, treatment with anti-Tac-H significantly prolonged the survival of cardiac allografts. In addition, all monkeys receiving anti-Tac-M developed a strong response to this mAb by day 11, whereas none of the animals receiving anti-Tac-H produced measurable anti-antibody responses until day 33, when 3 out of 5 monkeys elicited a detectable response. At present, the immunogenicity of humanized anti-Tac-H in humans is waiting evaluation in clinical trials.

C. Human Antibodies

The theoretical advantage of human mAbs is their lack of significant antigenicity in humans, and except for anti-idiotypic responses, human antibodies are not expected to be recognized as foreign by patients. Human mAbs, however, are difficult to produce. Conventional human-human or human-mouse fusions generated hybridomas are either unstable or possess very low secretory rates (65). Similarly, transformation of immune lymphocytes with Epstein-Barr virus has low transformation rates and tends to yield antibodies of the IgM class (66). New approaches, like the production of antibodies in Severe Combined Immuno Deficient (SCID) mice via repopulation of the mouse's lymphoid system with immunized human lymphocytes are currently at experimental stages (67). However, few human mAbs primarily directed against infectious agents, have been produced by different groups using immortalized lymphocytes from either patients recovering from infectious diseases or from immunized individuals. Two such human antibodies are now in clinical trials.

A human mAb that has been extensively studied is HA-1A, an IgM reacting with the lipid A domain of endotoxin (68). Endotoxin is a component of gram negative septic shock, a serious and often fatal consequence of bacterial infections. HA-1A mAb was generated by fusion of B lymphocytes from sensitized human spleens with heteromyeloma cells. When this antibody was tested in therapy trials of septic shock patients, an approximate 40% reduction of mortality was observed. All patients tolerated the administration of the antibody, and no anti-idiotypic response was reported (69).

Human antibodies against cytomegalovirus (CMV) have recently been raised by Ostberg et al. (7), as preventive therapy against CMV infections consequent to kidney, liver and allogeneic bone marrow transplantations. These antibodies have been tested in long-term studies in rhesus monkeys, and have recently completed Phase I clinical trials (71). Although these patients had varying degree of immunosuppression, no significant anti-idiotypic response has been reported. In addition, only one out of five immunocompetent monkeys injected with this antibody in preclinical trials elicited a detectable anti-idiotypic response. However, the long-term immunogenicity of these antibodies in non-immunosuppressed patients needs to be evaluated.

Lastly, the human IgM mAb 16.88 was recently tested in patients with advanced colorectal carcinomas (72). Contrary to the antibodies mentioned above, this mAb was developed from spontaneously transformed lymphocytes of a patient with colorectal carcinoma undergoing immunotherapy. Up to 1 gram of 16.88 mAb was administered in this clinical trial, with no reported anti-human antibody response (72).

D. Repertoire Cloning

Ultimately, cloning the entire human antibody genes and selection of their expression products for suitable antigen-binding characteristics may allow the identification of human antibodies of desired specificity. This approach, called "repertoire cloning" (73,74) involves the isolation of the heavy and light chain genes of the antibodies of desired specificity from the lymphocytes of an immunized donor. These genes are then amplified, inserted into bacteriophage and expressed in bacteria. With adequate screening, the relatively few combinations of the desired heavy and light chains genes can be isolated. These genes are then cloned, sequenced, and expressed in the appropriate system for widespread production. Repertoire cloning has the potential of providing new avenues toward the generation of antibodies with desired specificity and may render conventional immunizations, fusions and screening procedures obsolete.

However, this approach is still highly experimental, and antibodies generated with this method have not been tested in animal models or in clinical trials.

Acknowledgements

I wish to express my gratitude to Drs. Michele Pelligrino and Barry Wilson for their helpful comments, and to Ms. Alison Hagström for typing this manuscript.

References

1. Miller, R.A. and Levy. Response to cutaneous T cell lymphoma to therapy with hybridoma monoclonal antibody. Lancet. 1981; 2:226-230.
2. Dillman, R.O. Monoclonal antibodies in the treatment of cancer. Crit. Rev. Oncol. Hematol. 1984; 1:357-386.
3. Sears. H.F., Mattis, J., Herlyn, D., Häyry, P., Atkinson, B., Ernst, C., Steplewski, Z. and Koprowski, H. Phase I clinical trial of monoclonal antibody in treatment of gastrointestinal tumors. Lancet 1982; 2:762-765.
4. Cosini, A.B., Colvin, R.B., Burton, R.C., Rubin, R.H., Goldstein, G., Kung, P.C., Hausen, P., Delmonico, F.L. and Russell, P.S. Use of monoclonal antibodies to T-cell subsets for immunologic monitoring and treatment of renal allografts. N. Engl. J. Med. 1981; 305:308-311.
5. Ortho Multicenter Transplant Study Group. A randomized clinical trial of OKT3 monoclonal antibody for acute rejection of cadaveric renal transplant. N. Eng. J. Med. 1981; 305:308-311.
6. Takahashi, H., Okazaki, H., Terasaki, P.I. Yweaki, Y., Kinukawa, T., Taguchi, Y., Chia, D., Hadiwidjaja, S., Miura, K., Ishizaki, M. and Billing, R. Reversal of transplant rejection by monoclonal antibody. Lancet 1983; 2:1155-1157.
7. Herzog, C., Walker, C., Pichier, W., Aeschlimann, A., Wassmer, P., Stockinger, H., Knapp, W., Rieber, P. and Müller, W. Monoclonal anti-CD4 in arthritis. Lancet 1987; 2:1461-1462.
8. Hafler, D.A. and Weiner, H.L. Immunosuppression with monoclonal antibodies in multiple sclerosis. Neurology 1988; 38(2):42-47.
9. Dillman, R. Human antimouse and antiglobulin responses to monoclonal antibodies. Antibody Immunocon. Radiopharm. 1990; 3:1-15.
10. Reynolds, J.C., DelVecchio, S., Sakahara, H., Lora, M.E., Carrasquillo, J.A., Newman, R.D. and Larson, S.M. Anti-murine antibody response to mouse monoclonal antibodies: clinical findings and implications. Nucl. Med. Biol. 1989; 16:121-125.
11. Courtenay-Luck, N., Epenetos, A.A., Sivolapenko, G.B., Larche, M., Barkans, J.R. and Ritter, M.A. Development of anti-idiotypic antibodies against tumor

antigens and autoantigens in ovarian cancer patients treated intraperitoneally with mouse monoclonal antibodies. Lancet 1988; 2:894-897.

12. Parham, P. On the fragmentation of monoclonal IgG1, IgG2a and IgG2b from Balb/C mice. J. Immunol. 1983; 131:2859-2902.
13. Porter, R.R. The hydrolysis of rabbit γ-globulin and antibodies with chrystalline papain. Biochem. J. 1959; 73:119-123.
14. Hickey, A.R., Wenger, T.L., Carpenter, V.P., Tilson, H.H., Heatky, M.A., Furberg, E.D., Kirkpatrick, C.L., Strauss, H.C. and Smith, T.W. Digoxin immune Fab therapy in the management of digitalis intoxication: safety and results of an observational surveillance study. J. Am. Coll. Cardiol. 1991; 17:590-598.
15. Kalofonos, J.P., Sivolapenko, G.B., Courtenay-Luck, N.S., Snook, D.E., Hooker, G.R., Winter, R., McKenzie, C.G., Taylor-Papadimitriou, J.J., Lavender, P.J. and Epenetos, A.A. Antibody guided targeting of non-small cell lung cancer using ^{111}In-labelled HMFG1 F(ab')$_2$ fragments. Cancer Res. 1988; 48:1977-1984.
16. Bird, R.D., Hardinan, K.D., Jacobson, J.W., Johnson, S., Kaufman, R., Lee, S-M., Lee, T., Pope, S.H., Riordan, G.S. and Whitlow, M. Single-chain antigen binding proteins. Science 1988; 242:423-426.
17. Huston, J.S., Levinson, D., Mudgett-Hunter, M., Tai, M-S., Novotny, J., Margolis, M.N., Ridge, R.J., Bruccoleri, R.E., Haber, E., Crea, R. and Oppermann, H. Protein engineering of antibody binding sites: recovery of specific activity in an anti-digoxin single-chain FV analogue produced in Escherichia Coli. Proc. Natl. Acad. Sci. USA 1988; 85:5879-5883.
18. Colcher, D., Bird, R., Roselli, M., Hardman, K.D., Johnson, S., Pope, S., Dodd, S.W., Pantoliano, M.W., Milenic, D.E. and Schlom, J. In vivo tumor targeting of a recombinant single-chain antigen-binding protein. J. Natl. Cancer Inst. 1990; 82:1191-1197.
19. Milenic, D.R., Yokota, T., Filpula, D.R., Finkelman, M.A.J., Dodd, S.W., Wood, J.F., Whitlow, M., Snoy, P. and Schlom, J. Construction, binding properties, metabolism and tumor targeting of a single-chain Fv derived form of pancarcinoma monoclonal antibody CC 49. Cancer Res. 1991; 41:6363-6371.
20. Laroche, Y., Demaeyer, M., Stassen, J., Gansemans, Y., Demarsin, E., Matthyssens, G., Collen, D. and Holvoet, P. Characterization of a recombinant single-chain molecule comprising the variable domains of a monoclonal antibody specific for human fibrin fragment D-dimer. J. Biol. Chem. 1991; 266:16343-16349.
21. Traub, R., Gould, R.J., Garsky, V.M., Ciccarone, T.M., Hoxie, J., Friedman, P.A. and Shattil, S.J. A monoclonal antibody against the platelet fibronogen receptor contains a sequence that mimics a receptor recognition domain in fibronogen. J. Biol. Chem. 1989; 264:259-265.

22. Williams, W.V., Moss, D.A., Kieber-Emmons, T., Cohen, J.A., Meyers, J.N., Weiner, D.B. and Greene, M.I. Development of biologically active peptides based on antibody structure. Proc. Natl. Acad. Sci. USA. 1989; 86:5537-5541.
23. Williams, W.V., Kieber-Emmon, T., VonFeld, J., Green, M.I. and Weiner, D.B. Design of bioactive peptide based on antibody hypervariable region structure. Develop of comformationally costrained and chimeric peptides with enhanced affinity. J. Biol. Chem. 1991; 266:5182-5190.
24. Saragovi, H.V., Fitzpatrick, D., Raktabutr, A., Nakanishi, H., Kahn, M. and Greene, M.I. Design and synthesis of a mimetic from an antibody complementary-determining region. Science 1991; 253:792-795.
25. Abuchowski, A., Kazo, G.M., Verhoest, C.R., VanEs, T., Kafkewitz, D., Nucci, M.L., Viav, A. and Davis, F.F. Cancer therapy with chemically modified enzymes. I. Antitumor properties of polyethylene glycolasparaginase conjugates. Cancer Biochem. Biophys. 1984; 7:175-186.
26. Davis, S., Abuchowski, A., Park, K.K. and Davis, F.F. Alterations of the circulating life and antigenic properties of bovine adenosine deaminase in mice by attachment of polyethylene glycol. Clin. Exp. Immunol. 1981; 46:649-652.
27. Lee, W.Y. and Sehon, A.H. Abrogation of reaginic antibodies with modified allergens. Nature, Lond. 1977; 267:618-620.
28. Koide, A., Suzuki, S. and Kobayashi, S. Preparation of polyethylene glycol modified streptokinase with disappearance of binding ability toward anti-serum and retention of activity. FEBS Letters. 1982; 143:73-76.
29. Kimasaki, Y., Wada, H., Yagura, T., Matsushima, A. and Inada, Y. Reduction of immunogenicity and clearance rate of Escherichia Coli L-Asparaginase by modification with monometoxypolyethylene glycol. J. Pharm. Exp. Ther. 1981; 216:410-414.
30. Katre, N. Immunogenicity of recombinant IL-2 modified by covalent attachment of polyethylene glycol. J. Immunol. 1990; 144:209-213.
31. Wilkison, I., Jackson, C.J., Lang, G.M., Strevens-Holford, V. and Sehon, A.H. Tolerogenic polyethylene glycol derivatives of xenogeneic monoclonal immunoglobulins. Immunol. Lett. 1987; 15:17-22.
32. Rosemeyer, H., Körnig, E. and Sela, F. Adenosine deaminase covalently linked to soluble dextran. The effect of immobilization on thermodynamic and kinetic parameters. Eur. J. Biochem. 1982; 122:375-380.
33. Wileman, T.E., Foster, R.L. and Elliott, P.N.C. Soluble asparaginase-dextran conjugates show increased circulatory persistence and lowered antigen reactivity. J. Pharm. Pharmacol. 1986; 38:264-271.
34. Hurwitz, E. Specific ;and nonspecific macromolecule-drug conjugates for the improvement of cancer chemotherapy. Biopolymers. 1983; 22:557-567.
35. Tsukada, Y., Ohkawa, K. and Hibi, N. Therapeutic effect of treatment with polyclonal or monoclonal antibodies to α-fetoprotein that have been conjugated

to daunomycin via dextran bridge: studies with an α-fetoprotein-producing rat hepatoma tumor model. Cancer Res. 1987; 47:4293-4295.

36. Shih, L.B., Sharkey, R.M., Primus, F.J. and Goldenberg, D.M. Site-specific linkage of methotrexate to monoclonal antibodies using an intermediate carrier. Int. J. Cancer. 1988; 41:832-839.

37. Fagnani, R., Hagan, M.S. and Bartholomew, R. Reduction of immunogenicity by covalent modification of murine and rabbit immunoglobulins with oxidized dextrans of low molecular weight. Cancer Res. 1990; 50:3638-3645.

38. Royston, I., Majda, J.A., Bairds, S.M., Meserve, B.L. and Griffiths, J.C. Human T cell antigens defined by monoclonal antibodies: the 65,000 dalton antigen of T cells (T65) is also found on chronic lymphocytic leukemia cells bearing surface immunoglobulin. J. Immunol. 1980; 125:725-731.

39. Klein, J.L., Leichner, P.K., Callahan, K.M., Kopher, K.A. and Order, S.E. Effects of anti-antibodies on radiolabelled antibody therapy. Antibody Immunoconjugates Radiopharm. 1988; 1:55-64.

40. Shin, Su. Chimeric antibody: potential applications for drug delivery and immunotherapy. Biotherapy. 1991; 3:433.453.

41. Morrison, S.L., Johnson, M.J., Herzenberg, L.A. and Oi, V.T. Chimeric human antibody molecules: mouse antigen-binding domains with human constant regions domains. Proc. Natl. Acad. Sci. USA. 1984; 81:6851-6855.

42. Newberger, M.S., Williams, G.T. and Fox, R.O. Recombinant antibodies possessing novel effector functions. Nature. 1984; 312:604-608.

43. Boulianne, G.L., Hozumi, N. and Schulman, J.J. Production of functional chimeric mouse/human antibody. Nature. 1984; 312:643-646.

44. Sahagan, B.G., Sorai, H., Saltzgaber-Muller, J., Toneguzzo, F., Guidon, C.A., Lilly, S.P., McDonald, K.W., Morrissey, D.V., Stone, B.A., Davis, G.I., McIntosh, P.K. and Moore, G.I. A genetically engineered murine/human chimeric antibody retains specificity for human tumor-associated antigen. J. Immunol. 1986; 137:1066.

45. Brown, B.A., Davis, G.I., Saltzgaber-Muller, J., Simon, P., Ho, M.K., Shaw, P.S., Stone, B., Sands, H. and Moore, G.P. Tumor-specific genetically engineered murine/human chimeric monoclonal antibody. Cancer Res. 1987; 47:3577.

46. Hardman, N., Gill, L.L., DeWinter, R.F.J., Wagner, K., Hollis, M., Busunger, F., Ammaturo, D. Buchegger, F., Mach, J-P. and Heusser, C. Generation of recombinant mouse-human chimeric monoclonal antibody directed against human carcinoembryonic antigen. Int. J. Cancer. 1989; 44:424-433.

47. Bruggermann, M., Winter, G., Waldman, H. and Neuberger, M.S. The immunogenicity of chimeric antibodies. J. Exp. Med. 1989; 170:2153-2157.

48. Girardet, C., Vacca, A., Schmidt-Kessen, A., Schreyer, M., Carrol, S. and Mach, J. Immunochemical characterization of two antigens recognized by new

monoclonal antibodies against human colon carcinoma. J. Immunol. 1986; 136:1497-1503.

49. LoBuglio, A., Wheeler, R., Trang, J., Haynes, A., Rogers, K., Harvey, E.B., Sun, L., Gharayeb, J. and Khazaeli, M.B. Mouse/human chimeric monoclonal antibodies in man: kinetics and immune response. Proc. Natl. Acad. Sci. USA. 1989; 86:4220-4224.
50. Johnson, V. Schlom, J., Patterson, A., Bennett, J., Magnani, J. and Colcher, D. Analysis of a human tumor associated glycoprotein (TAG-72) identified by monoclonal antibody B72.3. Cancer Res. 1986; 46:850-857.
51. Whittle, N., Adair, J., Lloyd, C., Jenkins, L., Devin, J. Raubischak, A., Colcher, D. and Bodner, M. Protein Eng. 1987; 1:499-505.
52. Khazaeli, M.B., Saleh, M.N., Liu, T.P., Meredith, R.F., Wheeler, R.H., Baker, T.S., King, D., Secher, D., Allen, L., Rogers, K., Colcher, D., Schlom, J., Shochat, D. and LoBuglio, A.F. Pharmacokinetics and immune response of ^{131}I-chimeric mouse/human B72.3 monoclonal antibody in humans. Cancer Res. 1991; 51:5461-5466.
53. Carrasquillo, J., Sugerbaker, P., Colcher, D., Reynolds, J., Esteban, J., Begent, G., Keenan, A., Perentesis, P., Yoloyama, K., Simpson, D., Ferroni, P., Farkas, R., Schlom, J. and Larsen, S. Radioimmunoscintigraphy of colon cancer with iodine-131-labelled B72.3 monoclonal antibody. J. Nuc. Med. 1988; 29:1022-1030.
54. Reichmann, L., Clark, M., Waldmann, H. and Winter, G. Reshaping human antibodies for therapy. Nature. 1988; 332:323-327.
55. Kettleborough, C.A., Saldanha, J., Heath, V.J., Morrison, C.J. and Bendin, M.M. Humanization of a mouse monoclonal antibody by CDR-grafting: the importance of framework residues on loop confirmation. Protein Eng. 1991; 4:773-783.
56. Co, M.D., Deschamps, M., Whitley, R.J. and Queen, C. Humanized antibodies for antiviral activity. Proc. Natl. Sci. USA. 1991; 88:2869-2873.
57. Padlan, E.A. A possible procedure for reducing the immunogenicity of antibody variable domains while preserving their ligand-binding properties. Molecular Immunol. 1991; 28:489-498.
58. Benjamin, D.C., Berzofsky, J.A., Gurd, F.R.N., et al. The antigenic structure of proteins: A reappraisal. Ann. Rev. Immunol. 1984; 2:67-101.
59. Kurt-Jones, E.A., Lians, D., Hayglan, K.A., Benacerraf, B., Sy, M-S. and Abbas, A.K. The role of antigen-presenting B cells in T cells priming in vivo. Studies of B cell-deficient mice. J. Immunol. 1988; 140:3773-3778.
60. Hale, G., Dyer, M.J.S. and Clark, M.R. Remission in non-Hodgkins lymphoma with reshaped human monoclonal antibody Campath-1H. Lancet. 1988; 2:1394-1399.

61. Mathieson, P.W., Cobbold, S.P., Hale, G., Clark, M., Oliveira, D.B.G., Lockwood, C.M. and Waldman, H. Monoclonal antibody therapy in systemic vasculitis. New Eng. J. Med. 1990; 323:250-254.
62. Uchiyama, T., Broder, S. and Waldman, T.A. A monoclonal antibody (anti-Tac) reactive with activated and functionally mature human T cells. I. Production of anti-Tac monoclonal antibody and distribution of Tac (+) cells. J. Immunol. 1981; 126:1393-1397.
63. Queen, C., Schneider, W.P., Selick, H.E., Payne, P.W., Landolfi, N.F., Duncan, J.F., Avdalovic, N.M., Cevitt, M., Junghaus, R.P. and Waldman, T.A. A humanized antibody that binds to the interleukin-2 receptor. Proc. Natl. Acad. Sci. USA; 1989; 86:100290-10033.
64. Brown, S.P., Jr., Parenteau, G.L., Dirbas, F.M., Garsia, R.J., Goldman, C.K., Bukowski, M.A., Junghas, R.P., Queen, C., Hakini, J., Benjamin, W.R., Carl, R.E. and Waldman, T.A. Anti-Tac-H, a humanized antibody to interleukin-2 receptor, prolongs primate cardial allograft survival. Proc. Natl. Acad. Sci. USA. 1991; 88-2663-2667.
65. Glassy, M.C. and Dillman, R.O. Molecular biotherapy with human monoclonal antibodies. Mol. Biother. 1988; 1:7-12.
66. Casali, P. and Notkins, A.L. Probing the human B-cell repertoire with EBV: polyreactive antibodies and CD5 + B lymphocytes. Ann. Rev. Immunol. 1989; 7:513-536.
67. Bosma, M.J. and Carroll, A.M. The SCID mouse mutant: definition, characterization, and potential use. Ann. Rev. Immunol. 1991; 9:323-350.
68. Teng, N.N.H., Kaplan, H.S., Herbert, J.M., Moore, C., Douglas, H., Wunderlich, A. and Braude, A.I. Protection against gram-negative bacteremia and endotoxin with human monoclonal IgM antibodies. Proc. Natl. Acad. Sci. USA. 1985; 82:1790-1794.
69. Zeigler, E.J., Fisher, C.J., Sprung, C.L., Straube, R.C., et al. Treatment of gram-negative bacteriemia and septic shock with HA-1A human monoclonal antibody against endotoxin: a randomized, double-blind, placebo-controlled trial. New Engl. J. Med. 1991; 324:429-436.
70. Ostberg, L. and Pursch, E. Human x (mouse x human) hybridomas stably producing human antibodies. Hybridoma. 1988; 2:361-366.
71. Drobyski, W.R., Gottlieb, M., Carrigan, D., Ostberg, L., Grebenan, M., Schran, H., Magid, P. Ehrich, P., Nadler, P.I. and Ash, R.C. Phase I study of safety and pharmacokinetics of a human anti-cytomegalovirus monoclonal antibody in allogeneic bone marrow transplant recipients. Transplantation. 1991; 51:1190-1196.
72. Haisma, H.J., Pinedo, H.M., Kessell, M.A.P., vonMuijen, M., Roos, J.C., Plaizier, M.A.B.D., Martens, H.J.M., DeJager, R. and Boven, E. Human IgM monoclonal

antibody 16.88 pharmacokinetics and immunogenicity in colorectal cancer patients. J. Natl. Cancer Inst. 1991; 83:1813-1819.

73. Huse, W.D., Sastry, L., Iverson, S.A., Kaug, A.A., Alting-Mees, M., Burton, D.R., Benkovic, S.J. and Lerner, R.A. Science. 1989; 246:1275-1281.

74. Burton, D.R. Human and mouse monoclonal antibodies by repertoire cloning. Biotech. 1991; 9:169-175.

75. Blottiere, H.M., Diullard, J.T., Koprowski, H. and Steplewski, Z. Immunoglobulin G subclass analysis of human anti-mouse antibody response during monoclonal antibody treatment of cancer patients. Cancer Res. 1990; 50:1051s-1054s.

76. Courtenay-Luck, N.C., Epenetos, A.A., Moore, R., Larche, M., Pectasides, D., Dhokia, R. and Ritter, M.A. Development of primary and secondary immune responses to mouse monoclonal antibodies used in the diagnosis and therapy of malignant neoplasms. Cancer Res. 1986; 4:6489-6493.

77. Shawler, D.C., Bartholomew, R.M., Smith, L.M. and Dillman, R.O. Human immune responses to multiple injections of murine monoclonal IgG_1. J. Immunol. 1985; 135:1530-1535.

78. Jaffers, G.J., Fuller, T.C., Cosimi, B.A., Russell, P.S., Winn, H.J. and Colvin, R.B. Monoclonal antibody therapy. Anti-idiotypic and non-anti-idiotypic antibodies to OKT3 arising despite intense immunosuppression. Transplantation. 1986; 41:572-578.

79. Chatenoud, L., Baudrihaye, M.F., Chkoff, N., Kreis, H., Goldsten, G. and Back, J-F. Restriction of the human in vivo immune response against the mouse monoclonal antibody OKT3. J. Immunol. 1986; 137:830-838.

80. Khazaeli, M.B., Saleh, M.N., Wheeler, R.H., Huster, W.J., Holden, H., Carrano, R. and LoBuglio, A.F. Phase I trial of multiple large doses of murine monoclonal antibody CO 17-1A. II. Pharmacokinetics and immune responses. J. Natl. Cancer Inst. 1988; 80:937-942.

81. Hafler, D.A., Ritz, J., Schlossman, S. and Weiner, H.L. Anti-CD4 and anti-CD2 monoclonal antibody infusions in subjects with multiple sclerosis. Immunosuppressive effects and human anti-mouse responses. J. Immunol. 1988; 141:131-138.

82. Goldman-Leikin, R.E., Kaplan, E.H., Zimmer, A.M., Kazikiewicz, J., Manzel, L.J. and Rosen. S.T. Long-term persistence of human anti-murine antibody responses following radioimmunodetection and radioimmunotherapy of cutaneous T-cell lymphoma patients using 131I-T101. Exp. Hematol. 1988; 16:861-864.

83. Schroff, R.W., Foon, K.A., Beatty, S.M., Oldham, R.K. and Morgan, A.C. Human anti-murine immunoglobulin responses in patients receiving monoclonal antibody therapy. Cancer Res. 1985; 45:897-885.

84. Murray, J.L., Rosenblum, M.G., Lamki, L., Glen, H.J., Krizan, Z., Hersh, E.M., Plager, C.E., Bartholomew, R.M., Unger, M.W. and Carlo, D.J. Clinical

parameters related to optimal tumor localization of Indium-111-labelled mouse antimelanoma monoclonal antibody ZME018. J. Nuc. Med. 1987; 28:25-33.

85. Horneff, G., Minkler, T., Kalden, J.R., Emmrich, F. and Burmester, G. Human anti-mouse antibody response induced by anti-CD4 monoclonal antibody therapy in patients with rheumatoid arthritis. Clin. Immunol. Immunopath. 1991; 59:89-103.

86. Muto, M.G., Finkler, N.J., Kassis, A.I., Lepisto, E.M., Knapp, R.C. Human anti-murine antibody responses in ovarian cancer patients undergoing radioimmunotherapy with the murine monoclonal antibody OC-125. Gynecol. Oncol. 1991; 38:244-248.

87. Petersen, B.H., DeHerdt, S.V., Schneck, D.W. and Bumol, T.F. The human immune response to KS 1/4 - Desacetylvinblastine (Ly256787) and KS 1/4 - Desacetylvinblastine Hydrazide (LY203728) in single and multiple dose clinical studies. Cancer Res. 1991; 51:2286-2290.

88. Spitler, L.D., del Rio, M., Khentigan, A., Wedel, N.I., Brophy, N.A., Miller, L.L., Harkonen, W.S., Rosendorf, L.L., Lee, H.M., Mischack, R.P., Kawahata, R.T., Studemire, J.B., Fradkin, L.B., Bautista, E.E. and Scannon, P.J. Therapy of patients with malignant melanoma using a monoclonal antibody melanoma antibody - Ricin A immunotoxin. Cancer Res. 1987; 47:1717-1723.

89. Avner, B., Swindell, L., Sharp, E., Liao, S-K., Ogden, J., Avner, B.P. and Oldham, R. Evaluation and clinical relevance of patients immune responses to intravenous therapy with murine monoclonal antibodies conjugated to adriamycin. Mol. Biother. 1991; 3:14-21.

90. Hertler, A.A., Schlossman, D.M., Borowitz, M.J., Laurent, G., Jansen, F.K., Schmidt, C. and Frankel, E.A. A phase I study of T101-Ricin A chain immunotoxin in refractory chronic lymphocytic leukemia. J. Biol. Response Mod. 1987; 7:97-113.

2

Additive and Synergistic Interactions of Monoclonal Antibodies and Immunotoxins Reactive with Breast and Ovarian Cancer

Robert C. Bast, Jr.
Fengji Xu
Yinhua Yu
Jennie Crews
Yair Argon
Lisa Maier
Yaron Lidor
Andrew Berchuck
Cinda M. Boyer
Duke Comprehensive Cancer Center
Duke University Medical Center
Durham, North Carolina

I INTRODUCTION

Over the last several years our laboratory has studied the expression of human tumor associated antigens defined by murine monoclonal antibodies. Remarkable heterogeneity in the expression of different antigens has been observed within and between different carcinomas that arise from breast and ovarian epithelium.

II HETEROGENEITY OF ANTIGEN EXPRESSION BY BREAST AND OVARIAN CARCINOMAS

Expression of antigens has been studied using biotin avidin immunoperoxidase and a battery of 16 murine monoclonal antibodies that recognize 11 chemically defined epitopes.(1) When 18 breast cancers were examined in frozen sections, 16 had a distinct phenotype. Similarly, each of 14 epithelial ovarian carcinomas had a distinct antigenic phenotype. Heterogeneity has also been observed within breast cancers using individual monoclonal reagents. Intense staining of some cells was observed, whereas little, if any, binding could be detected to other cells within the some tumor nodule. Using a combination of five murine monoclonal antibodies, however, more than 95% of tumor cells could be stained intensely in more than 90% of tumors. Consequently, it is possible that more than one murine monoclonal antibody may need to be utilized if serotherapy is to be effective.

Research over the last 15 years indicates that there are few, if any, murine antibodies that are truly tumor specific.(2) Reactivity has generally been found with one or more normal tissues. Despite this lack of tumor specificity, a therapeutic advantage could still be attained if antigens can be identified that are coexpressed on tumor cells, but that are not coexpressed on normal tissues. For this strategy to succeed, additive or synergistic anti-tumor activity must be observed when two or more antibodies bind to the same tumor cell.

III ADDITIVE AND SYNERGISTIC CYTOTOXICITY OF IMMUNOTOXINS IN COMBINATION

In published reports, we have asked whether antigenic targets can be defined that will permit additive or synergistic cytotoxicity when different immunotoxins are used in combination.(3,4) For these studies, we have utilized immunotoxins prepared by conjugating recombinant ricin A chain (rRTA) to murine monoclonal antibodies using an imino-thiolane linkage. Ricin A chain has been conjugated with different murine monoclonal antibodies to examine the utility of different antigens as targets for the cytotoxicity of immunotoxins. Four toxin immunoconjugates were studied in depth including 317G5-rRTA, 260F9-rRTA, 454A12-rRTA, and 741F8-rRTA, which recognize antigens of 42 kd, 55 kd, 180 kd (transferrin receptor) and 185 kd (HER-2/neu). In the case of the 317G5-rRTA and 260F9-rRTA, each antibody individually can produce 1-2 logs of tumor cell killing, whereas the combination produces almost four logs of tumor cell killing. Whether this represents additive or synergistic cytotoxic activity can best be demonstrated by isobolographic analysis.

Isobolograms plot the combinations of concentrations of two agents required to kill a given fraction of cells. If the dose response curves for the individual agents are linear, additive values fall on a straight line cutting across both axis (Fig. 1 left panel). Subadditive combinations fall to the right and superadditive or synergistic combinations to its left. If the dose response curves are not linear, additivity can best be described as an envelope between two modes that assume the agents either interact or do not interact (Fig. 1 right panel). Subadditive combinations fall to the right of the envelope of additivity and synergistic combinations to its left. In the cases of 260F9-rRTA and 317G5-rRTA the individual dose response curves are markedly non-linear. When an envelope of additivity was calculated, the isobole for 1.6 logs of cytotoxicity fell within the limits of the envelope of additivity. In the case of 454A12-rRTA and 260F9-rRTA, true synergy was observed in killing 0.9 logs of clonogenic SKBr3 cells (Fig. 2). Studies were undertaken to determine the mechanism underlying this superadditive effect. Binding of the two antibodies to the tumor cell surface was only additive. When internalization was studied using acid conditions to strip antibody that had not been internalized, only additive effects were observed. When two color immunofluorescence analysis was undertaken, dells which expressed one, but not both antigens could readily be detected in the total tumor cell population. Subpopulations that lacked individual antigens would, however, only explain additive and not synergistic cytotoxicity.

Another more interesting possibility was posed by electron microscopy studies.(4) In these studies, 454A12 and 260F9 were labeled with 5 or 15 nanometer gold particles. 454A12 was promptly internalized through clathrin coated pits and vesicles, reaching deep endosomal structures within ten minutes. 260F9 was internalized more slowly reaching similar structures within 30 minutes. In double gold labeling experiments, 260F9 and 454A12 were cointernalized. The 260F9 markedly slowed the internalization of 454A12 permitting a longer dwell time in the endosomes. This may have permitted disassociation and translocation of a greater fraction of the immunotoxin.

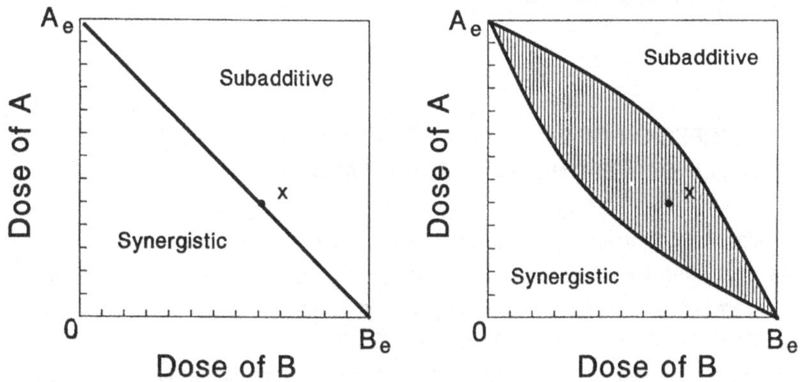

Fig.1 Isobolographic analysis of 2 agents (A and B) with linear dose response curves (left panel) or with nonlinear dose response curves (right panel)

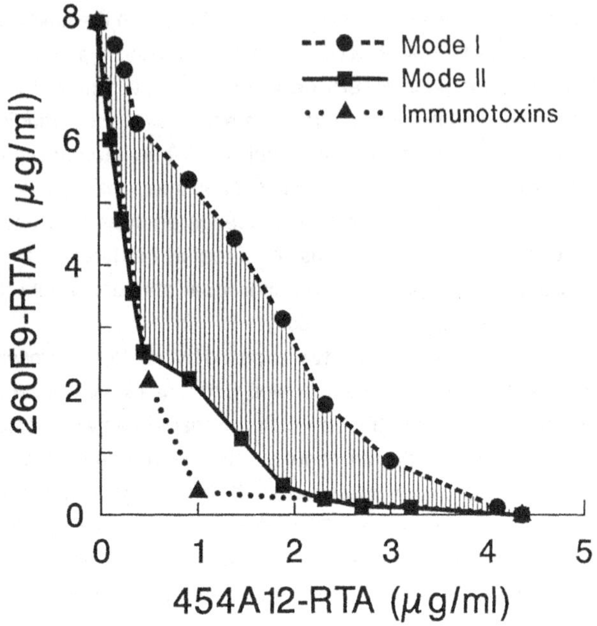

Fig.2 Isobolographic analysis of 0.9 log clonogenic elimination with 454A12-RTA and 260F9-RTA

260F9-rRTA and 454A12-rRTA have both been evaluated in clinical trials. The 260F9-rRTA produced a peripheral neuropathy due to binding of the immunotoxin to the myelin sheath cells.(5) The 454A12-rRTA produced central neurotoxicity.(6) Although their use in combination establishes the principal that synergy can be observed between different immunotoxins, they are not likely to be useful clinically.

IV EXPRESSION OF EPIDERMAL GROWTH FACTOR RECEPTOR (EGFR) AND C-ERBB-2 BY EPITHELIAL OVARIAN CANCERS

Other targets that have been explored include several integral membrane tyrosine kinases, including the EGFR and HER-2/neu. The EGFR has an extracellular ligand binding domain, a membrane spanning region, and an intracellular tyrosine kinase region. The viral erbB oncogene differs from EGFR in that the extracellular domain has been truncated permitting constitutive activation of the intracellular tyrosine kinase. The HER-2/neu (c-erbB-2) oncogene bears 40% homology to the entire EFGR and 80% homology to its intracellular domain. In the initial studies by Weinberg, et al,(7) the neu gene isolated from tumors of rat brain was shown to have a single point mutation within the membrane spanning domain. In human breast cancer, HER-2/neu has been amplified, but only the wild type sequence has been detected.(8) Overexpression of the p185 gene product has correlated with a poor prognosis. Recently, Lupu et al, have reported a ligand for p185 which is distinct from EGF.(9) The presence of such a ligand raises the possibility that autocrine growth regulatory control might exist in cells that express both p185 and the ligand.

When expression of EGFR was studied in human ovarian cancers, 77% of 87 neoplasms could be stained with the 225 antibody.(10) The prognosis of patients whose tumors expressed EGFR was worse than the prognosis of patient's whose tumors did not express EGFR. In subsequent studies of human ovarian cancer, expression of different oncogenes was studied by Northern transfers.(11) Expression of erbB could not be detected, but 86% of 22 cancers expressed HER-2/neu. In addition, HER-2/neu can be amplified and overexpressed in ovarian cancers.(12,13) When frozen sections of epithelial cancer were stained with the TA1 antibody against the extracellular domain of HER-2/neu, approximately two-thirds of cancers have low, but detectable, levels of antibody binding.(12) In one-third of ovarian cancers, there was marked overexpression of HER-2/neu and this was associated with a significantly worse prognosis. Patients whose tumors overexpressed p185 had a median survival of approximately 16 months, whereas those individuals whose tumors had normal p185 expression had a median survival of some 32 months.

V GROWTH INHIBITION OF TUMOR CELLS BY ANTIBODIES REACTIVE WITH HER-2/NEU

In subsequent studies we have explored the binding of multiple monoclonal antibodies to the extracellular domain of HER-2/neu.(14) Eleven different reagents were assembled from Applied Biotechnology and the Cetus Corporation. The relative positions of different epitopes on p185 have been defined by blocking the binding of ^{125}I labelled antibodies with nonlabelled monoclonal reagents. One of the 11 epitopes can be clearly separated, whereas the other ten epitopes are linked and fall in a roughly linear array.

Epitopes have been shown to be functionally distinct. In assays of anchorage independent growth performed in soft agar, seven of the 11 antibodies inhibited SKBr3 growth from 40-80%(14) Four of the 11 antibodies were inactive in this assay when used individually. When used in combination, however, each of these antibodies proved active for inhibiting clonogenic tumor growth. Two of the antibodies (BD5 and ID5) also significantly inhibited anchorage dependent growth. In collaborative studies with Dr. Ruth Lupu, the putative ligand for p185 inhibited the binding of each of BD5 and ID5 to the extracellular domain of p185, consistent with the possibility that these antibodies interact with epitopes near the ligand binding site.

Several models might explain these observations. First, if the natural ligand stimulated tumor growth, the antibodies might inhibit growth by displacing the ligand from its binding site. On the other hand, Lupu and her collaborators have shown that binding of ligand to SKBr3 cells that overexpress p185 inhibits, rather than stimulates, tumor growth.(9) Thus, binding of antibody to the extracellular domain of p185 might mimic ligand and inhibit tumor growth by interacting with the ligand binding site. Alternatively, for those antibodies that do not recognize epitopes near the ligand binding site, allosteric changes might decrease affinity for ligand or alter the kinase activity of the intracellular domain. Finally, bivalent antibodies could enhance dimerization according to the model of Schlessinger for activation of the EGFR. In the case of multiple monoclonal antibodies, aggregation of p185 could be achieved on the cell surface. Possible internalization of p185 once antibody is bound must also be taken into consideration. Against the possibility that antibodies simply enhance dimerization is the observation that Fab fragments are approximately as active as the intact immunoglobulin in down-regulating anchorage independent tumor growth.

Events downstream from binding of antibody to p185 have also been studied. Using two-dimensional gel electrophoresis, phosphorylation of several intracellular substrates has been shown to increase after binding of the ID5 antibody, which inhibits tumor growth, but not after binding of the TA1 antibody, which fails to affect tumor growth. The ID5 antibody also produces a marked decrease in the intracellular levels of diacylglycerol (DAG) after two hours of incubation, whereas TA1 does not effect DAG levels. As DAG is generated in part by phospholipase C, differential phosphorylation of this enzyme might contribute to changes in DAG levels.

VI GROWTH INHIBITION OF TUMOR CELLS BY IMMUNOTOXINS REACTIVE WITH HER-2/NEU

Immunotoxins are being prepared with each of the murine monoclonal antibodies that react with the extracellular domain of p185. For immunoconjugates that contain ricin A chain to be active, the immunotoxin must be taken up by endosomes and the ricin A chain translocated to the cytoplasmic compartment. Since internalization of the EGF receptor appears to depend, at least in part, on the presence of the tyrosine kinase, we examined whether internalization of HER-2/neu gene product p185 also required tyrosine kinase activity.(15) Several constructs were prepared that lacked the tyrosine kinase region or that lacked all but six amino acids of the intracellular domain. When these constructs were transfected into rat-1 cells, similar amounts of expression of the extracellular domain were obtained as judged by immunofluorescence and by Scatchard plots. Products encoded by each of these aberrant constructs served as effectively as products encoded by the wild type for internalization of radiolabelled TA1 or for tumor cell killing by immunotoxin. A similar pattern of internalization was confirmed by immunoelectron microscopy. In subsequent studies, combinations of TA1-rRTA AND ID5-

rRTA have demonstrated additive antitumor effects. These studies suggested that the anti-p185 immunotoxins did not require kinase activity for internalization, translocation of tumor cell killing. Growth inhibition by unconjugated anti-p185 has not been observed in transfectants that lack kinase activity. Consequently it seems likely that down-regulation of tumor growth by immunotoxin and by unconjugated antibody proceed by fundamentally different mechanisms. Further studies will be required to judge the significance of interactions between different immunotoxins against EGFR and p185 in nude mouse heterografts.

References

1. Boyer, C.M., Borowitz, M.J., McCarty, K.S., Jr., Kinney, R.B., Everitt, L., Dawson, D.V., Ring, D., and Bast, R.C., Jr.. Heterogeneity of antigen expression in benign and malignant breast and ovarian epithelial cells. Int. J. Cancer 1989; 43:55-60.
2. Boyer, C.M, Lidor, Y., Lottich, S.C., and Bast, R.C., Jr. Antigenic cell surface markers in human solid tumors. Antibody Immunocunj. Radiopharm. 1988; 1:105-162.
3. Yu, Y.H., Crews, J.R., Cooper, K., Ramakrishnan, S., Houston, L.L., Leslie, D.S., George, S.L., Lidor, Y., Boyer, C.M., Ring, D.B., and Bast, R.C., Jr. Use of immunotoxins in combination to inhibit clonogenic growth of human breast carcinoma cells. Cancer Res. 1990; 50:3231-3238.
4. K., George, S.L., Boyer, C.M., Argon, Y., and Bast, R.C., Jr. A combination of two cell lines. Submitted for publication.
5. Gould, B.J., Borowitz, M.J., Groves, E.S., Carter, P.W., Anthony, D., Weiner, L.M., and Frankel, A.E. Phase I study of an anti-breast cancer immunotoxin by continuous infusion: Report of a targeted toxic effect not predicted by animal studies. J. Natl. Cancer Inst. 1989; 81:775-781.
6. Bookman, M.A., Godfrey, S., Padavic, K. Griffin, T. Corda, J.P., Hamilton, T., Ozols, R.F., and Groves, E.S. Antitransferrin receptor immunotoxin (IT) therapy: Phase-I intraperitoneal (IP) trial. Proc. Am. Soc. Clin. Oncol. 1990; 9:A722.
7. Bargmann, C.I., Hung, M.C., and Weinberg, R.A. Multiple independent activations of the neu ongogene by a point mutation altering the transmembrane domain of p185. Cell 1986; 45:649-657.
8. Slamon, D.J., and Clark, G.M. Amplification of c-erbB-2 and aggressive human breast tumors. Science 1988; 240:1795-1798.
9. Lupu, R., Colomer, R., Zugmaier, G., Sarup, J., Shepard, M., Slamon, D., and Lippman, M.E. Direct interaction of a ligand for the erbB2 oncogene product with the EGF receptor and p185 erbB2. Science 1990; 249:1552-1555.
10. Berchuck, A., Rodriguez, G.C., Kamel, A., Dodge, R.K., Soper, J.T., Clarke-Pearson, D.L., and Bast, R.C., Jr. Epidermal growth factor receptor expression

in normal ovarian epithelium and ovarian cancer. I. Correlation of receptor expression with prognostic factors in patients with ovarian cancer. Am. J. Obstet. Gynecol. 1991; 164:669-674.

11. Tyson, F.L., Boyer, C.M., Kaufman, R., O'Briant, K., Cram, G., Crews, J.R., Soper, J.T., Daly, L., Fowler, W.C., Jr., Haskill, J.S., and Bast, R.C., Jr. Expression and amplification of the HER-2/neu (c-erbB-2) protooncogene in epithelial ovarian tumors and cell lines. Am. J. Obstet. Gynecol. 1991; 165:640-646.

12. Berchuck, A., Kamel, A., Whitaker, R., Kerns, B., Olt, G., Kinney, R., Soper, J.T., Dodge, R., Clarke-Pearson, D.L., Marks, P., McKenzie, S., Yin, S., and Bast, R.C., Jr. Overexpression of HER-2/neu is associated with poor survival in advanced epithelial ovarian cancer. Cancer Res. 1990; 50:4087-4091.

13. Slamon, D.J., Godolphin, W., Jones, L.A., Holt, J.A., Wong, S.G., Keith, D.E., Levin, W.J., Stuart, S.G., Udove, J., Ullrich, A., and Press, M.F. Studies of HER-2/neu proto-oncogene in human breast and ovarian cancer. Science 1989; 244:707-712.

14. Xu, F.J., Rodriguez, G.C., Whitaker, R., Boente, M., Berchuck, A., McKenzie, S., Houston, L., Boyer, C.M., and Bast, R.C., Jr. Antibodies against immunochemically distinct epitopes on the extracellular domain of HER-2/neu (c-erbB-2) inhibit growth of breast and ovarian cancer cell lines. Proc. Am. Assoc. Cancer Res. 1991; 32:A1544.

15. Maier, L.A., Xu, F.J., Hester, S., Boyer, C.M., McKenzie, S., Bruskin, A.M., Argon, Y. and Bast, R.C., Jr. Requirements for the internalization of a murine monoclonal antibody directed against the HER-2/neu gene product c-erbB-2. Cancer Res. 1991; 51:5361-5369.

3

Tumor Rejection Antigens of the Chemically Induced BALB/C Meth A Sarcoma

Albert B. DeLeo
Pittsburgh Cancer Institute and
University of Pittsburgh School of Medicine
Pittsburgh, Pennsylvania

I INTRODUCTION

The concept that tumors express tumor specific or associated antigens and the potential role of these antigens in the immunotherapy of cancer are based primarily on studies of the immunogenicity of chemically induced sarcomas of inbred mice. The murine tumor specific or associated antigens, which are functional in transplantation rejection of tumors in syngeneic mice, are commonly referred to as tumor specific transplantation antigens (TSTA) or tumor rejection antigens (TRA). Since cross-immunity among chemically induced tumors rarely occurs, the restricted immunogenicity of individual tumors has been attributed to their expression of highly restricted or unique TSTA (1).

Numerous hypotheses have been proposed to account for the antigenic diversity of chemically induced sarcomas. In general, these hypotheses tend to involve qualitative and/or quantitative changes in the expression of cellular components on tumor cells (2). A favored hypothesis is that the unique TSTA are the products of a polymorphic gene family. No relationship, however, has been established between TSTA and the products of the well- characterized polymorphic gene families, such as MHC, Ig or Murine leukemia virus(MuLV)-related antigens.

The molecular identification of the TSTA expressed on chemically induced sarcomas has been hampered by the lack of appropriate reagents for detecting them in in vitro assay systems. As a result of this, identification of these antigens has depended almost completely on their functional activities in the in vivo tumor rejection assay. Nonetheless, three TRA were isolated from the chemically induced BALB/c Meth A sarcoma, using conventional biochemical techniques and were shown to have restricted tumor rejection-inducing activities. These proteins are p82, heat shock related-protein

84/86 (HSP84/86) and gp96 (3-5). The HSP84/86 and gp96 are stress-related proteins: the HSP84/86 are murine equivalents of hsp90 (4), and gp96 is a hsp100-related gene product (6). Although these proteins have been purified from other tumors as well as Meth A sarcoma, and shown to have restricted tumor rejectioninducing activities, no genetic evidence can directly explain their restricted immunogenicities. The nucleotide sequences of the genes encoding these proteins in tumors appear to be identical to those of the genes found in normal tissues. Since heat shock-related proteins have now been shown to be involved in antigen processing and presentation, a possible explanation for the observed restricted immunogenicities of HSP84/96 and gp96 TRA might be the non-covalent binding of antigenic peptides to these proteins (7,8).

As a consequence of an effort to develop in vitro-based immunological assay systems to identify gp96, various approaches based on the ability of gp96 to induce tumor specific cytotoxic T lymphocytes (CTL) or proliferation of anti-tumor T cells, such as described by Naito et al. (9),have been utilized. It soon became apparent that when the purification of gp96 was extended beyond that which had been previously described for its isolation, such highly purified gp96 no longer had any significant tumor rejection inducing activity. Instead, the tumor rejection-inducing activity of the cytosol appeared to be associated with a Mr 110,000 glycoprotein, designated gp110.

II RESULTS

A. Isolation of the Meth A gp110 TRA (10).

Meth A gp96 was initially isolated by sequential lectin affinity and ion exchange chromatography of the cytosol fraction of Meth A sarcoma, using Con A Sepharose and Q Sepharose FF and/or Mono Q FPLC. Our procedure involved the isolation of the Con A Sepharose-binding cytosol protein fraction and its step-wise elution from Q Sepharose FF. Tumor rejection-inducing activity was, in general, associated with proteins which eluted between 0.2 to 0.4M NaCl, the P2 fraction, and contained gp96. Sequential Mono Q FPLC of the P2 fraction separated, however, tumor rejection-inducing activity from gp96, as indicated in Table 1. SDS-PAGE analysis of Mono Q FPLC-derived fractions with tumor rejection-inducing activity, such as P2D1 and P2D2, indicated that this activity was associated with the presence of a Mr 110,000 glycoprotein, designated gp110 (Table 1 and Figure 1).

B. Characterization of Meth A gp110 (10).

Immunoblot analysis of gp110 indicated that it was antigenically unrelated to any of the previously identified Meth A TRA. A panel of rabbit antisera prepared against intact gp96

Table 1
Tumor rejection-inducing activities of Mono Q FPLC fractions of the cytosol of Meth A sarcoma[a]

Fraction[b]	Antigen Content[c] gp96	gp110	MTD±SEM[d]	Incidence
CONTROL			7.0±1.1	12/18
Mono Q FPLC of P2 Fraction				
P2D	+	+	5.1±1.0	3/6
Mono Q FPLC of P2D Fraction				
P2D1	-	+	<u>0</u>[e]	0/6
P2D2	-	+	<u>1.0±0.6</u>	1/6
P2D3	+	±	3.9±1.6	3/6
Mono Q FPLC of P2D3 Fraction				
P2D3B	+	-	4.1±1.6	3/6

a Groups of BALB/c mice were immunized twice at a seven day interval with 5 ug of Meth A cytosol fractions. Mice were challenged with 5 x 10^4 Meth A sarcoma. This experiment was done by Dr. Lloyd W. Law at the National Cancer Institute.

b The Con A Sepharose binding fraction of Meth A cytosol was separated on Q Sepharose FF using a three-step NaCl gradient, 0.2 M, 0.4 M and 0.6 M NaCl. The 0.4 M fraction, P2, was then subjected to sequential Mono Q FPLC.

c Presence of gp96 and gp110 in cytosol fractions was based on immunoblot analysis using rabbit anti-gp96 N-terminal peptide serum and rat anti-gp110 serum.

d MTD= mean tumor diameter, mm. SEM= standard error of mean. Tumor measurements were taken 21 days after tumor challenge.

e <u>Underlined measurements indicate $p < 0.01$.</u>

or HSP84, or N-terminal and internal peptide sequences of gp96 and p82, respectively, failed to react with Meth A gp110, while monoclonal and conventionally prepared anti-gp110 antibodies of rat origin did not recognize gp96.

Highly enriched fractions of Meth A gp110 in addition to having restricted tumor rejection-inducing activity, also induced a cellular immune response detectable in an <u>in vitro</u> cell mediated cytotoxic (CMC) assay against Meth A target cells. These results are consistent with the possibility that gp110 might well be a polymorphic TSTA. A drawback to this conclusion, however, was the possibility that the immunogenicity of gp110 might be due to non-covalent bound peptides or proteins, an explanation which has also been proposed to account for the immunogenicity of stress-related Meth A TRA. The recognition of gp110 molecules separated by SDS-PAGE under denaturing conditions by antibodies present in the sera of Meth A-bearing mice, as well as Meth A-immune mice, suggests that gp110 molecules are inherently immunogenic in syngeneic mice (A. DeLeo, unpublished results).

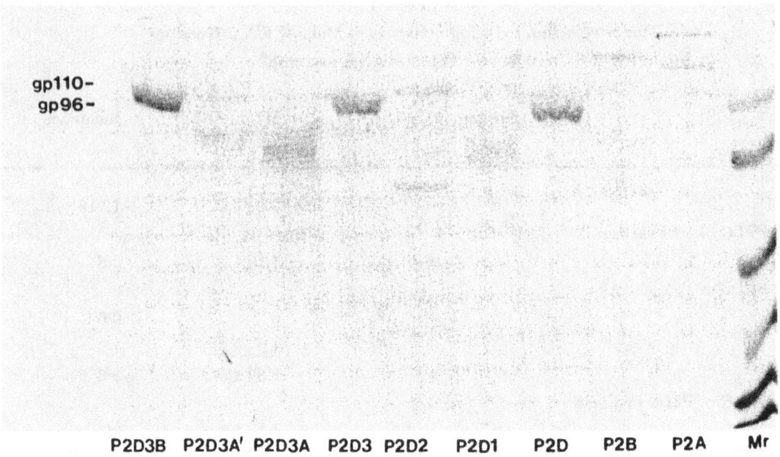

Mono Q FPLC of Meth A cytosol P2 Fraction

Fig. 1. Silver stain of SDS-PAGE analysis, under reducing conditions, of Mono Q FPLC fractions derived from the Meth A cytosol P2 fraction. Meth A gp96 is primarily detected in the P2D, P2D3 and P2D3B fractions, while gp110 is detected in the P2D1 and P2D2 fractions.

Figure reprinted with permission of the journal Cancer Research

recognition of gp110 molecules separated by SDS-PAGE under denaturing conditions by antibodies present in the sera of Meth A-bearing mice, as well as Meth A-immune mice, suggests that gp110 molecules are inherently immunogenic in syngeneic mice (A. DeLeo, unpublished results).

III CONCLUSION

A novel Mr 110,000 glycoprotein, designated gp110, has been isolated from the Meth A sarcoma and shown to have restricted tumor rejection-inducing activity, and induce effector cells that react with Meth A target cells in *in vitro* CMC assays. As part of our approach to developing an *in vitro*-based immunological assay system in which the inherent immunogenicity/ antigenicity of gp110 could be demonstrated, an anti-Meth A CTL cell line has been isolated from the peritoneal exudate cells of Meth A-immune mice using a modification of procedures initially described by Carbone et al (11). From these anti-Meth A effector cells, an anti-Meth A-specific CTL cell line, designated 9C, was cloned by limiting dilution. The anti-Meth A CTL cell line was found, in the presence of irradiated splenocytes acting as antigen- presenting cells, to proliferate in response to highly enriched fractions of Meth A gp110, but not by gp96 (DeLeo, et al. manuscript in preparation). This effect was tumor specific; enriched fractions of gp110 obtained from antigenically unrelated CMS3 and CMS4 sarcomas had no significant effect on the proliferation of anti-Meth A CTL cell line. Since the anti-Meth A CTL cell line was isolated from mice immunized with intact sarcoma cells, rather than a purified antigen, its specificity for gp110 provides further evidence that gp110 plays a role in defining the immunogenicity of Meth A sarcoma. Although Meth A gp110 has been the most extensively studied of the gp110 molecules isolated from chemically induced sarcomas, the analysis of the tumor rejection-inducing activity of other sarcomas also associates this activity with gp110-enriched fractions obtained from these tumors. The present level of characterization of gp110 supports our conclusion that this molecule represents a family of proteins expressed on a chemically induced sarcomas, in which variations in structure account for the antigenic diversity associated with these tumors.

Acknowledgement

The author gratefully acknowledges the generous advice and support of Dr. Ronald B. Herberman, Director, Pittsburgh Cancer Institute.

References

1. Old, L.J. Cancer immunology; The search for specificity. Cancer Research 1981; 41:361-375.
2. Srivastava PK, Old LJ. Individually distinct transplantation antigens of chemically induced mouse tumors. Immunol. Today 1988; 9:78-83.
3. Law, L.W. Tumour specific antigens. Cancer Surveys 1985; 4:3-17.
4. Ullrich, S.J., Robinson, E.A., Law, L.W., Willingham, M. and Appella, E.A. A mouse tumor-specific transplantation antigens is a heat shock-related protein. Proc. Natl. Acad. Sci. U.S.A. 1986; 83:3121-3125.
5. Srivastava, P.K., DeLeo, A.B. and Old, L.J. Tumor rejection antigens of chemically induced sarcomas of inbred mice.Proc. Natl. Acad. of Sci., U.S.A. 1986; 83:3407-3411.
6. Srivastava, P.K., Kozak, C.A. and Old, L.J. Chromosomal assignment of the gene encoding the mouse tumor rejection antigen gp96. Immunogenetics 1988; 28:205-208.

7. Ullrich, S.J., Moore, S.K., Law, L.W., Vieira, W.D., Sagaguchi, K. and Appella, E. The role of HSP90 in tumor specific immunity, In B. Maresca, and S. Lindquist, (eds.), Heat Shock, 1991; New York, Springer-Verlag, pp. 269-277.

8. Maki, R., Old, L.J. and Srivastava, P.K. Human homologue of murine tumor rejection antigen gp96: 5'regulatory and coding regions and relationship to stress-induced proteins. Proc. Natl. Acad. Sci. U.S.A. 1990; 87:5658-5662.

9. Naito, K., Pellis, N.R. and Kahan, B.H. Expansion of tumor-specific cytolytic T lymphocytes using in vitro stimulation with tumor specific transplantation antigen. Cell Immunol. 1987; 108:483-494.

10. DeLeo, A.B., Becker, M., Lu, L. and Law, L.W. Properties of a Mr 110,000 tumor rejection antigen of the chemically induced BALB/c Meth A sarcoma. Cancer Res. 1992; in press.

11. Carbone, G., Colombo, M.P., Sensi, M.L., Cernushi, A. and Parmiani, G. In vitro detection of cell mediated immunity to individual tumor specific antigens of chemically induced BALB/c fibrosarcomas. Int. J. Cancer 1983; 31:483-489.

PART II

Introduction

Antitumor Effector Cells: Characteristics and Basic Mechanisms

Rolf Kiessling
Karolinska Institute
Stockholm, Sweden

During the last two decades we have witnessed an amazing progress in the understanding of the effector cells active in immunity against experimental and human tumours. While in the early era the major focus of interest in tumour immunology was on mechanisms involved in specific, acquired immunity to tumours in experimental animals and cancer patients, it soon became apparent that a different type of "innate", non-adaptive immunity existed which was independent of mature T-cells or B-cells. The major cell type responsible in vitro for the killing of certain tumour cells without prior immunization and in vivo active in the rapid elimination of small numbers of tumour cells in non-immune experimental animals, was found to be the NK-cell (1, 2), one of the major topics during this meeting. Dr. Ronald B. Herberman was among the first to characterize NK cells as a defined lymphocyte subtype, distinct from previously described B-cells, T-cells or monocytes/macrophages. Early experimental work recognized NK-cells and T-cells as two distinct sets of lymphocytes, and while T-cells already before the advent of monoclonal antibodies were defined by the sheep red blood cell receptor and Thy-l antigen, NK cells were often characterised in negative terms as "null" cells. This difficulty of finding cell surface markers and receptors on NK cells, as compared to the explosive advancement in our understanding of the T-cell and its receptor, has resulted in that many immunologists tended to look upon NK cells as a curious "para-immunological" phenomenon rather than as a well-defined cell type. This has now at least partly changed, and we now have well-defined criteria for distinguishing NK cells from other cell types active in the lysis and elimination of tumour cells. It is clear, however, that the definition of various NK- and T-cell subsets must be based on several different markers, and that a considerable heterogeneity exists also among NK cells. Although NK cells were early defined as being $CD3^-$ (4), and expressing certain monoclonal markers such as CD56 (5), a substantial heterogeneity exists among $CD3^-$ $CD56^+$ NK cells. The morphological definition of NK cells as LGL (Large Granular Lymphocytes) was of crucial importance in the early definition of NK cells (6). Recently, however, this morphology-based distinction between NK cells and other anti-tumour effector cells has been found incomplete: not all large granular lymphocytes are NK cells– activated T-cells may have LGL morphology– and not all NK cells are LGLs- and in both mice and humans there are small-to medium sized agranular NK cells (7). LGL morphology may thus be a characteristic of an activated cytotoxic cell, not a general and unique property of NK cells. The difficulties in defining and distinguishing subsets of lymphocytes with anti-tumour activity becomes particularly obvious with IL-2 grown cell lines and clones from cancer patients. Thus, alpha/beta as well as gamma/delta T-cells often acquire "non-MHC restricted" cytotoxicity when cultured in IL-2, and such T-cells with "NK like" activity have often erroneously been classified as NK cells, leading to incorrect conclusions about the presence of T-cell receptors and markers on NK cells. A further issue of discussion relates to the classification of PBL from cancer patients grown in IL-2 as lymphokine activated killer cells (LAK). The therapeutical aspects of LAK cells will be dealt with later in this volume, but it should be stressed that LAK cells is an operational definition of the cytotoxic activity obtained after in vitro culture of PBL in IL-2, and <u>not</u> a definition of a lymphocyte subtype. Thus LAK cells are a heterogeneous mixture mainly composed of activated NK cells but also containing MHC-unrestricted T-cells (8), and the contribution of each lymphocyte subtype should be analysed for each particular LAK-tumor combination.

The rapid advancement in the understanding of the T-cell receptor and how cytotoxic T cells recognize MHC class I bound peptide on tumour targets has not yet been parallelled by a definition of the "NK-cell receptor". Already at the level of adhesion, as will be discussed by several contributors here, a multitude of different adhesion molecules and their ligands appear to be involved in the homing and infiltration of endothelial cells, extracellular matrix proteins and initial contact with the tumour cells. Thus, the NK specificity pattern may partly be determined by the selection of adhesion molecules used in a given situation. Apart from these adhesion molecules, none of which yet has been shown to be unique for NK cells, several candidates for more specific "NK cell receptors" exist, although it remains to be seen if any of these are unique for the NK cell lineage.

The possibility exists that the diversity and specificity among cellular effector mechanisms involved in the "non-adaptive" immunity against tumours does not necessarily depend on rearranging receptor genes and clonal selection, but that evolution may have fixed multiple receptors, enabling NK cells to perform "pattern recognition" of certain conserved structures of tumour cells (10). NK cells and cytolytic T cells share cytolytic mechanisms, and both types of lymphocytes express the signal-transducing zeta-chain polypeptide, in T-cells associated with the CD3 complex and in NK-cells associated with other moleculcs which may participate in a ligand recognition (9). It has therefore been argued that the "innate", non-specific NK part of the immune system is an evolutionary fore-runner to the specific T-cells (10). Cytotoxic T-cells are strictly dependent on MHC class I for the elimination of their target cells while NK cells on the contrary are triggered by the absence of these molecules (11). Thus "innate" NK mediated immunity complements the specific T cell system in the surveillance of tumours and other aberrant cells.

References

1. Herberman, R.B., Nunn, M.E., Lavrin, D.H. Natural cytotoxic reactivity of mouse lymphoid cells against syngeneic and allogeneic tumors. I. Distribution of reactivity and specificity. Int J Cancer 1975; 16:216.
2. Kiessling, R., Klein, E. and Wigzell, H. "Natural" killer cells in the mouse. I. Cytotoxic cells with specificity for mouse Moloney leukemia cells. Specificity and distribution according to genotype. Eur Immunol 1975; 5:112.
3. Trinchieri, G. Biology of Natural Killer cells. Advances in Immunology 1989; 47: 187.
4. Zarling, J.M., Kung, P. C. Monoclonal antibodies which distinguish between human NK cells and cytotoxic T lymphocytes. Nature 1980; 288:394.
5. Griffin, J. D., Hercend. T., Beveridge. R. and Schlossman S. F. Characterization of an antigen expressed on human Natural Killer cells. J. Immunol. 1983; 130:2947.
6. Saksela, E., Timonen, T., Ranki, A. and Hayry, P. Morphological and functional characterization of isolated effector cells responsible for human Natural Killer activity to fetal fibroblasts and to cultured cells line targets. Immunol Rev 1979; 44:71.
7. Karre, K., Hansson, M., Kiessling, R. Multiple interactions in Natural Killer Workshop. Immunology Today. 1991; 12:343.
8. Herberman, R.B. et al. Lymphokine-activated killer cell activity. Immunology Today. 1987; 8:6,178.
9. Lanier, L.L., Yu, G., Phillips, J.H. Co-association of CD3~ with a receptor (CD16) for EgG Fc on human Natural Killer cells. Nature 1989; 342:803.
10. Janeway, C. A. A primitive immune system. Nature 1989; 341:108.
11. Karre, K., Ljunggren, H.G., Piontekk, G., Kiessling, R. Selective rejection of H-2 deficient lymphoma variants suggests alternative immune defence strategy. Nature 1986; 319: 675.

4

Phenotypic and Functional Characteristics of Human A-NK Cells

Nikola L. Vujanovic
Theresa L. Whiteside
Pittsburgh Cancer Institute
and University of Pittsburgh School of Medicine
Pittsburgh, Pennsylvania

I INTRODUCTION

In 1986, an observation was made in our laboratories that a small proportion (1-5%) of rat spleen cells purified by the passage through nylon wool to remove adherent cells were able to adhere to the surface of plastic plates during 24 to 48h of incubation in the presence of 22 nM (1000 Cetus U/ml) of interleukin-2 (IL2) (1). The IL2-induced plastic-adherent spleen cells were shown to have the morphology of large granular lymphocytes (LGL) and phenotypic and functional characteristics of NK cells (1). Culture of these NK cells selected by adherence to plastic in the presence of IL2 resulted in their rapid expansion (100 folds in 3 days). The initially adherent population of NK cells grew as a single-cell suspension and had high levels of antitumor cytotoxicity against a large variety of NK-cell sensitive and resistant tumor cell targets but not against normal cells (1). These IL2-activated spleen NK cells have been named adherent lymphokine activated killer (A-LAK) cells based on their abilities to adhere to plastic after activation with IL2 and to mediate lysis of NK-cell resistent tumor cell targets <u>in vitro</u> (2). Plastic adherence of NK cells following IL2 activation appears to be a generalized phenomenon, which has been reproduced by Gungi and Gorelik in mice (3) and by Melder and colleagues, using human peripheral blood lymphocytes (2). These studies showed that exposure of human peripheral blood mononuclear cells or of spleen cells in rodents to high concentrations of IL2 induced adherence to plastic in a small proportion of cells, which had morphologic and functional characteristics of NK cells.

The next phase of our work focused on demonstrating that A-LAK cells in combination with IL2 were effective in elimination of established metastases in experimental animals (4). In a rat model of 3-day established lung or liver metastases, experimental animals (4). In a rat model of 3-day established lung or liver metastases, we compared therapeutic effectiveness of IL2 alone, standard LAK cells plus IL2 and A-LAK cells plus IL2. Both the reduction in the number of established metastases and survival were significantly better in animals treated with A-LAK cells and IL2 than those treated with IL2 alone or with LAK cells plus IL2 (4). In addition, in a xenograft model of human squamous cell carcinoma growing in immunosuppressed nude mice, A-LAK cells transferred peritumorally were found to be more effective in causing regression of established 7-day tumors than LAK cells or IL2 alone (5). These in vivo experiments showed that rodent or human A-LAK cells had a considerable therapeutic potential, and a phase I clinical trial was initiated in patients with advanced metastatic melanoma or renal cell carcinoma to determine toxicity and feasibility of adoptive immunotherapy with IL2 activated A-LAK cells and IL2. Among 15 patients treated, 14 were evaluable, and one complete, one partial, and two minimal responses were reported (6). At the same time, the phase I trial demonstrated the feasibility of generating large numbers of human A-LAK cells for therapy and indicated that these effector cells may be potentially useful for eliminating metastases.

The in vivo experiments in animal models and the phase I clinical trial in patients with advanced cancer described above focused attention on A-LAK cells and resulted in a more extensive evaluation of their functional characteristics. In both animals and humans, we observed that a proportion of A-LAK cells which adhered to plastic following IL2 activation never exceeded 1-10% of the NK cells present in the starting lymphoid population. These data suggested that IL2-induced adherence of NK cells is not a property of all NK cells but rather of a small NK cell subset. We have formulated a hypothesis that A-LAK cells represent a small distinct subset of NK cells. To test this hypothesis, human peripheral blood NK cells obtained by negative selection, using antibody-coated magnetic beads, were separated into plastic adherent and non-adherent cells in the presence of IL2. An in vitro model was developed in which both subsets of NK cells were examined immediately following adhesion to plastic as well as after in vitro expansion in the presence of IL2. In this chapter, we describe phenotypic and functional characteristics of a subset of human NK cells, which we designated as IL2-activated, adherent NK cells or "A-NK" cells. Because we used highly enriched $CD3^-CD56^+$ human NK cells in these experiments, A-LAK designation was not appropriate based on earlier studies indicating that LAK cells more represent mixtures of IL2 activated T and NK cells (7,8). The designation, "A-NK cells", indicates that these NK cells were both selected by adherence and activated by IL2.

II EXPERIMENTAL DESIGN

Purified human NK cells were obtained from peripheral blood of normal volunteers. First, PBLs recovered from nylon wool columns were incubated with anti-CD3 monoclonal antibody. T lymphocytes ($CD3^+$), which bound anti-CD3 mAb, were next removed using magnetic beads coated with goat anti-mouse IgG. The remaining cells were >90% $CD3^-CD56^+$ or $CD3^-CD16^+$, as determined by two-color flow cytometry. Purified NK cells were then exposed to IL2 for a brief period of time (3-5h) allowing for their adherence to plastic in culture flasks. A-NK cells were separated from non-adherent NK (NA-NK) cells (Figures 1b and 2a,c), and both cell subsets were analyzed phenotypically or functionally either immediately after plastic adherence or after 14 days of culture in the presence of 6,000 IU/ml of IL2 and allogeneic feeder cells (9,10,11; see Figures 1f and 2b,d). We have determined earlier that addition of allogeneic irradiated feeder cells (either EBV-transformed B cell lines or Concanavalin A-induced T cell blasts) were needed for optimal proliferation of A-NK cells (12). A-NK cells cultured in the presence of feeder cells tend to form aggregates, which appear to be necessary for their optimal proliferation in culture.

III ADHERENCE OF NK CELLS TO PLASTIC SURFACES

Initial experiments indicated that adherence of human NK cells to plastic was a very rapid phenomenon, which started within 5 min of exposure to IL2, peaking between 60 min and 5h and then declining significantly between 5 and 24h. At the peak of adherence, about 16% (range 4-30%) of human NK cells were adherent (Figs. 1a,b and 2a,c)9, indicating that only a proportion of NK cells was able to respond to IL2 by early adherence to plastic. In addition, these experiments confirmed that the optimal concentration of IL2 for adherence ranged from 2 to 22 nM. This concentration of IL2 is sufficient to saturate the intermediate-affinity IL2R (13,14). Furthermore, we observed that IL2-induced adherence of NK cells could not be blocked by anti-p55 IL2R monoclonal antibody but was efficiently and nearly completely blocked by anti p75 IL2R mAb, again indicating that the intermediate-affinity, but not high-affinity IL2R is involved in IL2-induced adherence of NK cells to plastic.

On the basis of these studies, we concluded that a subset of human NK cells is capable of a rapid response to IL2 which results in adherence to plastic. Using this property of early adherence, A-NK cells can be selected and separated from other, non-adherent, NK cells.

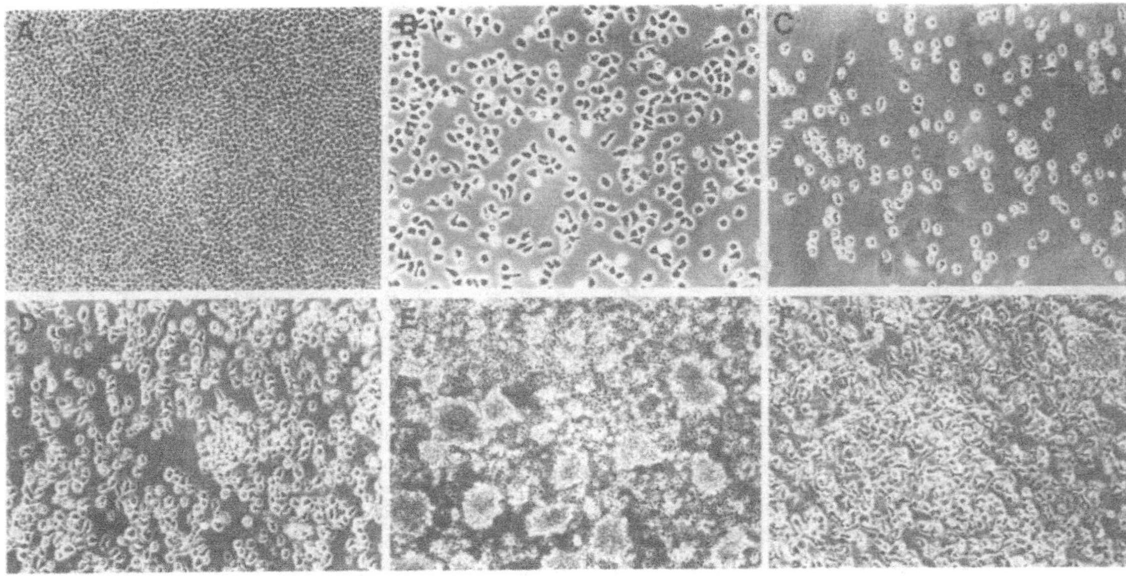

Fig. 1 Photomicrographs of human peripheral blood A-NK cells in culture. Resting NK (R-NK) cells were obtained by negative immunoselection of normal human PBMNC using OKT3 mAb and magnetic beads coated with goat anti-mouse Ig. R-NK cells ($1x10^6$/ml) were seeded in plastic 12-well tissue culture plates and incubated for 3h in the presence of 22 nM IL2. A. NK cells at the end of 3h incubation. After this incubation, A-NK and NA-NK cells were separated by washing NA-NK cells out with warm (37°C) medium; B. A-NK cells after 3h of IL2 induction; C. A-NK cells incubated in the presence of IL2 detached from plastic surface and by 24h on culture grew in suspension; D. A-NK cells in culture after addition of irradiated Con A-activated allogeneic lymphocytes as feeder cells (Day 1 of culture); E. A-NK cells and feeders form aggregates (Day 5 of culture); F. By day 14 of culture, A-NK cells grew as single-cell suspensions (x 200).

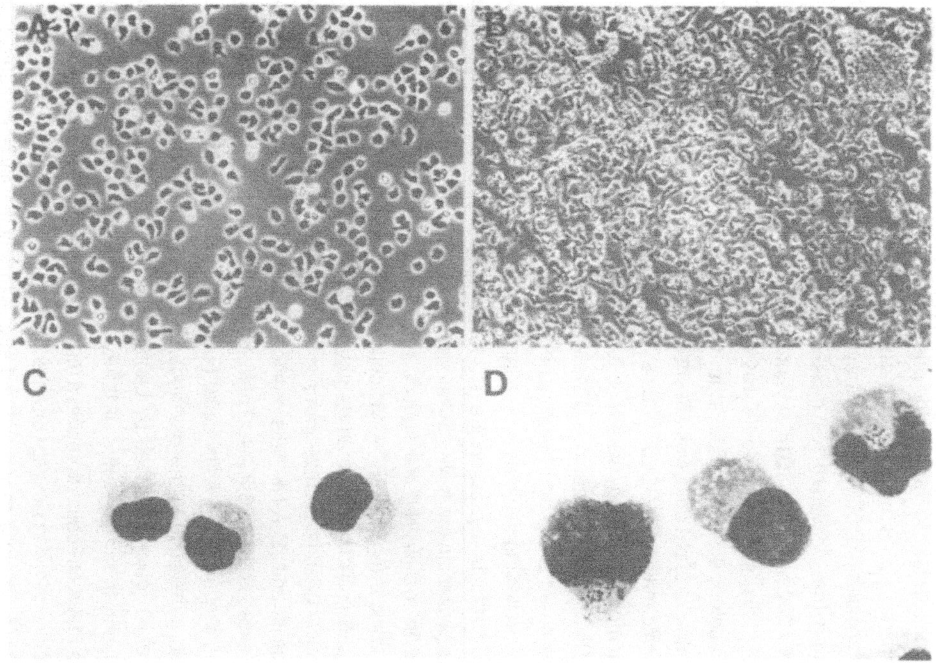

Fig. 2 Morphology of human A-NK cells: A. and C. Fresh A-NK cells selected by adherence to plastic after 3h IL2 induction; B. and D. A-NK cells after 14 days expansion in the presence of IL2 and feeder cells, as described in Figure 1.A. and B.; A-NK cells in culture (x 200); C. and D. Giemsa stain of cytospin preparations of A-NK cells shows that they are LGL both at the time of selection by adherence and at the end of 14-day culture (x 1000).

IV PHENOTYPE OF EARLY IL2-INDUCED A-NK CELLS

When freshly separated A-NK cells were studied by two-color flow cytometry for phenotypic characteristics, it appeared that they differed from NA-NK cells in expression of different surface molecules. In Table 1, the phenotypic profiles of A-NK cells and NA-NK cells are compared. It is clear from the data in Table 1 that A-NK cells are homogenous in expression of CD56 antigen: they are all CD56^{dim} cells. In contrast, NA-NK cells and resting (not activated) NK cells (R-NK) contain both CD56^{bright} and CD56^{dim} subsets of NK cells. The CD16 antigen is either expressed dimly or not at all expressed on A-NK cells, in contrast to NA-NK cells, which are mainly CD16^{bright} or CD16^{-}. By two-color flow cytometry, it was possible to demonstrate that CD56^{bright}CD16^{-} subset of NK cells was absent in the A-NK cell subset. Instead, A-NK cells contained significantly more CD56^{dim}CD16^{-} cells than either NA-NK or R-NK cells. One of the most important findings was that A-NK cells uniformly expressed the ß2 integrins CD18, CD11a and, particularly, CD11b at high levels of surface density. In contrast, NA-NK cells had a dim and heterogeneous expression of the ß2 integrins.

Immediately, following separation by adherence to plastic, a higher proportion of A-NK cells than NA-NK cells expressed the p75 IL2R. Similarly, a higher proportion of p55 IL2R^{+} cells was present in the A-NK cell subset than NA-NK or R-NK subpopulations. Furthermore, while in the NA-NK cell subset, the p55 IL2R was expressed only on CD56^{bright}CD16^{-} cells, the p55 IL2R^{+} A-NK cells were CD56^{dim}CD16^{+}. We also have evidence that the p55 IL2R is expressed de novo on A-NK cells following IL2-induced adherence to plastic (1) anti-p55 IL2R mAb was ineffective in blocking IL2-induced adherence of NK cells; and (2) A-NK cells preincubated with cycloheximide adhered to plastic but did not express the p55 IL2R. These data indicated that the α chain of IL2R is not expressed on precursors of A-NK cells, but is rapidly synthesized de novo at the time of early plastic adherence in the presence of IL2. Also, CD69 (Leu23), an early activation marker of NK cells described by Lanier et al (15), is rapidly expressed after only 1-5h of incubation in the presence of IL2 on A-NK cells but not on NA-NK cells. Again, suggestions that A-NK cells represent a distinct subset of NK cells poised for early and rapidly lactivation following delivery of an activation signal.

In summary, A-NK cells recovered and studied immediately after adherence to plastic are phenotypically distinct from NA-NK. These A-NK cells are CD56^{dim}, CD57^{+}, ß2 integrin^{bright}, CD69^{+}, CD16$^{dim\ or\ -}$ and largely IL2Rα^{+} and IL2Rß$^{+}$.

Table 1
Phenotypic properties of human A-NK and NA-NK cells

Marker Expression	A-NK	NA-NK
CD3	-	-
Granules	+	+ or -
N-CAMs CD56	Very dim	Dim or bright
8A2*	+ or -	+ or -
CD57	+	+ or -
Beta-1 Integrins (CD29, VLA-4, 5 and 6)	+	+
Beta-2 Integrins (CD18, CD11a and CD11b)	Bright	Dim
ICAM-1	Bright	Dim
CD16	Dim or -	Bright or -
IL2R p55*		- or +
p75	++	+
CD69*	+	-

* Markers which differentiate between A-NK and pre-A-NK cells. Pre-A-NK cells are positive for 8A2 marker but negative for the IL2Rp55 or CD69.

V FUNCTIONAL CHARACTERISTICS OF A-NK CELLS SEPARATED BY EARLY ADHERENCE TO PLASTIC

Early adherence to plastic of A-NK cells in the presence of IL2 was partially blocked by antibodies against various cell adhesion molecules except those against N-CAMs. This observation indicated that adhesion molecules participate in IL2-induced adherence of these cells to plastic. Using a single-cell cytotoxicity assay, we showed that early adherent A-NK cells were enriched in effector cells which bind and kill K562 targets. A-NK cells were significantly more active than NA-NK or R-NK cells in single-cell cytotoxicity assays. At the population level, using ^{51}Cr-release assays, A-NK cells also had higher cytotoxicity against K562 targets than NA-NK or R-NK. In addition, A-NK cells selected by adherence following incubation with cycloheximide, which blocks IL2-induced protein synthesis in these cells, still showed higher levels of NK activity than R-NK or NA-NK cells. These experiments suggest that A-NK cells constitutively express high levels of antitumor cytotoxicity and that they are a subset of mature cytotoxic cells. As expected from phenotypic studies, which showed low density of the CD16 antigens expression on A-NK cells, ADCC of freshly isolated A-NK cells was lower than that of NA-NK cells. Similarly, redirected killing in the presence of B73.1 (anti-CD16 mAb) was also lower in A-NK cells than NA-NK cells. Neither A-NK cells nor NA-NK cells showed any cytotoxic activity against Daudi targets at 3-5h following IL2 induction. After further 24h incubation in the presence of 22 nM of IL2, A-NK cells did not develop LAK activity, defined as the ability to lyse Daudi, while NA-NK cells did so. However, after 96h of incubation, A-NK cells developed the ability to lyse Daudi targets, although it was lower than that of NA-NK cells. Thus, A-NK cells appear to develop LAK activity at a slower rate than NA-NK cells in the presence of IL2. To determine the ability of A-NK cells to proliferate in the presence of IL2, autoradiography was performed following ^{3}[H]thymidine incorporation. A significantly higher proportion of A-NK than NA-NK cells proliferated after 48 to 96 hour culture in the presence of IL2 (9). This observation was confirmed by standard ^{3}[H]thymidine incorporation assays as well as by cell counts performed at different times during culture.

Proliferation in response to IL2 of all three NK cell preparations (A-NK, NA-NK and R-NK cells) incubated in the presence of 0.2 or 2.2 nM of IL2 were completely inhibited by preincubation of the cells with either anti-p55 or anti-p75 IL2R mAbs. This observation indicated that all NK cell subsets in the early phase of proliferation in response to IL2 used the high-affinity IL2R. Our data also suggested that the A-NK cell subset is able to proliferate better under these conditions than NA-NK or R-NK cells and, thus, these early adherent cells appear to be able to very rapidly upregulate expression of the high-affinity IL2R and more efficiently respond to activation by IL2. Summary of functional characteristics of A-NK cells following their separation by adherence in the presence of IL2 is shown in Table 2.

Table 2
Functional characteristics of human A-NK and NA-NK cells

Assays	A-NK	NA-NK
Adherence*	+ (rapid)	-
NK activity*	High	Low
ADCC*	Low	High
LAK development	Slow	Rapid
Proliferation	High	Low

* Similar functions in both A-NK and pre-A-NK cells.

VI EXPRESSION OF MRNA FOR THE IL2R AND CYTOKINES AND CYTOKINE PRODUCTION BY A-NK CELLS

As both phenotypic and functional studies of human NK cells exposed to IL2 indicated that A-NK cells were preferentially activated in the presence of high concentrations of IL2, we performed in situ hybridization (ISH) to look for mRNA for cytokine transcripts and IL2R gene expression in NK cells. Highly purified human NK cells were incubated in the presence of 22 nM of IL2. At different time points between 1 and 24h of activation, A-NK and NA-NK cells were selected by adherence and immediately assayed for mRNA expression of genes for IL2R or various cytokine transcripts using ^{35}S-labeled cDNA probes followed by autoradiography. Cells positive for the gene transcripts (i.e., containing >25 grains of radioactivity) were enumerated on cytocentrifuge smears. These experiments indicated that IL2 induced rapid expression of mRNA for IL2Rα or ß and for cytokines in both A-NK and NA-NK cells. However, at the peak of this mRNA expression at 3-5h post IL2 induction, mRNAs for IL2Rα and ß, IL2, TGFß, IL1, and IL6 were found in almost all A-NK cells, but in a significantly smaller proportion of NA-NK cells. On the other hand, mRNA transcripts for TNFα and IFNγ were expressed in a higher but similar proportion of both A-NK and NA-NK cells. These observations were confirmed by studies of cytokine or IL2R production and release into cell supernatants. As Table 3 indicates, cytokine production was uniformly higher in A-NK than NA-NK cells incubated under the same experimental conditions.

Table 3
Transcription of the IL2R and cytokine genes and production of cytokines by A-NK cells

IL2Rs/Cytokines	A-NK	NA-NK
	mRNA/protein expression	
p55	High/High	Low/Low
p75	High/High	Low/Low
IL2	High/High	Low/Low
TGFß	High/nt	Low/nt
IL1ß	High/High	Low/Low
IL6	High/nt	Low/nt
TNFα	High/High	High/High
IFNγ	High/Low	High/High

nt = not tested.

VII PHENOTYPIC AND FUNCTIONAL CHARACTERISTICS OF A-NK CELL IN 14-DAY CULTURES

To determine if the differences described above between A-NK cells and NA-NK cells were stable in culture, these two cell subsets were studied for phenotypic and functional characteristics after 14 days of growth in the presence of IL2 and irradiated allogenic feeder cells.

We observed that A-NK cells, which grow not as adherent monolayers but as a single-cell suspensions, proliferated 30-fold better than NA-NK cells in these cultures (Table 4). A-NK cells selected after 3 to 5h of IL2-induced adherence had the best proliferative activity as compared to, e.g., A-NK cells selected by 24h adherence. The optimal concentration of IL2 for proliferation of A-NK cells was 22 nM. In contrast, NA-NK cells proliferated optimally at lower IL2 concentrations, i.e., at 0.2 nM. Using ^{3}[H]thymidine incorporation assays, it was possible to determine that A-NK cells retained the ability to proliferate better than NA-NK cells even after 14 days of culture. However, at this time point in culture, both cell subsets depended on the presence of high IL2 concentrations (2 to 22 nM) for optimal growth.

Table 4
Phenotypic and functional characteristics of A-NK and NA-NK cells cultured for 14 days in the presence of IL2

Parameters	A-NK	NA-NK
CD56 (MFI)	Low	High
p55 expression (%)	High	Low
P55 (MFL)	High	Low
p75 (MFI)	High	Low
sp55 in supernatant (U/ml)	High	Low
Proliferation	High	Low
Cytotoxicity of K562 targets	High	Low
Cytotoxicity against Daudi targets	High	Variable

MFI = membrane fluorescence intensity

Both A-NK and NA-NK cells developed very high levels of cytotoxicity against tumor cell targets after 14 days of culture in the presence of IL2 (Table 4). However, while A-NK cells clearly needed 22 nM of IL2 for optimal development of LAK activity (measured against Daudi targets), NA-NK cells developed high levels of LAK activity at a lower IL2 concentration, of 0.2. After 14 days of culture, A-NK cells selected during 5h of IL2 induction, showed a significantly higher killing of K562 targets than NA-NK cells.

Phenotypically, both cell subsets substantially augmented expression of the CD56 antigen on the cell surface during 14-day cultures, and this expression was dependent on IL2 concentration. However, A-NK cells consistently showed relatively lower expression of CD56 antigen than NA-NK cells. Although membrane fluorescence intensity (MFI) of CD56 antigen on both A-NK and NA-NK cells after 14 days of culture was dependent on IL2 concentrations used, the difference in expression of this marker (i.e., dim on A-NK cells and bright on NA-NK cells) was consistently detectable at different concentrations of IL2. It is not clear at this time whether expression of this antigen is related to functional attributes of cultured NK cells. However, it may be that its differential expression on A-NK and NA-NK cells reflects functional differences between these two subsets of IL2-activated NK cells.

A higher proportion of A-NK than NA-NK cells express IL2Rα (43%$\pm$10 vs. 18%$\pm$6; n=6) in the log phase of growth (Table 4). In addition, A-NK cells had higher MFI of the IL2Rα (254$\pm$50 vs. 131$\pm$34; n=6) and of the IL2Rβ (209$\pm$27 vs. 98$\pm$5; n=6) than NA-NK cells in cultures in the log phase of growth. At this time, A-NK cells also produced and released 3 times higher levels of sIL2Rα (659$\pm$160 vs. 189$\pm$38; n = 3) than NA-NK cells (Table 4). Thus, A-NK cells appear to more actively modulate the α chain of the IL2R off their surface than NA-NK cells. We also found that IL2-dependent proliferation of A-NK or NA-NK cells in the log phase cultures was inhibited by antibodies to IL2Rβ but not to IL2Rα. This observation indicated that the intermediate-affinity receptor, rather than the high-affinity IL2R, was involved in proliferation of both NK cell subsets in culture. Other experiments, in which we compared the presence and MFI of the α and β chains of the IL2R on the cell surface of cultured NK cells in the presence of IL2 and, then, briefly incubated in the absence of IL2, showed that both NK-cell subsets expressed IL2R on fewer cells or had lower levels of expression of these receptors in cultures which have reached the stationary phase. At the same time, A-NK cells, which more rapidly shed IL2Rα but had higher density of the IL2Rβ, also appeared to express the IL2Rβ with somewhat lower affinity than that on NA-NK cells.

Our results show that the six main characteristics of A-NK cells which distinguished them from NA-NK cells before culture, that is, lower density of the CD56 epitope, a higher proportion of cells expressing IL2Rα, higher production of sIL2Rα, a requirement for higher concentrations of IL2 for optimal response, greater proliferation and higher level of cytotoxicity against K562 targets, were all maintained following 14 days of culture in the presence of 22 nM of IL2. Our conclusion is that the A-NK cell subset, originally identified by the ability to rapidly respond to high concentrations of IL2 by adherence to plastic surface, represent a distinct subset of mature NK cells which can be distinguished from other NK cells by several stable phenotypic and functional characteristics (Table 5).

Table 5
Summary of differences between A-NK and NA-NK cells

In comparison to NA-NK cells, A-NK cells have:	
1.	Distinct phenotype:$CD3^-CD56^{dim\ or\ -}$ β $integrins^{bright}CD16^{dim\ or\ -}p75^+$.
2.	Rapid response to high IL2 concentrations resulting in adherence to solid surfaces.
3.	Higher NK and lower ADCC activities.
4.	Slower development of LAK activity during culture in the presence of IL2.
5.	Greater fold proliferation in the presence of IL2.
6.	Rapid and high expression of IL2R mRNA and cytokine transcripts as well as higher levels of production of the corresponding proteins.
7.	In long-term cultures, lower expression of CD56, higher proportion of $p55^+$ cells, higher density of membrane p55 and p75 IL2R, higher production of soluble p55, better proliferation and higher cytotoxicity against K562 targets.

Acknowledgement

The authors wish to thank Professor Ronald B. Herberman for his encouragement and guidance during this work.

References

1. Vujanovic, N.L., Herberman, R.B., Maghazachi, A.A. and Hiserodt, J.C. Lymphokine-activated killer cells in rats. III. A simple method for the purification of large granular lymphocytes and their rapid expansion and conversion into lymphokine-activated killer cells. J. Exp. Med. 1988; 167: 15-29.
2. Melder, R.J., Whiteside, T.L., Vujanovic, N.L., Hiserodt, J.C. and Herberman, R.B. A new approach to generating antitumor effectors for adoptive immunotherapy using human adherent lymphokine-activated killer cells. Cancer Res. 1988; 48: 3461-3469.
3. Gunji, Y., Vujanovic, N.L., Hiserodt, J.C., Herberman, R.H. and Gorelik, E. Generation and characterization of purified adherent lymphokine-activated killer cells in mice. J. Immunol. 1989; 142: 1748-1754.
4. Schwarz, R.E., Vujanovic, N.L. and Hiserodt, J.C. Enhanced antimetastatic activity of lymphokine-activated killer cells purified and expanded by their adherence to plastic. Cancer Res. 1989; 49: 1441-1446.
5. Sacchi, M., Vitolo, D., Sedelmayer, P., Rabinowich, H., Johnson, J.T., Herberman, R.B. and Whiteside, T.L. Induction of tumor regression in experimental model of human head and neck cancer by human A-LAK cells and IL-2. Int. J. Cancer 1991, 47: 784-791.
6. Whiteside, T.L. and Herberman, R.B. Characteristics of natural killer cells and lymphokine-activated killer cells. Their role in the biology and treatment of human cancer. In Immunology and Allergy Clinics of North America. H.F. Oettgen, editor. W.B. Saunders Co., Philadelphia, PA. 1990; 10: 663-704.
7. Rosenberg, S.A. Adoptive immunotherapy of cancer using lymphokine-activated killer cells and recombinant interleukin-2. In: V.T. DeVita, S. Hellman, and S.A. Rosenberg (eds.), Important Advances in Oncology, New York: J.B. Lippincott. 1986; pp. 55-81.
8. Tiele, D.L. and Lipsky, P.E. The role of cell surface recognition structures in the initiation of MHC-unrestricted "promiscous" killing by T cells. Immunol. Today. 1989; 10: 375-381.
9. Vujanovic, N.L., Rabinowich, H., Lee, Y.J., Jost, L., Herberman, R.B. and Whiteside, T.L. Rapid interleukin 2-induced adherence of human natural killer

(NK) cells. I. Phenotypic and functional characteristics of adherent NK (A-NK) cells. J. Immunol. 1992; (Submitted).

10. Vitolo, D., Vujanovic, N.L., Rabinowich, H., Schlesinger, M., Herberman, R.B. and Whiteside, T.L. Rapid interleukin 2-induced adherence of human natural killer (NK) cells. II. Expression of mRNA for cytokines and IL2 receptors in adherent NK (A-NK) cells. J. Immunol. 1992; (Submitted).
11. Vujanovic, N.L., Herberman, R.B. and Whiteside, T.L. Rapid interleukin 2-induced adherence of human natural killer (NK) cells. III. Distinct phenotypic and functional properties of adherent NK (A-NK) cells in culture. J. Immunol. 1992; (Submitted).
12. Rabinowich, H., Sedelmayer, P., Herberman, R.B. and Whiteside, T.L. Increased proliferation, lytic activity, and purity of human natural killer cells cocultured with mitogen-activated feeder cells. Cellular Immunol. 1991; 135: 454-470.
13. Nagler, A., Lanier, L.L. and Phillips, J.H. Constitutive expression of high affinity interleukin 2 receptors on human CD16 natural killer cells in vivo. J. Exp. Med. 1990; 171: 1527-1533.
14. Caligiuri, M.A., Zmuidzinas, A., Manley, T.J., Levine, H., Smith, K.A. and Ritz, J. Functional consequences of interleukin 2 receptor expression on resting human lymphocytes. J. Exp. Med. 1990; 171: 1509-1526.
15. Lanier, L.L., Buck, D.W., Rhodes, L., Ding, A., Evans, E., Barney, C. and Phillips, J.H. Interleukin 2 activation of natural killer cells rapidly induces the expression and phosphorylation of the Leu-23 activation antigen. J. Exp. Med. 1988; 167: 1572-1585.

5

Regulation and Function of Fibronectin Receptors Expressed by Natural Killer Cells

Angela Santoni
Angela Gismondi
Fabrizio Mainiero[1]
Gabriella Palmieri
Stefania Morrone
Michele Milella
Mario Piccoli
Luigi Frati
University of Rome "La Sapienza"
Rome, Italy

I INTRODUCTION

Lymphocytes express a variety of cell surface receptors which mediate their adhesion to other cells and to extracellular matrix (ECM) and are critical for immunological recognition and lymphocyte recirculation and homing (1,2). The majority of these adhesion molecules belong to the three different supergene families of immunoglobulins, integrins and selectins and interact with an ECM ligand. Recently, an increasing number of lymphocyte ECM receptors has been identified in the integrin family (3). Their expression and function is highly regulated, thus allowing lymphocytes to rapidly pass from a nonadherent state to an adherent one. It has been shown that activation of T lymphocytes through the T cell receptor complex (TcR) (4-10) or via CD2 (5,6) surface antigen rapidly enhances the ability of LFA-1, VLA-4, VLA-5 and VLA-6 to bind their

---

[1] Recipient of an AIRC fellowship

specific ligands. The enhanced adhesiveness can be transient depending on the activation signal and on the integrin involved, and it is never associated with changes in the surface receptor expression.

The integrin receptors capable of binding fibronectin (FN) belong to the β1, β3 and β5 families. In the β1 (VLA) integrin subfamily there are at least four receptors: $\alpha v\beta 1$, $\alpha 3\beta 1$, $\alpha 4\beta 1$, $\alpha 5\beta 1$. The first antigen identified as FN receptor (FNr) is $\alpha 5\beta 1$ (11). It is a 150/110 heterodimer recognizing the RGD sequence in the central region of FN, mainly expressed among lymphoid cells by peripheral blood T lymphocytes and NK cells. Peripheral blood B cells and thymocytes bear none or only negligible levels of $\alpha 5\beta 1$. $\alpha 4\beta 1$ (150/110 Kd) is the receptor for the carboxy-terminal cell adhesion region containing the heparin II domain and the III connecting segment (IIICS) of FN (12,13). $\alpha 4\beta 1$ ligands include CS1 and CS5 regions in the IIICS of FN (with LDV being the sequence recognized in CS1), and VCAM-1, a vascular adhesion molecule induced on endothelial cells by inflammatory cytokines such as TNF-α and IL-1 (14). $\alpha 4\beta 1$ is expressed by peripheral blood T, B and NK cells, and by thymocytes. Interestingly, the apparent levels of α4 on B cells exceed those detected for β1. $\alpha 3\beta 1$ interacts with multiple ligands including FN, laminin, collagen and epiligrin (15). It binds FN in a RGD-dependent manner and is poorly expressed on lymphoid cells. $\alpha v\beta 1$ receptor too, recognizes the RGD sequence in the FN molecule (16,17) and has not been described on lymphocytes so far. No clear evidence is available on the expression of FN receptors belonging to β3 and β5 subfamilies on cells of lymphoid lineage. An integrin with high homology to the human $\alpha v\beta 3$ receptor has been shown to be expressed on mouse dendritic epidermal γ/δ T cell clones and LAK splenocytes (18,19). It displays a RGD-dependent binding to FN, but also interacts with vitronectin and fibrinogen and, like other β1 members ($\alpha 1\beta 1$, $\alpha 2\beta 1$), it is not expressed on resting lymphocytes, but can be induced after long term in vitro activation.

Natural killer (NK) cells are a heterogenous population of large granular lymphocytes (LGL) with the ability to mediate MHC-unrestricted and antibody-dependent cytotoxicity and to produce a large variety of cytokines (20). They are easily identified by the expression of surface antigens such as CD16 (FcRγIII) and CD56 (NCAM), with CD16 representing one of the most relevant structures on NK cells capable of triggering their functional program. NK cells mainly recirculate in the peripheral blood and localize in several nonlymphoid tissues under physiological and inflammatory conditions.

We have previously shown that fresh human peripheral blood NK cells express $\alpha 4\beta 1$ (VLA-4) and $\alpha 5\beta 1$ (VLA-5) integrin receptors which mediate their adhesion to FN (21). In this study we investigated the modulation of expression and function of these receptors following short term NK cell activation and the possible role of FN in the NK cell interaction with resting and activated endothelial cells.

II RESULTS AND DISCUSSION

A. NK cell activation by CD16 crosslinking or phorbol ester results in rapid enhancement of NK cell adhesion to FN, without affecting VLA-3, VLA-4, VLA-5 surface expression

It has previously been shown in several cell systems that integrin functions can be rapidly enhanced by cell activation (4,22). It was therefore of interest to investigate whether stimuli able to trigger several NK cell functions could affect their adhesion to FN. Treatment of ^{51}Cr-labeled NK cells for 10 min at 37°C with TPA (20 ng/ml) or with a mAb directed against the CD16 antigen (B73.1) resulted in a marked enhancement of adhesion to FN (Figure 1). Increased adhesion, like the constitutive one (21), was mediated by VLA-4 and VLA-5 as demonstrated by the ability of anti-β1, or anti-α4 plus anti-α5 mAb to completely inhibit binding (data not shown). In an attempt to understand the mechanisms responsible for the activation-dependent adhesion to FN, we evaluated whether expression of β1 FN receptors were affected upon NK cell activation. As shown by double immunofluorescence and cytofluorimetric analysis, treatment with TPA (10 ng/ml for 10 min at 37°C) did not induce any detectable changes in the levels of cell surface expression of β1, α3, α4 and α5 subunits (Figure 2). Similarly, β1 expression did not change after stimulation of NK cells with anti-CD16 mAb (data not shown).

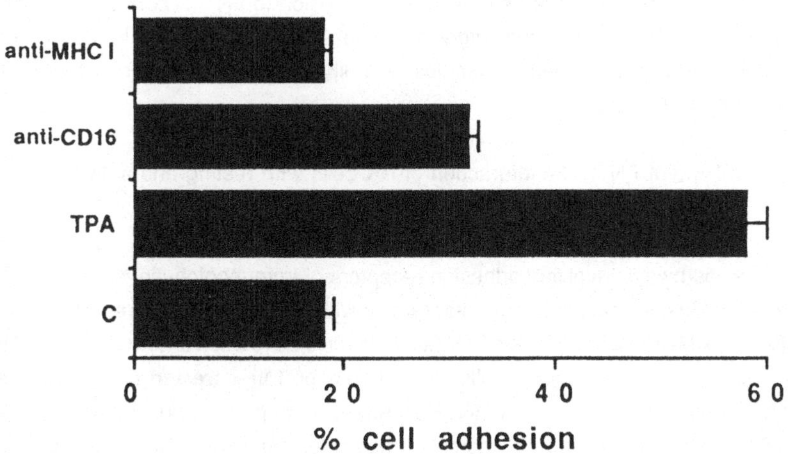

Fig. 1 NK cell activation by TPA or anti-CD16 mAb results in increased adhesion to FN. ^{51}Cr-labelled highly purified fresh human NK cells were treated with TPA (20 ng/ml) or anti-CD16 mAb for 10 min at 37°C and adhesion on FN-coated plates was measured after 2 h at 37°C. Anti-MHC class I mAb was used as control. Data are expressed as mean +/- standard deviation of quintuplicate determinations.

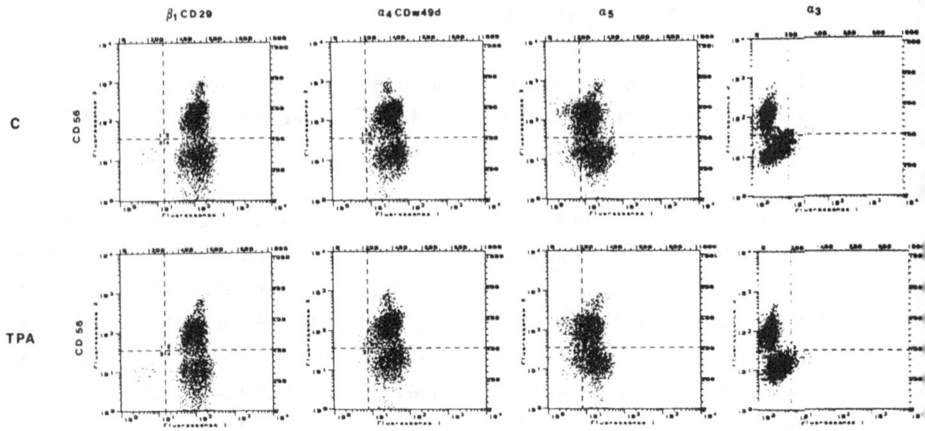

Fig. 2 TPA does not affect expression of β1 FNr on NK cells. Highly purified fresh human NK cells were treated or not (C) with TPA (20 ng/ml) for 10 min at 37°C and then doubly stained with mAb directed against β1, α4, α5 and α3 subunits plus a FITC-conjugated second-step reagent (fluorescence 1), and PE-conjugated anti-CD56 mAb (fluorescence 2). Samples were analyzed on a FACScan cytofluorimeter. Fluorescence intensity was measured on a logarithmic scale.

The rapid enhancement of NK cell adhesion to FN, associated with no changes in the levels of FNr expression, suggest that qualitative alterations of integrins themselves and/or integrin-associated molecules are likely responsible for activation-induced enhanced adhesiveness.

B. Role of FN in the interaction of NK cells with resting and activated endothelial cells

Several pathways involving adhesion receptors of immunoglobulin, selectin and integrin gene families control lymphocyte interaction with resting and activated endothelium (23). LFA-1/ICAM-1 as well as VLA-4/VCAM-1 pathways have been shown to mediate adhesion of resting or IL-2- stimulating NK cells to IL-1- or TNF-α-treated human umbilical vein endothelial cells (HUVEC) (24,25). Recently, CS1 peptide, representing the recognition site for LFA-4 within the FN molecule, has been found to inhibit the adhesion of rat lymphocytes to lymph node high endothelial venules (26). We have therefore investigated whether VLA-4/FN interaction, besides VLA-4/VCAM-1 adhesion pathway, might be implicated in the binding of NK cells to endothelial cells, particularly in the resting conditions when both ICAM-1 and VCAM-1 are barely expressed (23). CS1 peptide inhibited the adhesion of fresh human NK cells to untreated HUVEC, whereas it did not affect that to TNF-α (10 ng/ml, 18 hr at 37°C)-treated HUVEC. In these conditions a scrambled CS1 peptide had no inhibitory effect. CS1-mediated inhibition of NK cell adhesion to resting HUVEC was additive to that observed performing the adhesion assay in the presence of an anti-CD18 mAb (Figure 3).

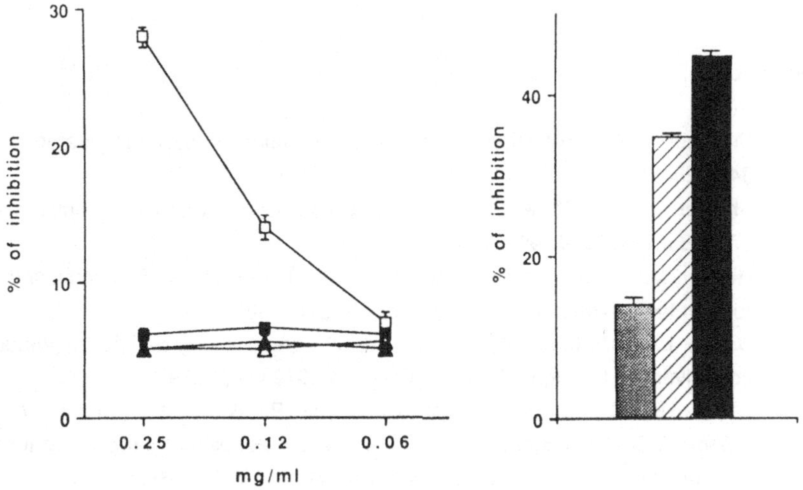

Fig. 3 VLA-4/FN adhesion pathway plays a role in NK cell adhesion to resting but not activated endothelial cells. Left panel: ^{51}Cr-labelled highly purified fresh human NK cells were assayed for adhesion to resting (open symbols) or TNFα (10 ng/ml, 18 h)-treated (closed symbols) human unbilical vein endothelial cells in the presence of different doses of CS1 (squares) or scrambled CS1 (triangles) peptides; Right panel: ^{51}Cr-labelled highly purified fresh human NK cells were assayed for adhesion to resting human umbilical vein endothelial cells in the presence of CS1 peptide (0.12 mg/ml) (▩), anti-CD18 mAb (▨), or CS1 peptide plus anti-CD18 mAb (■). Data are expressed as mean +/- standard deviation of quintuplicate determinations

Taken together, these results suggest that VLA-4/FN adhesion pathway plays a role in the interaction of NK cells with resting endothelium.

III CONCLUSIONS

The ability of NK cells to interact with an ECM component can be promptly regulated by their activation. This could result in a rapid change in their migration and tissue localization capacity during inflammatory or immune responses. In the extravasation process, interaction of NK cells with endothelium may imply the alternative usage of different receptor-counter receptor pairs consisting of the single integrin (VLA-4) capable of binding multiple ligands which can be differently expressed on endothelial cells, depending on their activation status.

References

1. Springer, T.A. Adhesion receptors of the immune system. Nature 1990, 346:425-434.
2. Shimizu, Y. and Shaw, S. Lymphocyte interactions with extracellular matrix. FASEB J. 5:2292-2299, 1991.
3. Hemler, M.E. VLA proteins in the integrin family: structures, functions, and their role on leukocytes. Ann. Rev. Immunol. 8:365-400, 1990.
4. Dustin, L.M., Springer, T.A. T cell receptor cross-linking transiently stimulates adhesiveness through LFA-1. Nature. 341:619-624, 1989.
5. vanKooyk, Y.P, van de Wiel-van Kemenade, P., Weder, P., Kuijpers, W. and Figdor, C.G. Enhancement of LFA-1-mediated adhesion by triggering through CD2 or CD3 on T lymphocytes. Nature. 342:811-813, 1989.
6. Shimizu, Y., van Seventer, G.A., Horgan, K.J. and Shaw, S. Regulated expression and binding of three VLA (β1) integrin receptors on T cells. Nature. 345:250-253, 1990.
7. Matsuyama, T., Yamada, A., Kay, J., Yamada, K.M., Akiyama, S.K., Schlossman, S.F. and Morimoto, C. Activation of CD4 cells by fibronectin and anti-CD3 antibody. A synergistic effect mediated by the VLA-5 fibronectin receptor complex. J. Exp. Med. 170:1133-1148, 1989.
8. Chan, B.M.C., Wong, J.G.P., Rao, A. and Hemler, M.E. T cell receptor-dependent; antigen-specific stimulation of a murine T cell clone induces a transient, VLA protein-mediated binding to extracellular matrix. J. Immunol. 147:398-404, 1991.
9. Damle, N.K. and Aruffo, A. Vascular cell adhesion molecule 1 induces T-cell antigen receptor-dependent activation of CD4+ T lymphocytes. Proc. Natl. Acad. Sci. 88:6403-6407.
10. Shimizu, Y., van Seventer, G.A., Horgan, K.J. and Shaw, S. Roles of adhesion molecules in T-cell recognition: fundamental similarities between four integrins

on resting human T cells (LFA-1, VLA-4, VLA-5, VLA-6) in expression, binding and costimulation. Immunol. Rev. 114:109-143, 1990.

11. Takada, Y, Huang, C. and Hemler, M.E. Fibronectin receptor structures in the VLA family of heterodimers. Nature. 326:607-609, 1987.

12. Wayner, E.A., Garcia-Pardo, A., Humphries, M.J., McDonald, J.A. and Carter, W.J. Identification and characterization of the lymphocyte adhesion receptor for an alternative cell attachment domain (CS-1) in plasma fibronectin. J. Cell Biol. 109:1321-1330, 1989.

13. Guan, J. and Hynes, R.O. Lymphoid cells recognize an alternatively spliced segment of fibronectin via the integrin receptor alpha4/beta1. Cell. 60:53-61, 1990.

14. Elices, M.J., Osborn, L., Takada, Y., Crouse, C., Luhowskyj, S., Hemler, M.J. and Lobb, R. VCAM-1 on activated endothelium interacts with the leukocyte integrin VLA-4 at a site distinct from the VLA-4/fibronectin binding site. Cell. 60:577-584, 1990.

15. Elices, M.J., Urry, L.A. and Hemler, M.E. Receptor functions for the integrin VLA-3: fibronectin, collagen, and laminin binding are differentially influenced by ARC-GLY-ASP peptide and by divalent cations. J. Cell Biol. 112:169-181, 1991.

16. Vogel, B.E., Tarone, G., Giancotti, F.G., Gailit, J. and Ruoslahti, E. A novel fibronectin receptor with an unexpected subunit composition ($\alpha v\beta 1$). J. Biol. Chem. 265:5934-5937, 1990.

17. Dedhar, S. and Gray, V. Isolation of a novel integrin receptor mediating ARC-GLY-ASP-directed cell adhesion to fibronectin and type I collagen from human neuroblastoma cells. Association of a novel $\beta 1$-related subunit with αv. J. Cell Biol. 110:2185-2193, 1990.

18. Moulder, K., Roberts, K., Shevach, E.M. and Coligan, J.E. The mouse vitronectin receptor is a T cell activation antigen. J. Exp. Med. 173:343-347, 1991.

19. Takahashi, K., Nakamura, T., Koyanagi, M., Kato, K., Hashimoto, Y., Yagita, H. and Okumura, K. A murine very late activation antigen-like extracellular matrix receptor involved in CD2- and lymphocyte function-associated antigen-1-independent killer-target cell interaction. J. Immunol. 145:4371-4379, 1990.

20. Trinchieri, G. Biology of natural killer cells. Adv. Immunol. 47:187-376, 1989.

21. Gismondi, A., Morrone, S., Humphries, M.J., Piccoli, M., Frati, L. and Santoni, A. Human natural killer cells express VLA-4 and VLA-5, which mediate their adhesion to fibronectin. J. Immunol. 146:384-392, 1991.

22. DeMinno, G., Thiagarajan, P., Perussia, B., Martinez, J., Shapiro, S., Trinchieri, G. and Murphy, S. Exposure of platelet fibrinogen-binding sites by collagen, arachidonic acid, and ADP: inhibition by a monoclonal antibody to the glycoprotein IIb-IIIa complex. Blood. 61:240-246, 1983.

23. Pober, J.S. and Cotran, R.S. Immunologic interactions of T lymphocytes with vascular endothelium. Adv. Immunol. 50:261-302, 1991.

24. Bender, J.R. Pardi, R., Karasek, M.A. and Engleman, E.G. Phenotypic and functional characterization of lymphocytes that bind human microvascular endothelial cells in vitro: evidence for preferential binding of natural killer cells. J. Clin. Invest. 79:1679-1688, 1987.
25. Allavena, P., Paganin, C., Martin-Padura, I., Peri, G., Gabili, M., Dejana, E., Marschisio, P.C. and Mantovani. A. Molecules and structures involved in the adhesion of natural killer cells to vascular endothelium. J. Exp. Med. 173:530-540, 1991.
26. Ager, A. and Humphries, M.J. Use of synthetic peptides to probe lymphocyte-high endothelial cell interactions. Lymphocytes recognize a ligand on the endothelial surface which contains the CS1 adhesion motif. Int. Immunol. 2:921-928, 1990.

6

Growth Requirements, Binding and Migration of Human Natural Killer Cells

Tuomo Timonen
Juha Jääskeläinen
Anna Mäenpää
Tuula Helander
Anatoly Malygin
Panu Kovanen
University of Helsinki
Helsinki, Finland

I INTRODUCTION

Very little is known about the growth and differentiation factors that induce the maturation and expansion of NK cells. NK cells express the p75 chain of IL-2 receptor (1), and, as T cells they upregulate the p55 chain of the IL-2 receptor upon activation (2-4). Thus IL-2 may be the major cyotkine in the expansion of NK cell pool. Yet it is not pivotal for NK cell maturation. NK cells are abundant in athymic nude rodents (5), and interferons induce indirectly NK cell proliferation in vivo by an IL-2-independent mechanism (6). NK cell proliferation and differentiation probably involve yet unidentified factors.

The cytolytic cascade in natural killer cell activity can be divided into four phases: binding of the effector cell to the target, triggering of the cytolytic machinery, secretion or liberation of cytolytic factors, and finally the killer cell independent phase of lysis. Some of the surface receptors involved in the binding and triggering phases have been identified, and it seems that natural killer cells recognize their targets via multiple parallel pathways instead of a single hypothetical NK cell receptor (1).

Endogenous circulating NK cells are relatively weakly cytotoxic and rarely kill autologous uncultured tumor cells. The cytotoxicity is strongly augmented by IL-2 (7,8), and also autologous malignant target cells are sensitive to IL-2-activated killer cells. The adhesion structures involved in the binding of endogenous and IL-2-activated NK cells are largely overlapping (9). However, IL-2-activated NK cells are more adherent, probably due to both quantitative and conformational changes of adhesion molecules induced by IL-2. Very little is known about the triggering receptors of IL-2-activated NK cells. It seems likely that they are at least partially different from those of endogenous NK cells.

A major emphasis on NK cell research is still in in vitro work, and only limited information on the in vivo distribution, infiltration, cytotoxicity and cytokine production of NK cells is available. IL-2-activated lymphocytes (LAK-cells) and purified IL-2-activated NK cells are known to selectively adhere to tumor vasculature in vivo (10,11). It is yet unclear whether any extravasation of NK cells in malignant tissue takes place, although it has been shown that NK cells infiltrate renal allografts (12).

In the following we describe some of our work concerning the growth requirements, adhesion molecule profile, and infiltration characteristics of human NK cells. We show that, in addition to IL-2 the optimal growth of NK cells requires cofactor cytokines. Some of the endogenously expressed adhesion molecules of NK cells are strongly upregulated by IL-2, some are downregulated, and a long-term stimulation by IL-2 also induces the expression of novel adhesion structures on NK cells. Finally, we describe our spheroid model for the studies of infiltration characteristics of IL-2-activated lymphocytes.

II GROWTH REQUIREMENTS

For the monitoring of NK cell growth, we use both thymidine incorporation and the morphology antibody chromosomes (MAC) technique, in which the phenotype of proliferating cells in metaphase is studied immunocytochemically in colchicine-arrested cultures (13). The MAC technique enables the follow-up of the proliferation of selected subpopulations of lymphocytes in culture. By this technique we have shown that in heterogenous peripheral blood lymphocyte cultures 10-15% of $CD16^+$ NK cells proliferate in response to IL-2 (13). However, NK cells stringently purified by FACS sorting respond to IL-2 poorly. The addition of purified $CD4^+$ T cells, either autologous or allogeneic, restores the proliferative capacity (Figure 1). $CD4^+$ T cells can be replaced by IL-4 or partially purified low molecular weight B cell growth factor (LMW-BCGF) (14), suggesting that helper T cell-derived cofactors are involved in the proliferation of NK cells (Fig. 1). LMW-BCGF has been a particularly efficient co-stimulating growth factor in our hands. However, the data have to be interpreted with caution because so far, the LMW-BCGF we have used is only partially purified.

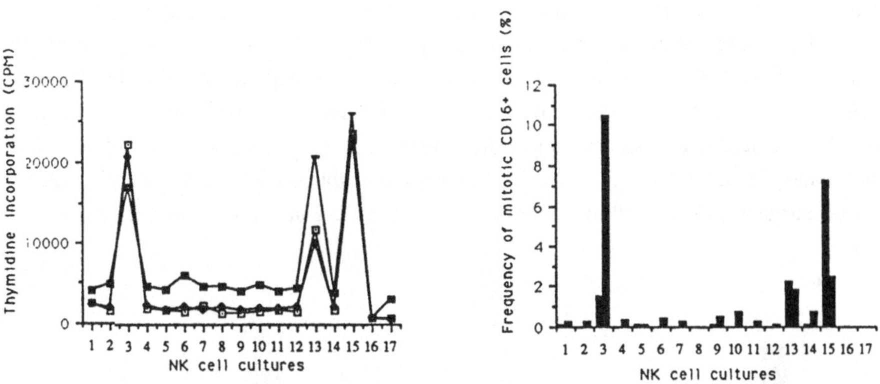

Fig. 1 Thymidine incorporation (cpm) and the frequency of mitotic CD16$^+$ cells in cultures of CD16$^+$ lymphocytes stimulated by rIL-2 (500 U/ml) for five days together with the following cytokines: 1. rIL-1 (100 U/ml); 2. rIL-3 (500 U/ml); 3. rIL-4 (500 U/ml); 4. rIL-6 (500 U/ml); 5. rIL-7 (50 ng/ml); 6. rIL-8 (50 ng/ml); 7. rTNFα (1000 U/ml); 8. rTNFß (1000 U/ml); 9. rIFNα (1000 U/ml); 10. rIFNγ (1000 U/ml); 11. rG-CSF (100 U/ml); 12. rGM-CSF (500 U/ml); 13. LMW-BCGF (5% of the total volume); 14. rIL-2 (500 U/ml); 15. Viable rIL-2-activated CD4$^+$ cells added to cultures at 10:1 ratio to CD16$^+$ cells; 16. Freeze-thawed unstimulated CD4$^+$ cells added to cultures as in 15; 17. Unstimulated PBL added to cultures as in 15

III ADHESION MOLECULES

We have recently analyzed the expression of adhesion molecules on NK cells with the aid of 19 different monoclonal antibodies. NK cells express adhesion structures abundantly, and the expression profile is profoundly affected by IL-2 (Table 1). Probably the changes in the expression of adhesion molecules during culture are due to both selective outgrowth of subpopulations within NK and T subsets, and the up- or down-regulation of surface molecule expression. The most apparent changes are the increases in the expression of $\alpha 1$, $\alpha 2$, $\alpha 3$ and ß3 integrin chains and CD2 as well as CD54. The alpha-chain associated with the ß3 has not yet been identified. We have previously shown that CD2, CD11a-c/CD18 and CD54 mediate the majority of NK cell binding to K562. Furthermore, our recent data indicate that mouse NK-resistant BW5147 become sensitive to IL-2-activated NK cells when hybridized with chromosome 6 (15). The cytotoxicity is completely abolished by monoclonal antibodies against CD18 (Fig. 2), further stressing the important role of ß2-integrins in IL-2 activated lymphocyte killing.

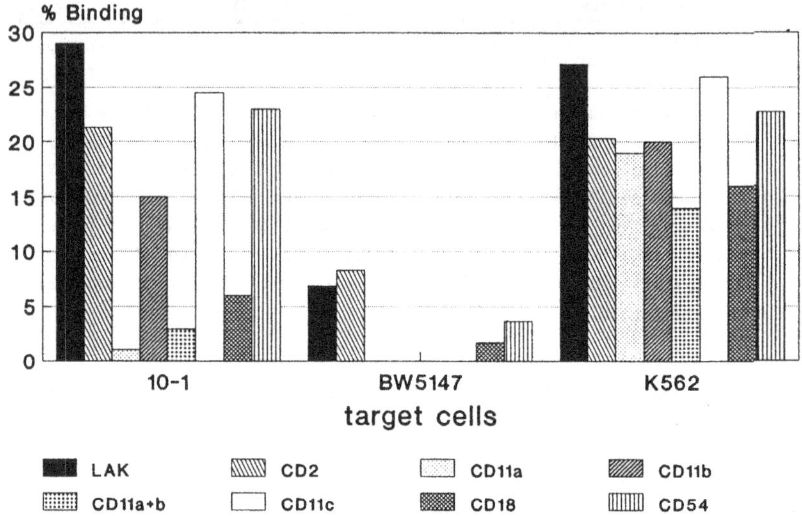

Fig. 2 Effect of mAb against adhesion molecules on the binding of rIL-2-activated peripheral blood lymphocytes to mouse BW 5147 thymoma cells, BW 5147 thymoma cell hybrids carrying human chromosome 6 (10-1), and K562

IV INFILTRATION OF IL-2-ACTIVATED LYMPHOCYTES INTO GLIOMA CELL SPHEROIDS

We have studied the infiltration of IL-2-activated PBMC into spheroids grown from a human H-2 glioblastoma cell line (16,17). Immunohistochemical staining of spheroids and flow cytometry of enzymatically dispersed spheroids show that all subtypes of lymphocytes are infiltrative (Table 2), but $CD56^+$ cells express the strongest infiltrative capacity and $CD4^+$ cells the weakest. The infiltration of all subtypes of lymphocytes was almost completely inhibited by anti-CD18 mAb (IB-4), and to some degree with synthetic peptides containing the RGD and CS-1 sequences of fibronectin (Table 3).

Immunohistological staining of the spheroids showed that the migrating lymphocytes clearly upregulated the expression of ICAM-1 on H-2 glioma cells in their vicinity, suggesting a positive feed-back loop in the regulation of lymphocyte adhesion to cells initially weakly expressing ICAM-1 (17). Altogether, the data indicate that both fibronectin receptors and ß2 integrins are involved in the migration of LAK cells into glioma spheroids.

Table 1

Impact of four-week IL-2 activation on the expression of adhesion molecules by T and NK cells

Adhesion molecule	Fresh T cells % pos	rIL-2 4wk T cells	Fresh NK cells % pos	rIL-2 4wk NK cells
VLA-1α	5	↑↑↑	0	↑↑
VLA-2α	0	↑↑↑	1	↑↑
VLA-3α	23	↑	24	↑
VLA-4α	67	↑↑	100	↓
VLA-5α	62	↑	93	↓
VLA-6α	63	↓	52	↓
CD29	96	↑	100	Ø
CD11a	100	↑↑	100	↑
CD11b	34	Ø	100	↓↓↓
CD11c	0	↑↑	67	↓↓
CD18	100	↑↑	100	Ø
CD41	0	Ø	4	Ø
CD51	0	Ø	0	Ø
CD61	0	↑↑	23	↑↑↑
CD2	100	↑↑	58	↑↑↑
CD44	100	↑	96	↑
CD54	25	↑↑↑	52	↑↑
CD58	42	↑	50	↑
L-selectin	74	↓↓↓	51	↓↓↓

↑↑↑ strong increase, ↑↑ moderate increase, ↑ small increase, Ø no effect, ↓ small decrease, ↓↓ moderate decrease, ↓↓↓ strong decrease.

Table 2

Phenotypic differences between infiltrative and non-infiltrative LAK cells in the H-2 glioma spheroid model[1]

Phenotype CD	Infiltrative LAK cells % at 72h	Non-infiltrative LAK cells % at 72h	Control LAK cell % at 0h
3+	96	95	94
4+	22	57	55
8+	60	46	48
16+	18	4	4
56+	37	11	14

1) A representative experiment using LAK cells from a single do PBMC were activated for three days with 500 U/ml of rIL-2. H- glioma spheroids were incubated with the activated lymphocyte for 72 hours, washed and dispersed enzymatically. The phenoty of infiltrative LAK cells harvested from dispersed spheroids, non infiltrative LAK cells remaining outside the spheroids, and contr LAK cells were analyzed by flow cytometry.

Table 3

Density of lymphocyte infiltration in H-2 glioma spheroids incubated for 72 hours with LAK cells and adhesion blocking mAb or peptides

Adhesion molecule subunit	Blocking reagent	Median number of lymphocytes/ 10^4 μm^2 in the depth of 0-100 μm	100-200 μm	200-300 μm
α4	HP2/1	9.5	3.1	2.2
	GPEILDVPST	7.4	3.1	0.6
α5	BIIG2	14.5	3.4	1.8
	GRGDS	9.5	3.7*	2.2**
	GRGES	14.8	3.7	4.9
α4 + α5	GPEILDVPST+GRGDS	6.0*	1.2*	0.6*
	GPSVDPTLIE+GRGES	16.3	7.2	2.2
β1	4B4	15.8	4.9	3.5
	AIIB2	12.3	8.0	3.4
β2	IB4	2.8***	0.3***	0***
CD54	LB-2	3.7***	0.9*	0.3
CD44	Hermes 3	8.0	2.5	1.7
CD56	Leu-19	11.1	4.3	2.5
VCAM-1	E1/6	14.5	10.5	4.5
Control spheroids		12.3	5.5	3.1

The density of lymphocytes in each three depths is compared to the density in control spheroids incubated with LAK cells from corresponding donors. p values in the Mann-Whitney U-test: <0.01*, <0.001**, <0.001***.

V CONCLUSIONS

Although it has been well established that NK cells proliferate in response to IL-2, it is evident that the generation of NK cell clones is rather difficult, and requires the use of feeder cells in addition to IL-2. NK cell proliferation is stimulated by lymphoblastoid cell lines, and the presence of helper T cells is required for this phenomenon (18). It is not clear whether the contact of NK cells and T cells is needed for the helping effect. Our data suggest that the contact is not needed and cofactor lymphokines together with IL-2 are sufficient to induce NK cell proliferation. NK cell stimulatory factor may be one of the costimulants (19), and our data suggest that LMW-BCGF and IL-4 may also be effective.

It is possible that LMW-BCGF does not affect NK cell proliferation directly, but instead first activates B cells, which then stimulate NK cells as the lymphoblastoid cell lines do. However, our experiments with NK cell preparations very stringently depleted of B cells suggest that the participation of B cells is unlikely. Furthermore, LMW-BCGF activates B-cells only after an initial stimulation by antigen or anti-IgM antibodies (20), not present in our growth medium. Clearly, the major concern in our LMW-BCGF data is the purity of the cytokine preparation, and the lack of recombinant BCGF as well as neutralizing antibodies against it.

The costimulation of NK-cell proliferation by IL-4 is somewhat surprising, because the inhibition of NK-cell proliferation by IL-4 has been reported (21). As our purification method of NK cells is negative selection (depletion of non-NK cells by immunomagnetic beads or by sheep erythrocyte rosetting), it is possible that IL-4 affects NK cell precursors, which may not be recovered in the populations purified by positive selection using anti-CD16-antibodies. This is supported by our preliminary data showing that IL-4 does not act as a costimulatory cytokine for NK cells purified by positive selection in flow cytometry. Another possibility is that some $CD3^-$ $CD16^-$ co-operative cells are required for the IL-4 effect on NK cell proliferation.

NK cells express many adhesion molecules (9) which makes them the most adhesive lymphocyte subpopulation in peripheral blood. The panel of the mAb we used in the present study covers most of the known adhesion receptors. Our results show that the already abundant expression of adhesion structures is on the average upregulated by IL-2. An interesting novel finding is that CD61, the ß3 chain of integrins, is strongly upregulated by IL-2. We have not yet been able to identify the alpha chain associated to ß3, although immunoprecipitations with anti-ß3 antibodies clearly reveal a band that fits to the molecular weight range of the integrin alpha chains. We are currently using alpha chain oligonucleotide probes in order to identify the molecule by Northern blotting.

The changes in the adhesion molecule profile of NK cells after the incubation with IL-2 probably profoundly affect their distribution characteristics _in vivo_. This may be important in the immunotherapy of cancer by cytokines. The major problem in these treatments is the side effects that at least partially may be caused by undesired homing of activated cells to normal tissues. It will be of interest to study which adhesion molecules are involved in the binding of NK cells to normal and malignant tissues. If differences exist, an obvious goal would be to reduce the undesired binding and homing.

It is yet uncertain to what extent NK cells actually infiltrate tumor tissue in vivo. NK cells have been recovered from tumor- and renal allograft-infiltrating lymphocyte populations, but the possibility of blood contamination cannot be entirely ruled out. Our in vitro data indicate that $CD56^+$ cells preferentially infiltrate three-dimensional tumor tissue, suggesting that tumor-infiltrating lymphocytes really include NK cells. ß2-integrins are pivotal for the infiltration, but it is improbable that they employ the infiltrative capacity of lymphocytes maximally. Nitta and co-workers recently showed that the use of bispecific mAb against CD3 and NCAM strongly improved the results of local adoptive immunotherapy of glioma patients (22). A suitable bispecific antibody, in addition to enchancing the cytotoxicity, may also increase the infiltrative capacity of injected LAK cells. Our spheroid model helps in choosing the proper reagents for maximal destruction of three-dimensional tumor tissue by activated NK cells.

References

1. Trinchieri, G. Biology of natural killer cells. Adv Immunol 1989; 47:187-376.
2. London, L., Perussia, B., Trinchieri, G. Induction of proliferation in vitro of resting human natural killer cells: expression of surface activation antigens. J Immunol 1985; 134:718-727.
3. Kehri, J.E., Dukovich, M., Whalen, G., Kaatz, P., Fauci, A.S., Green, W.C. Novel interleukin-2(IL-2) receptor appears to mediate IL-2-induced activation of natural killer cells. J Clin Invest 1988; 81:200-205.
4. Siegel, J.P., Sharon, M., Smith, P.L., Leonard, W.J. The IL-2 receptor beta chain (p70): Role in mediating signals for LAK, NK, and proliferative activities. Science 1987; 238:75-77.
5. Reynolds, C.W., Timonen, T., Holden, H.T., Hansen, C.T. and Herberman, R.B. Natural killer cell activity in the rat. Analysis of effector cell morphology and effect of interferon on natural killer cell function in the athymic (nude) rat. Eur J Immunol 1982; 12:577-582.
6. Biron, C.A., Sonnenfeld, G. and Welsh, A.M. Interferon induces natural killer cell blastogenesis in vivo. J Leukocyte Biol 1984; 35:31-37.
7. Grimm, E.A., Mazumder, A., Zhang, H.Z. and Rosenberg, S.A. Lymphokine-activated killer cell phenomenon. Lysis of natural killer-resistant fresh solid tumor cells by interleukin 2- activated autologous human peripheral blood lymphocytes. J Exp Med 1982; 155:1823-1841.
8. Phillips, J.H. and Lanier, L.L. Discussion of the lymphokine-activated killer phenomenon: relative contribution of peripheral blood natural killer cells and T lymphocytes to cytolysis. J Exp Med 1986; 164:814-825.

9. Timonen, T., Gahmberg, C.G. and Patarroyo, M. Participation of CD11a-c, CD18, CD2 and RGD-inding receptors in endogenous and interleukin-2-stimulated NK activity of CD3-negative large granular lymphocytes. Int J Cancer 1990; 46:1035-1040.
10. Timonen, T., Lehtovirta, P., Gripengerg, J., Sarlomo-Rikala, M., Jääskeläinen, J. and Saksela, E. Interleukin-2 activated lymphocytes (LAK cells) as potential tumor tracers. Acta Radiol Suppl 1990; 374:109-111.
11. Basse, P., Herberman, R.B., Nannmark, U., Johansson, B.R., Hokland, M., Wasserman, K., Goldfarb, R.H. Accumulation of adoptively transferred adherent, lymphokine-activated killer cells in murine metastases. J Exp Med 1991; 174:479-488.
12. Nemlander, A., Saksela, E. and Hayry, P. Are NK cells involved in allograft rejection? Eur J Immunol 1983; 13:348-350.
13. Kovanen, P.E., Timonen, T., Seppälä, I. and Knuutila, S. MAC-technique (Morphology Antibody Chromosomes) in phenotypic identification of NK and T cells in interleukin-2-stimulated lymphocyte cultures. Clin Exp Immunol 1989; 75:407-413.
14. Sharma, S., Mehta, S., Morgan, J. and Maizel, A. Molecular cloning and expression of a human B cell growth factor gene in Eschericia coli. Science 1987; 238:1144-1146.
15. Helander, T., Timonen, T., Kalliomäki, P. and Schröder, J. Recognition of chromosome 6-associated target structures by human lymphokine-activated killer cells. J Immunol 1991; 147:2063-2067.
16. Jääskeläinen, J., Kalliomäki, P., Paetau, A. and Timonen, T. Effect of LAK cells against three dimensional tumor tissue: in vitro study using human glioma spheroids as targets. J Immunol 1989; 142:1036-1045.
17. Jääskeläinen, J., Mäenpää, A., Gahmberg, C.G., Somersalo, K., Tarkkanen, J., Kallio, M., Patarroyo, M. and Timonen. T. Migration of rIL-2 activated T and NK cells in the intercellular space of human H-2 glioma spheroids in vitro: a study of adhesion molecules involved. J Immunol 1992; 149 (1):260-268.
18. Perussia, B., Ramoni, C., Anegón, I., Cuturi, M.C., Faust, J. and Trinchieri, G. Preferential proliferation of natural killer cells among peripheral blood mononuclear cells cocultured with B lymphoblastoid cell lines. Nat Immun Cell Growth Regul 1987; 6:171-188.
19. Kobayashi, M., Fitz, L., Ryan, M., Hewick, R.M., Clark, S.C., Chan, S., Loudon, R., Sherman, F., Perussia, B. and Trinchieri G. Identification and purification of natural killer cell stimulatory factor (NKSF). J Exp Med 1989; 170:827-845.
20. Shields, J.G. and Bonnefoy, J-Y. B cell assays for growth and differentiation factors. In: Cytokines and B lymphocytes. London: Academic Press 1990; 253-265.

21. Nagler, A., Lanier, L.L. and Phillips, J.H. The effects of IL-4 on human natural killer cells. A potent regulator of IL-2 activation and proliferation. J Immunol 1988; 141:2349-2351.

22. Nitta, T., Kato, K., Yagita, H., Okumura, K. and Ishii, S. Preliminary trial of specific targeting therapy against malignant glioma. Lancet 1990; 335:368-371.

7

Regulation of Monocytes by IL-2-Activated Killer Cells

Julie Y. Djeu
Sheng Wei
D. Kay Blanchard
H. Lee Moffitt Cancer Center and Research Institute
University of South Florida College of Medicine
Tampa, Florida

I INTRODUCTION

It is now widely accepted that the majority of IL-2 activated killer (LAK) cells are derived from the large granular lymphocyte (LGL) natural killer (NK) subset of lymphocytes with a CD3-CD16+CD56+ phenotype (1,2). These cytotoxic cells have demonstrated broad lytic reactivity against a large variety of tumor cell lines and fresh, surgically-obtained human tumors (3). This property has been exploited in the clinic for the treatment of cancer by immunotherapy with IL-2 and/or LAK cells (4). However, severe toxicities associated with this type of therapy prompted us to investigate other effects of IL-2 that might contribute to in vivo toxicity. In addition to their tumoricidal activity, we have found that IL-2 induced human LAK cells can also acquire the ability to lyse autologous and allogeneic monocytes in vitro (5). The recognition of cultured monocytes is selective, with monocytes differentiated by granulocyte-macrophage colony stimulating factor (GM-CSF) being the most susceptible to lysis (6). On the other hand, monocytes cultured in interferon (IFN)-g (gamma) are resistant to LAK lysis (7). Monocytes play an important role in all aspects of host defense, beginning with phagocytosis of invading microbes, continuing with antigen processing and release of key cytokines that trigger T cell and B cell activation. The biological significance of LAK recognition of monocytes is unknown. It is possible that LAK cells may be involved in immunohomeostasis by downregulation of an immune function that is no longer required. The control may be at the level of monocytes at certain stages of differentiation and activation or at a stage when they have accomplished their functions and need to be inactivated. This chapter will explore evidence that was generated to support this hypothesis.

II LYSIS OF MONOCYTES BY LAK CELLS

Our early studies indicated that normal human LGL, isolated by Percoll density gradient centrifugation, when cultured 3 - 4 days with recombinant IL-2, expressed optimal lytic activity against autologous and allogeneic monocytes (5). Freshly isolated LGL, in the absence of IL-2, had no ability to lyse monocytes and required IL-2 activation to acquire this function. The phenotype of the anti-monocyte LAK cell fitted that of the primary anti-tumor LAK cell in being CD2+CD16+CD56+CDllb+CD4-CD8-. The ability to recognize normal monocytes appeared to be specific because granulocytes and lymphocytes did not serve as LAK targets, whether they were freshly isolated or cultured for 3 - 7 days in vitro. On the other hand, monocytes became increasingly sensitive to LAK lysis with in vitro culture (Table 1). The increased susceptibility could be due to culture-induced differentation of monocytes to macrophages and sensitivity of monocytes may therefore depend on their differentiation state. This was confirmed by the use of monocyte activating and differentiating factors. Differentation with GM-CSF or IL-3 was found to further enhance monocyte susceptibilility (6,8). Peak susceptibility occurred after 4 - 6 days of culture in GM-CSF or IL-3 and was increased 4-fold over that of monocytes cultured in medium alone, which already displayed significant LAK sensitivity. Again, the effects of GM-CSF and IL-3 were specific for monocytes because neutrophils that are also activated by these cytokines remained resistant to LAK lysis after similar treatments (Table I). GM-CSF and IL-3 appeared to differentiate monocytes along the same pathway as assessed by FACS analysis and SDS-polyacrylamide gel electophoresis of cell surface antigens. This was also confirmed by cold target inhibition of lysis with reciprocal labelled GM-CSF-treated monocytes with unlabelled IL-3 activated monocytes and vice versa.

Human monocytes, neutrophils or nonadherent lymphocytes were incubated with either medium, GM-CSF, IL-3 or IFN-g for 4 days at 37°C and then labelled ^{51}Cr and used as targets for lysis with either fresh LGL or IL-2-activated LAK cells.

Analysis of other monocyte differentiating factors indicated that IFNg had the opposite effect on monocytes (7). Treatment with IFNg rendered monocytes resistant to LAK lysis (Table 1). Kinetic studies showed that as little as 2 hr incubation with IFNg was sufficient for the protective effect to take hold. In analyzing the mechanism of protection, we found that IFNg treatment appeared to inhibit a postbinding event rather than initial recognition of the binding ligand on monocytes by LAK cells. This conclusion was derived from evidence that IFNg-treated monocytes could compete efficiently with untreated monocytes for binding to LAK cells, as measured by cold target inhibition. In addition, both IFN-g-treated and untreated monocytes could equally form conjugates with LAK cells. Therefore, IFN-g may modulate a LAK-triggering target ligand separate from the LAK binding epitope.

Table 1
Lysis of Monocytes by LAK cells

Targets	Treatment of targets	Lysis mediated by:	
		fresh LGL	IL-2-activated LAK cells
Monocytes	medium	-	++
Neutrophils	medium	-	-
Lymphocytes	medium	-	-
Monocytes	GM-CSF	-	++++
Neutrophils	GM-CSF	-	-
Monocytes	IL-3	-	++++
Neutrophils	IL-3	-	-
Monocytes	IFN-g	-	-

III INHIBITION OF MONOCYTE FUNCTION AGAINST CANDIDA ALBICANS BY LAK CELLS

The biological significance of the interaction of LAK cells with monocytes is unknown. To determine if LAK lysis of monocytes was associated with loss of monocyte function, we assessed the functional activity of monocytes against Candida albicans, subsequent to brief incubation with autologous LAK cells (9). Monocyte function against this opportunistic fungal pathogen was assessed by measuring the amount of ^{3}H-glucose incorporated into residual Candida after I8 hr incubation with monocytes (10). Cultured monocytes, after 2 - 6 hr incubation with LAK cells were found to be substantially suppressed in their ability to control fungal growth. Moreover, monocytes cultured in the presence of GM-CSF or IL-3 were even more diminished in their capacity to inhibit fungal growth after a brief exposure to LAK cells. The effect of GM-CSF was both time and dose dependent. Peak susceptibility was induced in monocytes after 4 days of culture in GM-CSF and as little as I0 units/ml of GM-CSF was sufficient to produce this effect. On the contrary, freshly-isolated monocytes were relatively resistant to LAK lysis and displayed normal antifungal activity, equivalent to that in fresh monocytes untreated with LAK cells. Similarly, IFNg-treated monocytes showed significant functional integrity, after exposure to LAK cells. In fact, IFN-g protected cultured monocytes from LAK inhibition (Table I). The usual functional inhibition of medium-cultured monocytes by LAK cells was reversed when the monocytes were cultured in IFN-g. Thus, the opposing effect of GM-

CSF and IFN-g on monocyte functional susceptibility was similar to that found earlier on monocyte lysis by LAK cells. It appears, however, that functional suppression, as assessed by growth inhibition of C. albicans, was even more sensitive as a measurement of the effect of LAK cells on monocytes than membrane damage as assessed by ^{51}Cr release, because antifungal activity in GM-CSF-treated monocytes was suppressed within l hr of exposure to LAK cells whereas ^{51}Cr release from monocytes was not detected until 4 hr incubation with LAK cells (Table 2). Therefore LAK inhibition of monocyte function may occur rapidly, even before the occurrence of cell membrane damage in monocytes, leading to lysis.

Table 2
Kinetics of Inhibition of Monocyte Function
or lysis by LAK cells

Time of incubation with LAK cells	Level of inhibition of function or lysis of monocytes by LAK cells		
	Anti-fungal[a]	Antigen presentation[b]	Lysis[c]
1 hr	++	++	-
2 hr	+++	+++	-
4 hr	++++	++++	++

a GM-CSF-treated monocytes were incubated with autologous LAK cells for 1 - 4 hr at 37°Cr; then the nonadherent LAK cells were washed off and the adherent monocytes were incubated C. albicans to test for antifungal activity.

b GM-CSF-treated monocytes were incubated with autologous LAK cells for 1 - 4 hr at 37°C; then the nonadherent LAK cells were washed off and the adherent monocytes were pre-pulsed with Candidal antigen for 2 hr before mixing with autologous T cells for a further 6 day incubation. The cultures were then pulsed with ^{3}H-thymidine to assess Candida-specific T cell proliferation.

c GM-CSF-treated monocytes were labelled with ^{51}Cr and added to autologous LAK cells for 1 - 4 hr at 37°C to assess specific lysis of the monocytes.

It is to be noted that fresh LGL, without activation by IL-2, had no deleterious effect on monocytes, whether they were cultured with or without GM-CSF. Therefore, IL-2 is required to induce in LGL the capacity to recognize and inhibit monocytes, particularly if they are differentiated by GM-CSF. We have earlier found differences in the activation state of monocytes cultured in GM-CSF and IFN-g (l0). While GM-CSF could maintain monocytes in a highly activated form for up to 5 days in culture, as assessed by functional activity against C. albicans, IFN-g could only activate and maintain high antifungal activity in monocytes for a very brief period, confined to the first 24 hr. Additionally, monocytes cultured in medium alone, tended to progressively lose their antifungal capacity, but 5 day-old aged monocytes could be treated with GM-CSF to recover function while IFN-g could not. Therefore, there are major differences in the differentiation and activation of monocytes by GM-CSF and IFN-g. These differences are apparently recognized by LAK cells. It is possible that the LAK cells are involved in the regulation of monocytes and the state of differentiation dictates the level of control by LAK cells.

IV INHIBITION OF MONOCYTE ANTIGEN PRESENTATION TO T CELLS BY LAK CELLS

To further define the role of LAK cells in regulation of monocyte function, we next investigated the effect of LAK cells on monocyte antigen presentation to autologous T cells. Antigen presentation is a sentinel process in the immune system (11,12). Monocytes take up antigen, process it and display an immunogenic fragment to lymphocytes, thereby delivering an activation signal to T cells. In addition, monocytes secrete costimulatory molecules, e.g. IL-1, which are important to initiate the immune cascade. We therefore examined whether LAK cells might be involved in downregulation of antigen presentation in GM-CSF-treated but not IFN-treated monocytes. Monocytes after treatment with autologous LAK cells were assessed for presentation of C. albicans antigen to autologous T cells, as measured by ^{3}H-thymidine incorporation during antigen-specific lymphocyte proliferation (13).

Adherent monocytes cultured 4 days in medium or GM-CSF were found to equally process and present Candidal antigen to autologous Percoll-purified T cells. However, only the GM-CSF-treated monocytes were functionally inhibited by autologous 4-day IL-2 induced LAK cells. Pretreatment of these monocytes with LAK cells for l hr, followed by subsequent removal of the nonadherent LAK cells, was sufficient to cause inhibition of antigen presentation (Table 2). Inhibition of antigen presentation was thus significantly earlier than cell membrane damage in LAK-treated monocytes. In addition to inhibition of antigen presentation, short exposure to LAK cells also downregulated monocyte production of IL-lα and IL-lß (13). Northern blot analysis indicated that the inhibition of IL-l production occurred not only at the protein level but also at the mRNA level.

As with earlier findings, the effect of IFN-g was opposite that of GM-CSF and IL-3 (13). IFN-g protected monocytes from LAK suppression of antigen presentation. The opposing effects of GM-CSF and IFN-g on monocyte functional susceptibility, as measured by antigen presentation, were equivalent to those seen on monocyte antifungal activity. Thus, both antimicrobial activity and antigen presentation function are susceptible to inhibition by LAK cells, depending on the state of differentation of the monocytes.

V CONCLUSIONS

Monocytes are important accessory cells in the activation of T cells for specific antigen recognition and for control of microbial invasion; yet little is known of their regulation. Our studies have provided insight into the contribution of LAK cells to this process. Several levels of LAK cell-monocyte interactions were identified: l) freshly-isolated LGL and LGL cultured without IL-2 were relatively inert against monocytes and had no capacity to lyse monocytes or inhibit their function, whether they were cultured with GM-CSF or IFN-g; 2) IL-2 could induce LAK cells to acquire lytic and inhibitory functions against cultured monocytes, particularly those differentiated with GM-CSF and IL-3; 3) IL-2 activated LAK cells could not lyse or inhibit freshly-isolated monocytes or monocytes differentiated with IFN-g. These results taken together suggest that the regulation of LAK cells on monocyte function may represent a system specifically to eliminate accessory cells at certain stages of differentation for the control of immune homeostasis. It is possible that the response to a foreign antigen leads to IL-2 induction, followed by LAK activation, which might coincide with the induction of GM-CSF at a time that monocytes have already presented antigen to T cells and are no longer required.

On the other hand, the inhibitory effect on monocytes by LAK cells may serve other purposes. One positive effect is in the elimination of monocytes infected with intracellular pathogens (14-17). On the negative side, suppression of monocytes by LAK cells could contribute to toxic side effects seen in immunotherapy with IL-2/LAK cells. Further work is warranted to clearly define what key roles the interaction of LAK cells and monocytes play in the overall host defense system.

Acknowledgements

The work conducted in our laboratory was supported by USPHS Grants CA46820 and AI24699. We thank Ann Taylor, Susan McMillen, and Carolyn Pearson for their technical support.

References

1. Ortaldo, J.R. and Herberman, R.B. Heterogeneity of natural killer cells. Ann. Rev. Immunol. l984; 2:359-94.
2. Trinchieri, G. Biology of natural killer cells. Adv. Immunol. l989; 47:187-303.
3. Grimm, E.A., Mazumder, A., Chang, A. and Rosenberg, S.A. Lymphokine-activated killer cell phenomenon. Lysis of natural killer-resistant fresh solid tumor cells by interleukin-2-activated autologous human peripheral blood lymphocytes. J. Exp. Med. l982; 155:1823-41.
4. Rosenberg, S.A., Lotze, M.T., Muul, L.M., Chang, A.E., Avis, F.P., Leitman, S., Linehan, W.M., Robertson, C.N., Lee, R.E., Rubin, J.T., Seipp, C.A., Simpson, C.G. and White, D.E. A progress report on the treatment of l57 patients with advanced cancer using lymphokine-activated killer cells and interleukin-2 or high-dose interleukin-2 alone. N. Engl. J. Med. l987; 316:889-97.
5. Djeu, J.Y. and Blanchard, D.K. Lysis of human monocytes by lymphokine-activated killer cells. Cell. Immunol. l988; lll:55-65.
6. Blanchard, D.K., Serbousek, D. and Djeu, J.Y. Induction of human monocyte susceptibility to lymphokine-activated killer cell lysis by granulocyte-macrophage colony stimulating factor. Cancer Res. l989; 49:5037-43.
7. Blanchard, D.K. and Djeu, J.Y. Protection of cultured human monocytes from lymphokine-activated killer-mediated lysis by IFN-g. J. Immunol. l988; 141:4067-73.
8. Djeu, J.Y., Widen, R. and Blanchard, D.K. Enhanced lysis of monocytes by LAK cells after treatment with GM-CSF or IL-3. Blood l989; 1264-74.
9. Wei, S., Serbousek, D., McMillen, S., Blanchard, D.K. and Djeu, J.Y. Suppression of human monocyte function against Candida albicans by autologous IL-2 induced lymphokine-activated killer cells. J. Immunol. l99l; 146:337-42.
10. Wang, M., Friedman, H. and Djeu, J.Y. Enhancement of human monocyte function against Candida albicans by the colony stimulating factors (CSF): IL-3, granulocyte-macrophage-CSF, and macrophage-CSF. J. Immunol. l989; 143:671-77.
11. Weaver, C.T. and Unanue, E.R. The costimulatory function of antigen-presenting cells. Immunol. Today l990; ll:49-55.
12. Dinarello, C.A. Biology of interleukin l. FASEB J. l988; 2:l08-l5.
13. Wei, S., Blanchard, D.K., McMillen, S. and Djeu, J.Y. Lymphokine-activated killer cell regulation of T-cell mediated immunity to Candida albicans. Inf. Immun. l992; 60:3586-95.
14. Blanchard, D.K., Stewart, W.E., Klein, T.W., Friedman, H. and Djeu, J.Y. Cytolytic activity of human peripheral lymphocytes against Legionella pneumophila-infected

monocytes: characterization of the effector cell and augmention by interleukin-2. J. Immunol. l987; 139:55l-56.

15. Blanchard, D.K., Michelini-Norris, M.B., Friedman, H. and Djeu, J.Y. Lysis of mycobacteria-infected monocytes by IL-2-activated killer cells: role of LFA-l. Cell. Immunol. l989; ll9:402-ll.

16. Carl, M. and Dasch, G.A. Characterization of human cytotoxic lymphocytes directed against cells infected with typhus group rickettsiae: evidence for lymphokine activation of effectors. J. Immunol. l986; 136:2654-61.

17. Steinhoff, U., Wand-Wurttenberger, A., Bremerich, A. and Kaufmann, S.H.E. Mycobacterium leprae renders Schwann cells and mononuclear phagocytes susceptible or resistant to killers. Inf. Immunol. l99l; 59:684-88.

8

Identification and Enrichment of Proteolytic Enzymes of IL-2 Activated Rat Natural Killer (A-NK) Cells: Potential Physiological Roles in NK Cell Function

Richard P. Kitson
Ken Wasserman
Ronald H. Goldfarb
Pittsburgh Cancer Institute and
University of Pittsburgh School of Medicine
Pittsburgh, Pennsylvania

I INTRODUCTION

Significant progress has been made in recent years towards elucidating the cellular and molecular mechanism(s) underlying cell-mediated cytotoxicity (1-5). Striking advances have been made in the investigation of these events for natural killer (NK) cells (2-5). Proteolytic enzymes, implicated to play a potential role in cell-mediated cytotoxicity, have been focused upon in many investigations of cytolytic T cells (CTL), NK cells and lymphokine activated killer (LAK) cells (reviewed in 5). The observation that large granular lymphocytes (LGL) account for NK cell activity and that LGL can be enriched to a high degree of purity have allowed for the use of enriched populations of NK cells for biochemical studies (5).

The granzymes are the most widely examined of these killer cell-associated proteolytic enzymes. This collection of neutral, serine proteases are found in the cytolytic granules of cloned, murine cytolytic T-cells (6-9). Proteases have also been discovered in the lytic granules of human, cloned cytotoxic T-cells (10,11), rat NK cells lines (12,13), and LAK cells from both human (14) and murine (15,16) sources. To date, seven granzymes (A-G) have been isolated from the granules of cloned mouse CTLs. Only granzymes A, B and D have known synthetic substrates (11,17). That the other granzymes are serine proteases has been inferred by their sequence homology to each other and to model serine proteases including trypsin and chymotrypsin (18). Granzyme A (i.e., Hanukah (H) factor, cytotoxic T-lymphocyte-associated protease (CTLA-3), or BLT esterase) is a 69 kD protein which cleaves benzoxycarbonyl lysine thiobenzyl ester (BLT). This enzyme has also been discovered in the cytolytic granules of rat NK and LAK cells (15). In addition to granzyme A, proteases with aromatic specificity have also been identified in erythrolytic granules isolated from a rat LGL tumor cell line (RNK-16) (13). A genomic sequence (RNKP-1) has been isolated from this cell line which yields 80% homology with CCPI (granzyme B) from mouse CTL (19). Rat natural killer cell protease 1 (RNKP-1) has been isolated as a cDNA clone from an RNK-16-λgt11 library. This clone codes for a protein containing 248 amino acids. Although it shares eighty per cent sequence homology with the murine CTL cDNA clone CCPI (granzyme B), there are significant differences in the substrate binding region in the number of cysteine residues indicating that these clones code for functionally distinct proteins. The protease encoded by this gene has recently been purified from the RNK cell line (20). This molecule, with an Mr of 29-31 kDa, appears to have a substrate specificity similar to that for granzyme B. In addition, another recently described granular protein, fragmentin, has also been isolated from the same cell line and has been found to have tryptic peptides which are highly analogous to RNKP-1.

Although recent studies have identified some candidate substrate for granzymes in target cell DNA degradation or dissolution of basement membrane type IV collagen (23,24) it remains unknown what actual physiological functions these enzymes contribute to the physiologic function of NK cells. Moreover, several studies have suggested that granzymes/BLT esterase may not contribute to cell-mediated cytotoxicity; e.g., some freshly isolated CTL have been found to have essentially no BLT esterase activity (25,26).

The inhibition of cell-mediated cytotoxicity by protease inhibitors has suggested a role for proteases in the killing process. Studies in both CTL and NK cells indicated that preincubation of effector cells with either tryptic (TLCK, benzamidine, p-aminobenzamidine, or leupeptin) or chymotryptic (TPCK, α-1-antichyrmotrypsin, or chymostatin) protease inhibitors could inhibit cell-mediated cytotoxicity (27,28). Nevertheless, the targets for these protease inhibitors remain unknown.

To date the major focus in the study of proteolytic enzymes of killer cells by most investigators has been almost exclusively directed towards only the proteases of cytolytic granules. While several reports have documented that killer cell proteases can exist in: cell surface-associated, non lytic -cytoplasmic granules, and extracellular, secreted forms (e.g., 29-32), the role of such enzymes and possibly other proteases in killer cell function have largely been ignored in favor of proteases found only in cytolytic granules. In recent years, a major emphasis in our laboratory has been to further investigate the subcellular localization and release of NK/LAK proteolytic enzymes that might contribute to cell-mediated functions including cytotoxicity. In these investigations we have employed rat A-NK cells as a source of NK/LGL since these cells may be easily expanded and enriched to yield large number of cells appropriate for biochemical investigation (34). We have recently determined that: 1) rat A-NK cells have, in addition to the granzymes, a number of previously undescribed proteases which cleave with tryptic, chymotryptic and collagenolytic specificity; 2) these enzymes exist extracellularly or in cytosolic, or cytoplasmic granule subcellular domains; 3) some of these proteases are predominantly located in the cytosol and, by isopycnic sucrose and Percoll gradient centrifugation, are distinct from granzymes of cytolytic granules; 4) our studies suggest that some, but not all, of these proteases are involved in A-NK cell-mediated cytotoxicity; 5) two of these proteolytic activities exist as part of a macromolecular complex which we have purified to apparent homogeneity; 6) biochemical, immunochemical, biophysical and structural/EM studies have shown that this complex is molecularly related to the multicatalytic proteinase complex (MCP); 7) rat A-NK MCP specific activities are increased substantially following IL-2 activation of NK cells (R.P. Kitson and R.H. Goldfarb, unpublished observations).

A. Proteases of Rat A-NK Cells

We have tested a number of substrates with rat A-NK postnuclear supernatants (PNS) as a source of cell-associated enzymes. Of these the most active are those which cleave Boc-FSR-7-amino-4-methyl coumarin (AMC), (trypsin-like, A-NKP 1); Suc-AAF AMC (chymotrypsin-like A-NKP-2); SucGPLGP AMC (collagenase-like, A-NKP-3); and Z-FR AMC (another trypsin-like enzyme, A-NKP-4). Activity was also detected against Suc-AFK AMC (plasmin-like) and Z-GGR AMC (plasminogen activator-like). To determine the subcellular location of these activities, rat A-NK cell PNS were separated on isopycnic sucrose gradients. The Suc-GPLGP AMCase was clearly located in the cytosol; whereas the Z-FR AMCase activity appeared to co-migrate with the lysosomal marker aryl sulfatase. The Boc-FSR AMCase, Suc-AAF AMCase and Z-GGR AMCase co-migrated with each other at a position which did not correspond to either plasma membrane (5'-nucleotidase) or Golgi complex (galactosyl transferase). Suc-AFK AMCase could not be detected in gradients. To resolve the location of the complex with three activities, the PNS was treated with digitonin since this detergent is known to shift cholesterol-containing membranes to higher densities. Although the two marker enzymes, 5' nucleotidase and galactosyl transferase showed characteristic digitonin-induced shifts, there was no change in the position of the protease complex. Increasing the force-time integral (w^2t) gave no change in the position of 5' nucleotidase in the gradient, whereas the median density of both Boc-FSR AMCase and Suc-AAF AMCase increased. To eliminate the possibility that our findings were merely a potential artifact of sucrose gradients, we also employed Percoll gradients. Separation of A-NK PNS on a Percoll gradient showed that these activities appeared at the top of the gradient. In sum, all of these results indicated that these proteolytic activities (e.g., A-NKP-1 and A-NKP-2) resided in the cytoplasm of rat A-NK cells and that they probably existed as a high molecular weight complex (35).

B. Identification and partial purification of rat A-NK multicatalytic protease complex (MCP)

The aggregate Boc-FSR AMCase and Suc-AAF AMCase activity from sucrose gradients were further purified over Sephacryl S-300 and heparin-Sepharose columns with a final purification factor of 6000-8000 fold. We believe that we have purified this enzyme to apparent homogeneity since even though three bands are visible on a modified SDS PAGE gel, all three bands are catalytically active. The molecular weight of the complex was determined to be approximately 7000,000 by molecular sieving on a calibrated column of Sephaccryl S-300. The biochemical properties of A-NKP-1 and A-NKP-2, including sensitivities to salts and detergents, coupled with their high molecular weight and substrate specificities led us to consider that this protease complex was related to MCP. The possibility that our enzymes were possibly related to MCP was intriguing since this complex appears to be ubiquitous in all cells examined to date, but which had heretofore been ignored in studies of lymphocyte proteases. We were therefore excited by the possibility that a "ubiquitous" enzyme system might play a specialized role in a specialized cell type: A-NK cells. In collaboration with Dr. A.J. Rivett (U. of Leicester), we have determined that our highly purified rat A-NK protease complex is cross reactive with anti-rat liver (MCP on a Western blot (35). Ultrastructural analysis by EM indicated that this enzyme had the typical barrel-shaped morphology described for MCP (36).

C. Proteases in A-NK cells from other species

We have looked at the protease profile in mouse A-NK cells and cells from human A-NK leukemia (supplied by Dr. Carol Dahl.) The sucrose density gradient protease profile in mouse cells is almost identical to that of the rat. The profile of the human A-NK leukemia does have some differences. Although the MCP activities are in the same position, the Suc-AAF AmCase appears to be split. In addition the Z-GGR AMCase activity is now not associated with the Boc-FSR AMCase, but comigrates with Suc-GPLGP AMCase.

D. Role of proteases in A-NK cytotoxicity

To determine whether a correlation existed between the inhibition of cytotoxicity and the inhibition of proteolytic activity, rat A-NK cells were treated with protease inhibitors, washed thoroughly and analyzed for protease activity and cytotoxicity. The results show a correlation between the inhibition of the chymotrypsin-like protease, A-NKP-2 and the inhibition of rat A-NK cell-mediated cytotoxicity against both YAC-1 and P815 targets (36). As an example of the effect of protease inhibitors on the lysis of YAC-1 and P815 target cells by rat A-NK cells is shown in Figure 1.

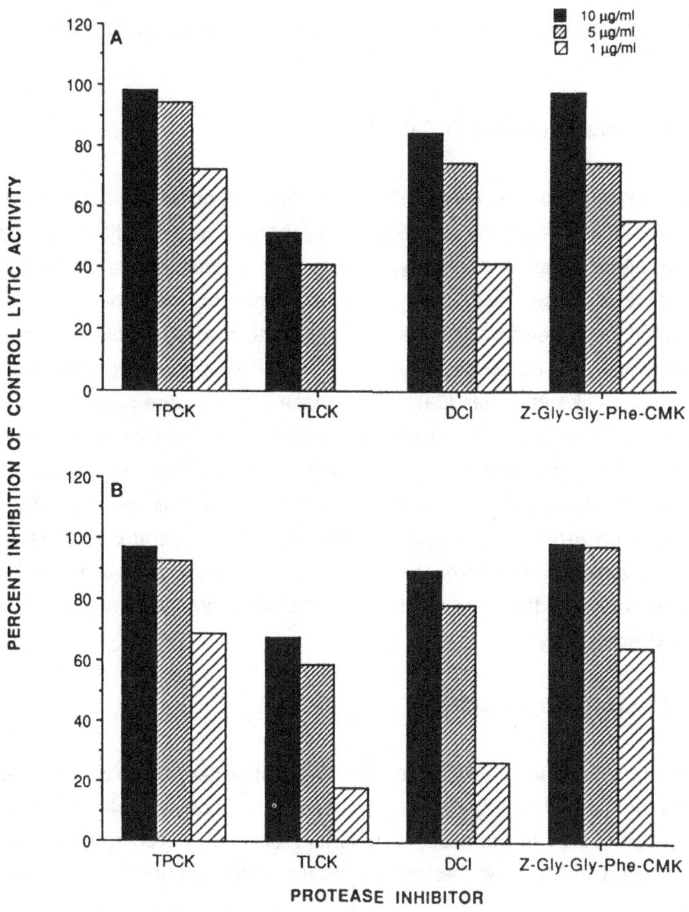

Fig. 1 Effect of protease inhibitors on the lysis of YAC-1 (Panel A) and P815 (Panel B) target cells by rat A-NK cells. Rat A-NK cells (2 x 10^6/2 ml) were incubated with protease inhibitors (TPCK, TLCK, 3,4-dichloroisocoumarin (DCI), and Z-Gly-Gly-Phe chloromethyl ketone at the indicated concentrations for 30 min. at room temperature and then assayed for lytic activity. Results are expressed as percent inhibition of control lytic units (30%)

E. Induction of Rat A-NK MCP by IL-2

We have observed that highly enriched NK cells, obtained by panning using the NK-specific MoAb 3.2.3 (37), upon stimulation with IL-2 show an increased specific activity of MCP (R.P. Kitson and R.H. Goldfarb, unpublished observations.) This enhancement is of particular interest since under the identical conditions we have noted no consistent enhancement of BLT esterase activity. These preliminary studies, which need to be confirmed and extended, further support the view that MCP plays an important role in A-NK lytic function.

F. Extracellular Proteases of Rat A-NK Cells

Culture supernatants from rat A-NK cells were collected, concentrated and subjected to batch affinity chromatography on ligated sepharose previously characterized for selectivity for either tryptic or chymotryptic enzymes: p-aminobenzamidine agarose and D-phenylalanine agarose, respectively (38). Zymographic patterns for each of these enzymes which have been partially purified by these methods indicated that rat A-NK cells produce relatively low Mr proteolytic enzymes (< 30 Kd) with specificities of trypsin- and chymotrypsin- like enzymes (38). Since these enzymes are produced in very small quantities we have not as yet been able to purify them to homogeneity nor to compare them to the rat A-NK cell-associated MCP. While the extracellular enzymes were able to cleave casein and casein plus plasminogen in SDS-PAGE zymography gels, it is noteworthy that rat A-NK MCP is unable to cleave these substrates suggesting that the extracellular enzymes have different cleavage specificities and are not derived from MCP. We therefore conclude that it is not likely that MCP proteases are either released or functional extracellularly.

II CONCLUDING COMMENTS

In summary our studies indicate that an assortment of proteolytic enzymes exists in A-NK cells that are not limited to only cytolysin-containing cytolytic granules or described only by the cleavage of BLT in the assay of esterase activity. Moreover, these studies add additional support to the view that proteolytic enzymes contribute to NK cell-mediated cytotoxicity.

Our finding and purification to homogeneity of MCP in NK cells now allows for the comparison of MCP to granzymes of lytic granules with respect to their physiologic, functional role in NK cell-mediated cytotoxicity. Our identification and enrichment of proteases released by A-NK cells might be highly relevant to a potential role for these degradative enzymes in cellular migration in vivo. We have recently observed that adoptively transferred A-NK cells infiltrate and accumulate within established experimental tumor metastases (39,40). We hypothesize that it is possible that proteases released by A-NK cells might contribute to detachment from microvascular endothelial cells, invasion through extracellular matrices, and/or cellular migration, analogous to degradative proteases of invasive tumor cells such as type IV collagenase and urokinase (41).

Further experimentation is clearly required to fully define the physiologic role of these newly described and purified cell-associated and released proteolytic enzymes in NK cell physiologic and regulatory functions.

References

1. Young, D.E. Killing of target cells by lymphocytes: A mechanistic view. Physiologic. Rev. 1989. 69:250-314.
2. Henkart, P.A. Mechanism of lymphocyte-mediate cytotoxicity. Ann. Rev. Immunol. 1985. 3:31-58.
3. Herberman, R.B., Reynolds, C.W. and Ortaldo, J.R.. Mechanisms of cytotoxicity by natural killer (NK) cells. Ann. Rev. Immunol. 1986. 4:651-680.
4. Tschopp J. and Jongeneel, C.V.. Cytotoxic T lymphocyte mediated cytolysis. Biochem. 1988. 27:2641-2646.
5. Goldfarb R.H. Cell-mediated cytotoxic reactions. Human Path. 1986. 17:138-145.
6. Masson, D., Nabholz, M., Estrude, C. and Tschopp, J. Granules of cytolytic cells contain two serine esterases EMBO J. 1986. 5:1595-1600.
7. Pasternak, M.S., Verret, C.R., Liu, M.A. and Eisen, H.N. "Serine esterase in cytolytic T lymphocytes" Nature 1986. 322:740-743.
8. Fruth, U., Prester, M., Golecki, J.R., Hengartner, H., Simon, H.G., Kramer, M.D. and Simon, M.M. The T cell-specific serine protease TSP-1 is associated with cytoplasmic granules of cytolytic T lymphocytes Eur. J. Immunol. 1987. 17:613-621.
9. Young, J.D.-E., Leong, L.G., Liu, C-C., Damiano, A., Wall, D.A. and Cohn, Z.A. Isolation and characterization of a serine esterase from cytolytic T cell granules Cell 1986. 47:183-194.
10. Poe, M., C.D., Bennett, W.E., Biddison, J.T., Blake, G.P., Norton, J.A., Rodkey, N.H., Sigel, R.V., Turner, J.K., Wu, and Zweerink, H.J. Human cytotoic lymphocyte tryptase. J. Biol. Chem. 1988. 263:13215-13222.
11. Poe, M., J.T. Blake, D.A., Boulton, M., Gammon, N.H., Sigel, R.V., Turner, J.K., Wu, and Zweerink, H.J. Human cytotoxic lymphocyte granzyme B. J. Biol. Chem. 1991. 266:98-103.
12. Goldfarb, R.H. Role of proteases in NK activity. In. Mechanisms of Cytotoxicity by NK Cells. Edit. R.B. Herberman and D.M. Callewaert, Academic Press, New York, 1985. pp 205-211.
13. Zunino, S.J., Allison, N.J., Kam, C-M., Powers, J.C. and Hudig, D. Localization, implications for function, and gene expression of chymotrypsin-like proteinases of cytotoxic RNK-16 lymphocytes. Biochim. Biophys. Acta 1988. 967:331-340.
14. Hameed, A., Lowrey, D.L., Lichtenheld, M. and Podack, E. Characterization of three serine esterases isolated from human IL-2 activated killer cells. J. Immunol. 1988. 141:3142-3147.
15. Velotti, F., Palmieri, G., Morrone, S., Piccoli, M., Frati, L. and Santoni, A. Granzyme A expression by normal rat natural killer (NK) cells <u>in vivo</u> and by interleukin 2-activated NK cells <u>in vitro</u>. Eur. J. Immunol. 1989. 19:575-578.

16. Lin, T.-H. and Chu, T.M. Enhancement of murine lymphokine-activated killer cell activity by retinoic acid. Cancer Res. 1990. 50:3013-3018.
17. Jenne, D.J. and Tschopp, J. Granzymes, a family of serine proteases released from granules of cytolytic T lymphocytes upon T cell receptor stimulation. Immunological Rev. 1988. 103:53-71.
18. Masson, D. and Tschopp, J. A family of serine esterases in lytic granules of cytolytic T lymphocytes. Cell 1987. 49:679-685.
19. Zunino, S.J., Bleackley, R.C., Martinez, J. and Hudig, D. RNKP-1, a novel natural killer-associated serine protease gene cloned from RNK-16 cytotoxic lymphocytes. J. Immunol. 1990. 144:2001-2009.
20. Sayers, T.J., Wiltrout, R.A., Sowder R., Munger, W.L., Smyth, M.J. and Henderson, L.E. Purification of a factor from the granules of a rat natural killer cell line (RNK) that reduces tumor cell growth and changes tumor morphology. Molecular identity with a granule serine protease (RNKP-1). J. Immunol. 1992. 148:292-300.
21. Shi, L., Kraut, R.P., Aebersold, R. and Greenberg, A.H. A natural killer cell granule protein that induces DNA fragmentation and apoptosis. J. Exp. Med. 1992. 175:553-66.
22. Pasternack, M.S., Bleier, K.J. and McInerney, T.N. Granzyme A binding to target cell proteins. Granzyme A binds to and cleaves nucleolic in vitro. J. Biol. Chem. 1991. 266:14703-14708.*
23. Hayes, M.P., Berrebi, G.A. and Henkart, P.A. Induction of target cell DNA release by the cytotoxic T lymphocyte granule protease granzyme A. J. Exp. Med. 1989. 170:933-946.
24. Simon, M.M., Kramer, M.D., Prester, M. and Gay, S. Mouse T-cell associated serine proteinase 1 degrades collagen type IV: a structural basis for the migration of lymphocytes through vascular basement membranes. Immunology 1991. 73:117-119.
25. Dennert, G, Anderson, C.G. and Prochazka, G. High activity of N-α-benzyloxycarbonyl-L-lysine thiobenzyl ester serine esterase and cytolytic perforin in cloned cell lines is not demonstrable in in vivo-induced cytotoxic effector cells. Proc. Natl. Acad. Sci. USA. 1987. 84:5004-5008.
26. Somersalo, K. and Saksela, E. N-α-Benzyloxycarbonyl-L-lysine thiobenzyl ester (BLT) serine esterase in human cytolytic effector cells and cell line targets. Scand. J. Immunol. 1989. 29:459-467.
27. Hudig, D., Redelman, F. and Minning, L.L. The requirement for proteinase activity for human lymphocyte-mediated natural cytotoxicity (NK): evidence that the proteinase is serine dependent and has aromatic amino acid specificity of cleavage. J. Immunol. 1984. 133:2647-2654.

28. Goldfarb, R.H., Timonen, T.T. and Herberman, R.B. In NK Cells and Other Natural Effector Cells. R.B. Herberman, (Ed.) Academic Press, New York, NY. 1982. p. 931-938.
29. Zucker-Franklin, D., Yang, J. and Fuks. A. Different enzyme classes associated with human natural killer cells may mediate disparate functions. J. Immunol. 1984. 132:1451-1455.
30. Lavie, G., Leib, Z. and Servadio, C. The mechanism of human NK cell-mediated cytotoxicity. Mode of action of surface-associated proteases in the early stages of the lytic reaction. J. Immunol. 1985. 135:1470-1476.
31. Goldfarb, R.H., Timonen, T. and Herberman, R.B. Production of plasminogen activator by human natural killer cells, large granular lymphocytes. J. Exp. Med. 1984. 159:935-951.
32. Carpen, O., Saksela, O. and Saksela, E. Identification and localization of urokinase - type plasminogen activator in human NK cells. Int. J. Cancer 1986. 38:355-360.
33. Jiang, S.C., Hassalkus-Light, S., Ojcius, D.M. and Young, J.D.-E. Purification of a membrane-associated serine esterase from murine cytotoxic T lymphocytes by a single reverse-phase column. Prot. Expr. Purif. 1990. 1:77-82.
34. Vujanovic, N.L., Herberman, R.B., Maghazachi, A.A. and Hiserodt, J.C. Lymphokine-activated killer cells in rats. III. A simple method for the purification of large granular lymphocytes and their rapid expansion and conversion into lymphokine-activated killer cells. J. Exp. Med. 1988. 167:15-29.
35. Goldfarb, R.H., Wasserman, K., Herberman, R.B. and Kitson, R.P. Non-granular proteolytic enzymes of rat interleukin-2 activated natural killer cells. I. Subcellular localization and functional role. J. Immunol. 1992. In press.
36. Wasserman, K., Rivett, A.J., Sweeney, S.T., Gabauer, M.D., Herberman, R.B., Kitson, R.P., Watkins, S. and Goldfarb, R.H. Purification and characterization of multicatalytic proteinase complexes from rat interleukin-2 activated natural killer cells and rat CRNK-16 leukemia. Submitted for publication. 1992.
37. Chambers, W.H., Vujanovic, N.L., DeLeo, A.B., Olszowy, M.W., Herberman, R.B. and Hiserodt, J.C. Monoclonal antibody to a triggering structure expressed on rat natural killer cells and adherent lymphokine-activated killer cells. J. Exp. Med. 1989. 169:1373-1389.
38. Wasserman, K., Kitson, R.P., Gabauer, M.D., Miller, C.A., Herberman, R.B., and Goldfarb, R.H. Zymographic analysis of cell-associated and released proteases of rat interleukin 2-activated natural killer (A-NK) cells. Submitted for publication. 1992.
39. Basse, P., Herberman, R.B., Nannmark, U., Johansson, B.R., Hokland, M., Wasserman, K. and Goldfarb, R.H. Accumulation of adoptively transferred adherent, lymphokine-activated killer cells in murine metastases. J. Exp. Med. 1991. 174:479-488.

40. Basse, P., Nannmark, U., Johansson, B.R., Herberman, R.B. and Goldfarb, R.H. Establishment of cell-to-cell contact by adoptively transferred adherent lymphokine-activated killer cells with metastatic murine melanoma cells. J. Natl. Cancer Inst. 1991. 83:944-950.

41. Goldfarb, R.H. and Liotta, L.A. Proteolytic enzymes in cancer invasion and metastasis. Sem. Thromb. and Hemosta. 1986. 12:294-307.

PART III

Introduction

Animal Models in Tumor Biology

Craig Reynolds
Biologic Resources Branch
National Cancer Institute
Bethesda, Maryland USA

Most scientists agree that preclinical animal models provide information that is useful in the study of tumor biology and in the design of new immunotherapeutic approaches for human cancers. This type of information is not available from any in vitro or artificial model system and is critical to the development of cancer therapeutics. Although there is still some disagreement as to the types of preclinical information that can be extrapolated to humans, and the extent to which this information will be predictive for responses, animal tumor models continue to provide the best source of data for the study of tumor biology and immunotherapeutics.

Each of the papers in this section contributed by the laboratories of Drs. Wiltrout, Gorelik and Riccardi use specific animal models to address an important topic in the field of tumor immunology and its potential prospects for therapy. The first manuscript from Robert Wiltrout's group describes his pioneering work on the antitumor and hematological effects of recombinant cytokines, especially interleukin-7 (IL-7). This work in a murine tumor transplant model has demonstrated an important potential therapeutic value for IL-7 in promoting lymphopoiesis and in stimulating the host's immune response against tumors. The manuscript from Dr. Riccardi and his associates on their studies of NK cell precursors in mice provides an essential basis for understanding the development of this important component of the immune system. Their data suggest that a number of different cytokines are critical for the maturation of NK cells, and that stimulating precursors to develop into mature effector cells should be considered when studying immunomodulatory agents with antitumor activity. Dr. Gorelik and associates describe a series of results indicating an increase in sensitivity to NK cells in murine tumor cells exposed to ultraviolet (UV) irradiation. These data suggest an increase in tumor cell immunogenicity following UV irradiation and are directly relevant to our understanding of the experimental and clinical data on the antitumor effects of photodynamic therapy.

These experimental studies in animals provide an important insight into the potential prospects of using immunotherapy for the treatment of human malignancy. In view of the potential of immunotherapy as a novel treatment modality for cancer, many other aspects of the immune system response to tumor need to be investigated. The use of these and other animal model systems for such studies may lead to the identification of new therapeutic strategies and should provide valuable information regarding the problems and solutions that may be encountered during future clinical trials.

9

Effects of rhIL-7 on Leukocyte Subsets in Mice: Implications for Antitumor Activity

Kristin L. Komschlies
Timothy T. Back
Program Resources, Inc./DynCorp
Frederick, Maryland

Theresa A. Gregorio
M. Eilene Gruys
Giovanna Damia
Robert H. Wiltrout
Laboratory of Experimental Immunology
National Cancer Institute-Frederick Cancer Research and Development Center
Frederick, Maryland

Connie R. Faltynek
Sterling Drug, Inc.
Malvern, Pennsylvania

I INTRODUCTION

Interleukin 7 (IL-7) is a 25 kDa glycoprotein that was originally characterized as a product of a cloned murine bone marrow stromal cell line (1,2), and its mRNA has since been detected in the thymus and spleen (3). Human and murine recombinant IL-7 exhibit a 60% homology at the protein level with all of the six cysteine residues conserved (4). In vitro, IL-7 induces the proliferation of pro-B and pre-B lymphocytes (1,2) and affects the growth of both immature and mature cells of the T lymphocyte lineage (5-9). IL-7 also induces lymphokine-activated killer (LAK) activity from human peripheral blood leukocytes (9,10) and mouse splenocytes (11) in vitro. Further, IL-7 affects myeloid progenitor cells since splenic granulocyte-macrophage colony forming units (CFU-GM) and multipotential

CFU (CFU-GEMM) are increased in mice treated with IL-7 (12,13), while total bone marrow CFU are decreased (13). Thus, the aims of these studies are to 1) determine whether the administration of recombinant human IL-7 (rhIL-7) to mice could alter the incidence or total number of various lymphoid subsets and 2) assess the effects of IL-7 against experimental metastases in mice.

II EFFECTS OF rhIL-7 ON LYMPHOID SUBSETS IN MICE

A. Effects of rhIL-7 on Early B Cells

The twice daily ip injection of C57B1/6 mice with increasing doses of rhIL-7 (generously provided by Sterling Drug, Malvern, PA; specific biological activity of 2-5x10^7 U/mg) for various periods of time has been shown to induce a dose and time dependent lymphocytosis in the peripheral blood (13) and in several lymphoid organs (14). When the surface phenotype of the splenic lymphocytes was studied, a large increase in the percentage and number of $B220^+$ lymphocytes was detected.

Specifically, the percentage of cells bearing B220 (detected using the monoclonal antibody 14.8) increased by about 50% and the total numbers of $B220^+$ cells increased by about 8-fold following 7 days of IL-7 administration (14). The increase in the percentage of $B220^+$ cells was due mainly to an increase in the $B220^+$ cells of intermediate brightness (Figure 1) which are most likely pre-B cells. The increase in these cells become evident by 4 days (Figure 1C) of IL-7 administration and is most prominent following 7 days of rhIL-7 treatment (Figure 1D). Subsequent studies (14) have directly shown that the increase in $B220^+$ cells was due to an increase in pre-B cells (i.e. $B220^+$, surface IgM^- cells). These results demonstrate that rhIL-7 has the predicted biological effects in mice, and is thus active across species <u>in vivo</u>.

B. rhIL-7 Alters T Cell Subsets and These Effects are Reversible

C57BL/6-Ly 5.1 mice were injected ip with 5 μg/injection of rhIL-7 twice a day for 7 days. Mice were sacrificed 1 to 14 days following cessation of rhIL-7 treatment and their spleens and lymph nodes were examined for cellularity and surface phenotype. There was a 3-fold increase in cellularity of the spleen and lymph nodes following rhIL-7 treatment, and this increase was undiminished 3 days after treatment with the cellularity of the spleen and lymph nodes decreasing to control levels by day 14 (14). As can be seen in the spleen (Figure 2), the percentage of $B220^+$ cells (Figure 2A) is elevated approximately 2-fold, 1 day after treatment, and by 3 days after treatment the percentage of these cells begins to drop and returns to control levels by day 6. Conversely, the percentage of $CD4^+$ cells is decreased 3-fold at 1 day after treatment compared to controls and returns to control levels by day 6 after treatment and to 50% above control levels 14 days after cessation of rhIL-7 treatment (Figure 2B). $CD8^+$ cells (Figure 2C) however, do not change in their percentage until greater than 3 days after treatment, thus by 6 days until at least 14 days after cessation of treatment their percentage is double that of control mice. Further examination of the relative numbers of leukocytes bearing each marker shows that the initial increase in the number of $B220^+$ cells has declined to control levels by day 14 after cessation of treatment (Figure 2D). The numbers of $CD4^+$ cells however, remain near control levels 1 day after cessation of treatment and increase in number until 6 days after treatment when their numbers are 27.7 x 10^6 in control mice (Figure 2E). By day 14 after treatment the number of $CD4^+$ cells begins to drop. Perhaps most interestingly, $CD8^+$ cell numbers are strikingly elevated immediately (1 day) after completion of rhIL-7 treatment and reach the maximum number of 32.1 x10^5 cells per mouse at 3 days after treatment compared to 7.7 x 10^6 for control mice (Figure 2F). The numbers of $CD8^+$ cells drop thereafter but remains 2-fold higher than controls even by day 14 after cessation of rhIL-7 treatment. Thus, the repeated administration of rhIL-7 to mice dramatically increases the total numbers of both $CD4^+$ and $CD8^+$ T cells, with the $CD8^+$ cells being increased disproportionately. These changes result in a decrease in the CD4 to CD8 ratio from about 1.5:1 to 1:2.

III ANTI-METASTATIC EFFECTS OF rhIL-7

Previous studies from our laboratory have demonstrated that various combinations of flavonoid components and rhIL-2 (15-19) or the combination of IL-2 and IFNα (20) have potent antitumor effects against murine renal cancer (Renca) and other murine tumors, and that these effects are at least partially T cell-mediated. Because administration of 5-10 μg of rhIL-7 twice per day to mice profoundly perturbs hematopoiesis (13) as well as lymphopoiesis and/or expansion of T cell numbers (14, Figure 2), studies were performed to determine whether the administration of rhIL-7 could cause regression of pre-existent experimental metastases.

C57BL/6 mice that had been injected iv with 2x10^5 MCA-38 tumor cells on day 0, were treated on days 3 to 12 with HBSS, rhIL-7 (25 μg/injection) or rhIL-2 (10μg/injection) as a positive control. The mice were euthanized on day 14 and their lungs were removed to determine the antimetastatic effects of rhIL-7 (Figure 3). The results of this study demonstrate that the administration of rhIL-7 to mice bearing pulmonary MCA-38 metastases reduced the median number of metastases by about 70%, an effect similar to that induced by an optimal regimen of rhIL-2. Subsequent studies have revealed a similar antimetastatic effect in mice bearing murine renal cancer (14).

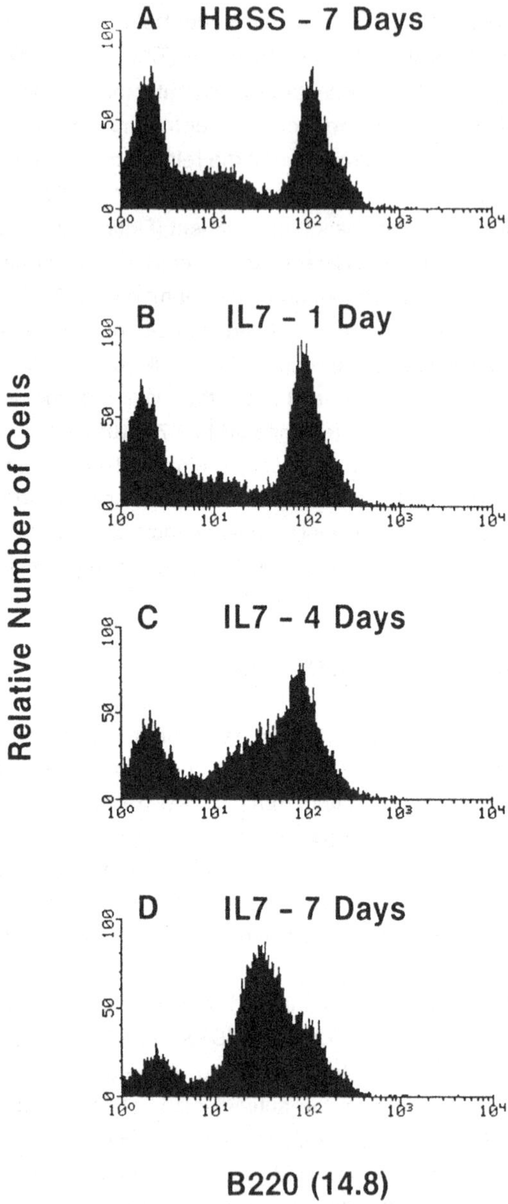

Fig. 1 B220 (14.8) positive cells of intermediate density in the spleen increase in proportion following in vivo IL-7 treatment. C57BL/6 mice were injected ip twice a day for 1 to 7 days with HBSS or 10 μg of IL-7 per injection. Following treatment, the spleens of 5 mice/group were removed, pooled, and labeled by immunofluorescence for detection of B220 expression using the monoclonal antibody clone 14.8. Results are displayed as histograms of the intensity of B220 expression versus the relative number of cells expressing that density

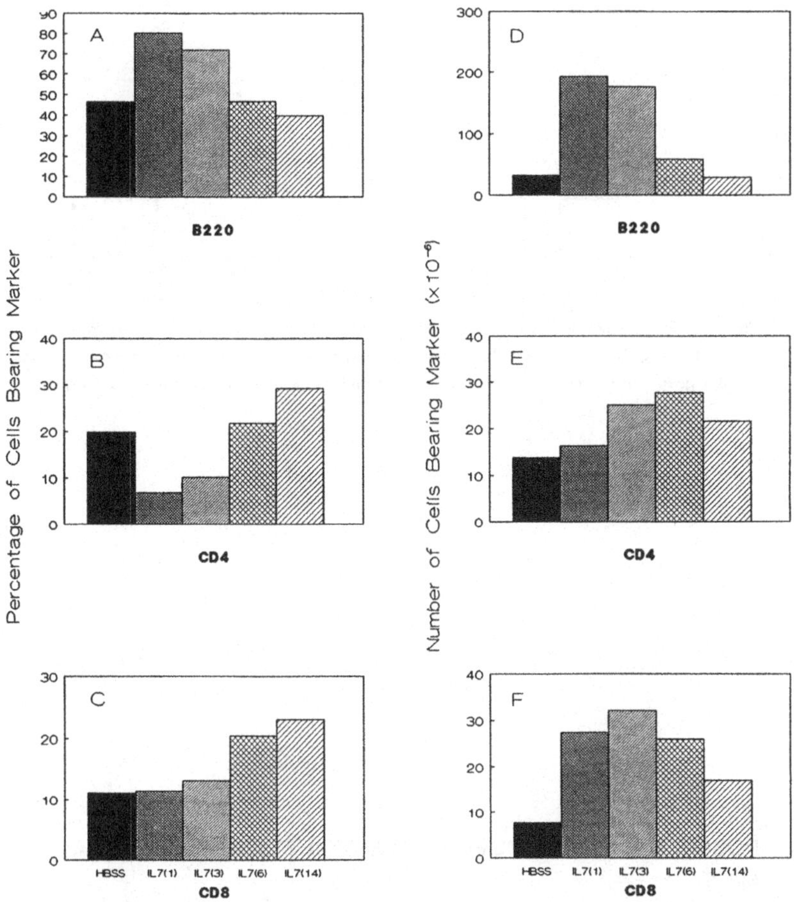

Fig. 2 The effects of <u>in vivo</u> administration of IL-7 on percentages and numbers of B220, CD4, and CD8, bearing cells are reversible. C57BL/6-Ly 5.1 mice were injected ip twice a day for 7 days with HBSS or 5 μg of IL-7/injection. On 1, 3, 6 or 14 days following treatment, the spleens of 4 mice/group were pooled and assayed using immunofluorescence analysis to determine the percentage (Panels A-C) and number (Panels D-F) of cells bearing B220, CD4 or CD8, respectively

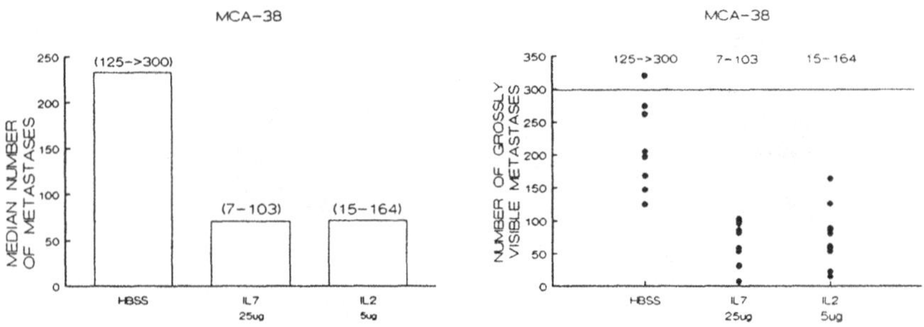

Fig. 3 There is a 70% reduction in the number of experimental lung metastases in MCA-38 tumor-bearing mice following IL-7 treatment. The results are expressed as the median number of metastases in the bar graph and as the individual number of metastases per animal in the scattergram

IV POSSIBLE MECHANISMS FOR THE ANTIMETASTATIC EFFECTS OF rhIL-7 IN MICE

Because rhIL-7 has significant antimetastatic effects against both the Renca and MCA-38 tumors, studies were initiated to determine which host effector mechanisms might be contributing to this effect. Initially, extensive studies were performed in normal mice to determine whether rhIL-7 could induce NK/LAK activity in vivo. A representative experiment is presented in Figure 4. In this experiment, C57BL/6 mice were treated twice daily for 7 days with a 50 μg/injection of rhIL-2 vs. rhIL-7. Levels of NK activity were then assessed in the spleen, liver and lungs. The results showed that as expected, rhIL-2 induced significant increases in NK activity. Specifically, the twice daily administration of 50 μg rhIL-2 significantly increased the NK activity as defined by LU/10^7 cells in the spleen (from 43LU to 157 LU, $p<0.01$), liver (from 58LU to 752LU, $p<0.001$), and especially in the lungs (from 6LU to 87LU, $p<0.001$). The NK-augmenting effect of rhIL-2 was even more evident when the data was calculated as total LU/organ, where NK in spleen (389LU to 548LU, $p<0.001$), liver (108LU to 13,122LU, $p<0.001$), and lungs

(<1LU to 100LU, $p<0.001$) was significantly increased. In contrast, rhIL-7 at a twice daily dose of 50 μg/injection did not significantly augment NK activity/10^7 cells in any organ. The only significant increase in NK activity ($p<0.01$) induced by rhIL-7 occurred when the data was calculated as total LU/organ, in the spleen, with a twice daily dose of 50 μg. Under these conditions, there was a 2-3 fold increase in total splenic NK activity, and this increase appears to be due to the general leukocytosis induced by rhIL-7. Subsequently, we also found that rhIL-7 also did not significantly increase NK/LAK activities in tumor-bearing mice (data not shown).

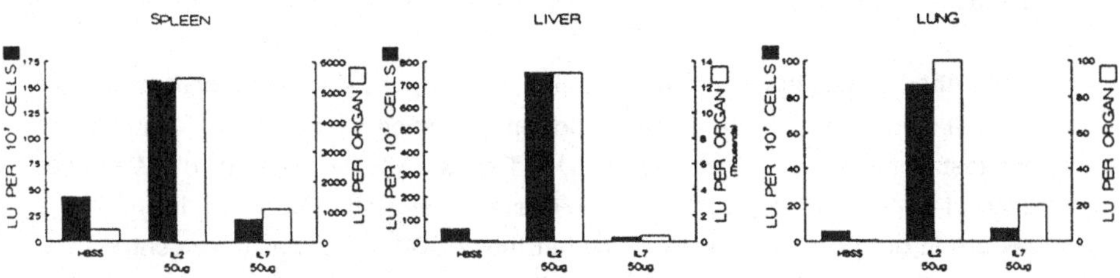

Fig. 4 NK activity is not significantly altered by in vivo IL-7 administration. Normal C57BL/6 mice were injected ip twice a day for 7 days with HBSS or 50 μg/injection of IL-2 or IL-7. Leukocytes from spleen, liver and lung were tested for NK activity against YAC-1 target cells using a ^{51}Cr-release assay. The results are expressed as the number of lytic units (LU) per 10^7 cells (black bars) and as the number of LU per organ (white bars)

Further studies were performed to determine whether rhIL-7 could induce LAK activity in vitro or in vivo. The overall results from these experiments indicated that rhIL-7 at doses as high as 1000 U/ml did not induce LAK in vitro within 7 days, while the twice daily administration of up to 100 μg/day rhIL-7 did not significantly augment LAK activity in vivo (data not shown). Thus, while rhIL-7 induces LAK from human cells (9,10), and recombinant mouse IL-7 (rmIL-7) induces LAK from mouse cells, (11), rhIL-7 does not efficiently induce LAK from mouse cells. Therefore, the generation of LAK activity is unlikely to be critical in the rhIL-7-induced reduction in the number of metastases in mice bearing Renca or MCA-38. Although NK/LAK cells may not be involved in the antimetastatic activity, subsequent studies have shown that there is a 9-fold increase in the number of $CD8^+$ lymphocytes that infiltrate into the lungs during the rhIL-7-induced regression of day 3 MCA-38 pulmonary metastases (14). These results demonstrate that rhIL-7 can have antimetastatic activity against early established metastases and suggest that T lymphocytes may play some role in those effects.

VI CONCLUSIONS AND IMPLICATIONS

The studies presented herein demonstrate that the twice daily administration of IL-7 preferentially stimulated a splenic lymphocytosis, largely through an increase in $B220^+$ cells. This effect has also been observed in the bone marrow (14). Perhaps, more interesting, was the ability of rhIL-7 to also alter the composition of T cell subsets. While there is a dose and time dependent increase in both $CD4^+$ and $CD8^+$ cells, the $CD8^+$ cells are increased to a greater degree resulting in decrease in the CD4/CD8 ratio from 1.5:1 to 1:2.

Further studies have shown that rhIL-7 also induces significant increases in the $CD8^+$ T cell subset in both the spleens and lungs of mice bearing MCA-38 pulmonary micrometastases (14). This increase in $CD8^+$ T cells coincides with an rhIL-7 induced decrease of 40-90% in the number of pre-existent metastatic tumor foci in the lungs. Studies are in progress to determine whether these $CD8^+$ T cells actually contribute to therapeutic efficacy of rhIL-7. In contrast rhIL-7 has no demonstrable effects on either NK or LAK activities in either normal or tumor-bearing mice suggesting that this mechanism may not be involved in the antimetastatic effects of rhIL-7, and demonstrating that rhIL-7 may not fully cross species since rhIL-7 induces human LAK (9,10) and rmIL-7 increases mouse LAK (11).

Thus, IL-7 may be of therapeutic value in promoting lymphopoietic recovery in immunodeficient states, either iatrogenic in origin (e.g., bone marrow transplantation, chemotherapy and/or radiation therapy treatment of cancer) or disease-related (e.g., AIDS). Preliminary studies from our laboratory suggest that rhIL-7 may accelerate the regeneration of the lymphoid compartment in both chemotherapy-treated (13) and irradiated (Boerman et al., manuscript in preparation) mice. It also may be of value in stimulating the host response against tumors.

References

1. Namen, A.E., Lupton, S., Hjerrild, K., Wignall, J., Mochizuki, D.Y., Schmierer, A., Mosley B., March, C.J., Urdal, D., Gillis, S., Cosman, D., and Goodwin, R.G. Stimulation of B-cell progenitors by cloned murine interleukin-7, Nature. 1988; 233:571.
2. Namen, A.E., Schmierer A.E., March, C.J., Overell, R.W., Park, L.S., Urdal, D.L. and Mochizuki, D.Y. B cell precursor growth-promoting activity - Purification and characterization of a growth factor active on lymphocyte precursors, J. Exp. Med. 1988; 167:988.
3. Goodwin, R.G., and Namen, A.E. The cloning and characterization of interleukin-7, The Year in Immunology, Vol. 6 (J.M. Cruse and R.E. Lewis Jr., eds.), Karger, Basel, Switzerland, 1990; p. 127.
4. Goodwin R.G., Lupton S., Schmierer, A., Hjerrild, K.J., Jerzy, R., Clevenger, W., Gillis, S., Cosman, D., and Namen, A.E. Human interleukin 7: Molecular cloning and growth factor activity on human and murine B-lineage cells. Proc. Natl. Acad. Sci. USA. 1989; 86, 302.
5. Watson, J.D., Morrissey, P.J., Namen, A.E., Conlon, P.J., and Widmer, M.B. Effect of IL-7 on the growth of fetal thymocytes in culture, J. Immunol. 1989; 143:1215.
6. Conlon, P.J., Morrissey, P.J., Nordan, R.R., Grabstein, K.H., Prickett, K.S., Reed, S.G., Goodwin, R., Cosman, D., and Namen, A.E. Murine thymocytes proliferate in direct response to interleukin 7. Blood 1989; 74:1368.
7. Morrissey, P.J., Goodwin, R.G., Nordan, R.P., Anderson, D., Grabstein, K.H., Cosman, D., Sims, J., Lupton, S., Acres, B., Reed, S.G., Mochizuki, D., Eisenman, J., Conlon, P.J. and Namen, A.E. Recombinant interleukin 7, pre-B cell growth factor, has costimulatory activity on purified mature T cells. J. Exp. Med. 1989; 169:707.
8. Welch, P.A., Namen, A.E., Goodwin, R.G., Armitage, R., and Cooper, M.D. Human IL-7: A novel T cell growth factor. J. Immunol. 1989; 143:3562.
9. Alderson, M.R.,, Sassenfeld, H.M., and Widmer, M.D. (1990). Interleukin 7 enhances cytolytic T lymphocyte generation and induces lymphokine-activated killer cells from human peripheral blood. J. Exp. Med. 1989; 172:577.
10. Stötter, H., Custer, M.C., Bolton, E.S., Guedez, L., and Lotze, M.T. IL-7 induces human lymphokine-activated killer cell activity and is regulated by IL-4. J. Immunol. 1991; 146:150.
11. Lynch, D.H., and Miller, R.E. Induction of murine lymphokine-activated killer cells by recombinant IL-7, J. Immunol. 1990; 145:1983.
12. Namen, A.E., Williams, D.E., and Goodwin, R.G. A new hematopoietic growth factor. Hematopoietic Growth Factors in Transfusion Medicine (J. Spivak, W. Drohan, an D. Donley, eds.), Wiley-Liss, Inc., New York, 1990; p.65.

13. Damia, G., Komschlies, K.L., Faltynek, C.R., Ruscetti, F.W., and Wiltrout, R.H. Administration of recombinant human interleukin 7 alters the frequency and number of myeloid progenitor cells in the bone marrow and spleen of mice. Blood, in press.

14. Komschlies, K.L., Gregorio, T.a., Gruys, E., Back, T., Faltynek, C.R., and Wiltrout, R.H. Administration of recombinant human interleukin 7 to mice alters the composition of B lineage cells and T cell subsets, and induces regression of established metastases. In press.

15. Salup, R.R., Back, T.A., and Wiltrout, R.H. Successful treatment of advanced murine renal cell cancer by bicompartmental adoptive chemoimmunotherapy. J. Immunol. 1987; 138:641.

16. Wiltrout, R.H., Boyd, M.R., Back, T.C., Salup, R.R., Arthur, J.A., and Hornung, R.L. Flavone-8-acetic acid augments systemic natural killer cell activity and synergizes with interleukin 2 for treatment of murine renal cancer. J. Immunol. 1988; 140:3261.

17. Hornung, R.L., Back, T.C., Zaharko, D.S., Urba, W.J., Longo, D.L., and Wiltrout, R.H. Augmentation of natural killer (NK) activity, induction of interferon and development of tumor immunity during the successful treatment of established murine renal cancer using flavone acetic acid (FAA) and interleukin 1. J. Immunol. 1988; 141:3671.

18. Mace, K.F., Hornung, R.C., Wiltrout, R.H., and Young, H.A. Induction of cytokine gene expression in vivo by flavone acetic acid: Strict dose dependency and correlation with therapeutic efficacy against murine renal cancer. Cancer Res. 1990; 50:1742.

19. Futami, H., Hornung, R.L., Back, T.T., Gruys, M.E., and Wiltrout, R.H. Effect of systemic alkalinization on biologic response modification and therapeutic antitumor efficacy of flavone acetic acid plus recombinant interleukin 2. Cancer Res. 1990; 50:7926.

20. Sayers, T.J., Wiltrout, T.A., McCormick, K., Husted, C., and Wiltrout, R.H. Antitumor effects of the IFNα and IFN on a murine renal cancer (Renca) in vitro and in vivo. Cancer Res. 1990; 50:5414.

10

Up-Regulation of Tumor Cell Sensitivity to Natural Cell-Mediated Cytotoxicity by UV Light Irradiation

Mirsada Begovic
Ronald B. Herberman
Elieser Gorelik
Pittsburgh Cancer Institute and
University of Pittsburgh School of Medicine
Pittsburgh, Pennsylvania

I INTRODUCTION

Natural cell-mediated immunity could play an important role in the elimination of tumor cells and preventing their local and metastatic growth (1). Indeed, stimulation or inhibition of natural effector cells resulted in inhibition or stimulation, respectively, of local tumor growth and metastatic spread (1). The efficiency of natural immunity was found also to be dependent on the level of tumor cell sensitivity to natural effector cells. It is considered that natural effector cells are mostly comprised of natural killer (NK) and natural cytotoxic (NC) cells. NK cells are a morphologically and phenotypically distinct population of large granular lymphocytes (LGLs) that are $CD3^-$, $CD16^+$ and $CD56^+$. In mice NK cells express asialo GM1 and NK1.1 or NK1.2 determinants. NK cells are capable of killing tumor cells after short (4h) exposure, probably via the exocytosis of cytoplasmic cytolytic granules (2). In contrast, NC cells lyse tumor cells after prolonged (12-18h) incubation (3-6) and their cytotoxicity appears to be mediated by the release of TNF, since NC activity can be blocked in the presence of anti-TNF antibodies (7-9). NC activity has been associated with a variety of cell types, including T-, B-, NK, and mast cells, that are capable of generation of TNF (6-9).

The sensitivity of tumor cells to NK- and NC cell cytotoxicity varies in a broad range. Some tumor cells have preferential sensitivity to NK and other to NC cells. YAC-1 cells are highly sensitive to NK and resistant to NC cells, whereas WEHI-164 and L929 cells are resistant to NK and sensitive to NC cells. Some tumor cells can be lysed by both NK and NC cells (4). Mechanisms responsible for tumor cell resistance to lysis by

NK and NC cells are mostly unknown. It is considered that NK resistance of tumor cells could be due to a lack of appropriate NK recognizable determinants, failure to switch on the lytic program, or their resistance to lytic machinery (2). Resistance of tumor cells to NC activity is probably due to their resistance to NC cell generated TNF. Indeed, tumor cell lines that manifested sensitivity to cytotoxic action of TNF were also sensitive to NC cell-mediated lysis (7-9).

It is believed that the antitumor efficiency of natural immunity could be substantially increased by stimulation of the immune system with various biological response modifiers. On the other hand, its antitumor activity would be higher if to find an appropriate approach of converting tumor cell resistance to NK and NC cell-mediated cytotoxicity. Several agents have been found to modify tumor cell sensitivity to natural cell-mediated cytotoxicity (NCMC), but this effect is rather transient. In addition, the same treatment could differently affect tumor cells sensitivity to NK or NC cells. For example, interferon treatment of tumor cells makes them more sensitive to NC/TNF-mediated cytotoxicity (10) and more resistant to NK cell lysis (2).

It is desirable to develop an approach of stable increase of tumor cells sensitivity to natural effector cells. It might help to clarify the mechanisms of regulation of tumor cell sensitivity to NCMC and might be useful for potentiation of the antitumor activity of natural cell-mediated immunity.

We previously demonstrated that treatment of murine BL6 melanoma cells with N-methyl-N-nitro-nitroso-guanidine (MNNG) resulted in a significant increase in tumor cell immunogenicity (11) as well as a stable increase in sensitivity to both NK and NC cells (12). We also found that similar to MNNG, in vitro treatment of murine 3LL Lewis lung carcinoma, MCA102 and MCA105 fibrosarcomas with short-wave UVC light also resulted in substantial increase in their immunogenicity (13-15). UV-treated tumor cells grew progressively in immunosuppressed (x-irradiated) C57BL/6 mice or T-cell deficient nude mice (13-15). Immunocompetent C57BL/6 mice were capable of rejecting $1X10^6$-$2X10^6$ inoculated UV-treated MCA102UV and MCA105UV tumor cells. However, at lower doses ($1X10^5$-$5X10^5$) MCA102UV tumor cells were also rejected in 80-60% of nude mice, respectively.

Failure of the lower doses of tumor cells to grow in nude mice might have been due to natural antitumor resistance since pretreatment of nude mice with NK-depressive antibodies resulted in tumor development. These observations might indicate that UV irradiation not only increased tumor cell immunogenicity but also increased their sensitivity to natural cell-mediated immunity.

In this article we will summarize our studies of the effect of UV light irradiation of tumor cell sensitivity to NCMC (14-16).

II CHARACTERIZATION OF EFFECTOR CELLS INVOLVED IN LYSIS OF UV-TREATED AND UNTREATED TUMOR CELLS

First, we tested whether UV treatment of MCA102 and MCA105 fibrosarcoma cells render them sensitive to lysis by normal spleen cells of C57BL/6 mice. Tumor cells were irradiated in vitro with UVC light from a germicidal lamp under laminar flow hood with a total dose of 610 J/m^2, the incident-dose rate was 1.2 W/m^2 as measured by a Black-Ray UV meter, Model J225 (American UV Co., Chatham, NJ) (14-16).

Irradiated cells initially adhered to the plastic but after 1-2, days the vast majority of cells died and detached and only a very small proportion of cells remained adherent to the flask and slowly expanded. It took 3-4 weeks for surviving cells to saturate a T-75 culture flask. Expanded tumor cells were harvested and irradiated again. However a second course of UV irradiation with 610 J/m^2 was lethal for all cells. Therefore, tumor cells were irradiated for a second time with 457 J/m^2. Cells that survived two courses of UV irradiation were expanded and termed MCA102UV and MCA105UV cells.

UV irradiation induced some morphological changes in tumor cells. The original MCA102 and MCA105 tumor cells were strongly adherent fibroblast-like cells. MCA102UV and MCA105UV cells were less plastic adherent, more round without long processes, and they grew in close contact. In addition, UV-treated tumor cells demonstrated in vitro 3-4 times higher rate of proliferation in comparison to the parental cells.

When cytotoxic activity of normal spleen cells was tested in a 4h ^{51}Cr release assay, very low levels of tumor cell lysis were observed. Cytotoxic activity of normal spleen cells became apparent after 18h incubation with the radiolabeled tumor cells and was substantially higher against UV-treated tumor cells than the parental nontreated tumor cells (Figure 1).

Cytotoxic activity of normal spleen cells after prolonged (18h) incubation with tumor cells could be a result of action of NK as well as NC cells.

Therefore, we analyzed the involvement of NK and NC cells in lysis of UV-treated and nontreated tumor cells by comparing the cytotoxic activity of normal spleen cells containing both NK and NC cell activity (NK+, NC+) with: a) normal spleen cells in the presence of anti-TNF Abs (NK+, NC-); b) NK-depleted or NK-deficient spleen cells (NK-, NC+); and c) NK-depleted or -deficient spleen cells which NC activity was blocked by anti-TNF Abs (NK-, NC-) (15,16).

To block NC activity, serum of New Zealand rabbits immunized with recombinant murine TNF-alpha (IP-400, Genzyme, Boston, MA) was used. One ml of this serum contains 1x10^6 TNF-alpha neutralizing units (nu). Our titration experiments show that 2000-3000 nu of the anti-TNF antibodies were sufficient to block NC activity of spleen cells against NC sensitive WEHI-164 or L929 cells.

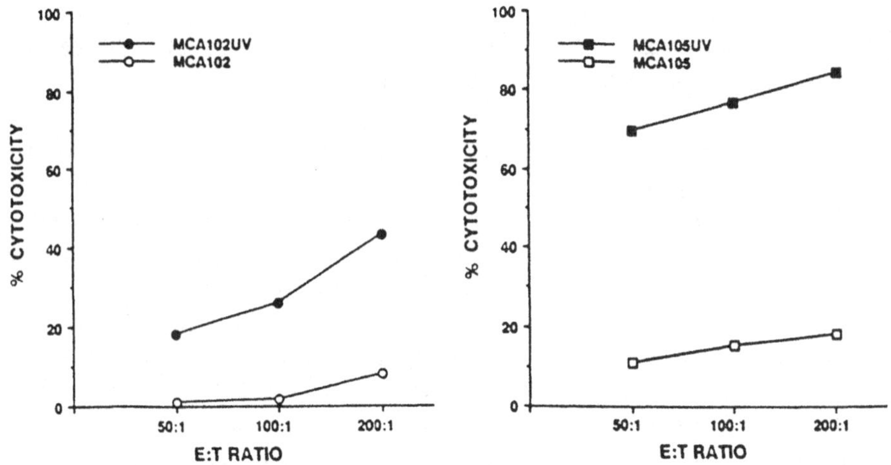

Fig. 1 Effect of UV light treatment on sensitivity of tumor cells to natural cell-mediated cytotoxicity. UV-treated and untreated MCA102 and MCA105 fibrosarcoma cells were labeled with ^{51}Cr. Radiolabeled tumor cells were mixed with spleen cells of C57BL/6 mice and incubated for 18 h. The levels of ^{51}Cr released into supernatants were determined and percent cytotoxicity was calculated.Differences in mean cytotoxicity above 10% were significant ($p < 0.05$) according to the Student's t test

To evaluate the level of NK and NC activity of the tested effector cells, NK-sensitive, NC-resistant YAC-1 and NK-resistant, NC-sensitive L929 and WEHI-164 target cells were included as selective positive controls and tested in parallel with UV-treated and nontreated MCA102 and MCA105.

The results presented in Table 1 demonstrate that MCA102UV and MCA105UV cells are more sensitive to normal spleen cell cytotoxicity than the parental MCA102 and MCA105 cells. Anti-TNF Abs partially inhibited the cytotoxic activity of normal spleen cells against UV-treated and nontreated MCA105 and MCA102 tumor cells. The same antibody treatment completely inhibited lysis of L929 cells (NC/TNF sensitive, NK-resistant) and had no effect on lysis of YAC-1 cells (NK-sensitive and NC/TNF-resistant). These results indicated the involvement of NC cells in lysis of the UV-treated and nontreated MCA105 and MCA102 tumor cells. Since anti-TNF antibodies had partial inhibitory effect, it might indicate that the residual cytotoxicity is mediated by other effector cells. Selective NK cell inhibition by pretreatment of the spleen donors with anti-asialo GM1 serum completely abrogated lysis of YAC-1 cells and had no effect on lysis of L929 cells. Such NK cell inhibition significantly reduced lysis of MCA102UV cells but had no significant effect on lysis of MCA105 and MCA102 cells. In the presence of anti-TNF Abs, the cytotoxicity of NK-depleted spleen cells against MCA105 and MCA102 cells was completely abrogated and was further reduced against MCA102UV and MCA105UV cells (Table 1).

Table 1
Analysis of the effector cells involved in lysis of
UV-treated and nontreated MCA102 and MCA105 tumor cells

Spleen cells[a]	Anti-TNF Abs	% Cytotoxicity, E:T 100:1					
		YAC-1	L929	MCA102	MCA102UV	MCA105	MCA105UV
Normal							
(NK^+, NC^+)	--	40.1	48.2	21.2	68.2	29.8	56.4
(NK^+, NC^+)	+	38.6	4.7	10.5	56.2	11.0	32.9
asGM1⁻ treated							
(NK^-, NC^+)	--	3.9	49.3	16.0	54.4	29.8	42.0
(NK^-, NC^-)	+	4.7	5.2	4.8	34.8	5.4	22.4
Beige							
(NK^-, NC^+)	--	6.6	NT	17.4	51.5	26.2	40.2
(NK^-, NC^-)	+	7.1	NT	2.9	28.9	-7.7	23.6

[a]Radiolabeled tumor cells were mixed with spleen cells of normal, anti-asialo GM1-treated C57BL/6 or beige mice and incubated for 18 h. The ability of anti-TNF Abs (3000 nu/well) to block the cytotoxicity of these spleen cells was also tested. In parenthesis are indicated the expected natural effector cell profile for each experimental group.

Differences in mean cytotoxicity above 10% were significant ($p < 0.05$) according to the Student's t test

A similar pattern of results was obtained when the targets were tested for susceptibility to lysis by spleen cells from beige mice (Table 1). Spleen cells of beige mice showed very low levels of cytotoxicity against YAC-1 cells but they were as efficient as spleen cells of C57BL/6 mice in lysis of MCA102 and MCA105 cells. However, lysis of MCA102UV and MCA105UV cells by spleen cells of beige mice was significantly lower than by normal spleen cells. In the presence of anti-TNF Abs, the cytotoxic activity of spleen cells of beige mice against MCA102 and MCA105 cells was completely neutralized, and there was substantial but not complete reduction of MCA102UV and MCA105UV cell lysis (Table 1).

Thus, these data indicate that the relatively low levels of lysis of the original MCA102 and MCA105 cells were mostly mediated by NC cells with very low contribution of NK cells. UV treatment of these tumor cells substantially increased their sensitivity to NC- and NK cell-mediated cytotoxicity.

To test dose-effect of UV irradiation on tumor cell sensitivity to NCMC, MCA102 tumor cells were irradiated with various doses of UV light (76-610 J/m^2). Surviving cells were expanded and their sensitivity to lysis by normal spleen cells was tested 6 weeks after irradiation (Figure 2). A significant increase in tumor cell sensitivity to NCMC was observed with cells irradiated with 152 J/m^2. With a further increase in dose of irradiation, no significant increase in tumor cell sensitivity to lysis by normal spleen cells was found (Figure 2). When irradiation with the same dose of UV light was repeated no further increase in their lysability was observed.

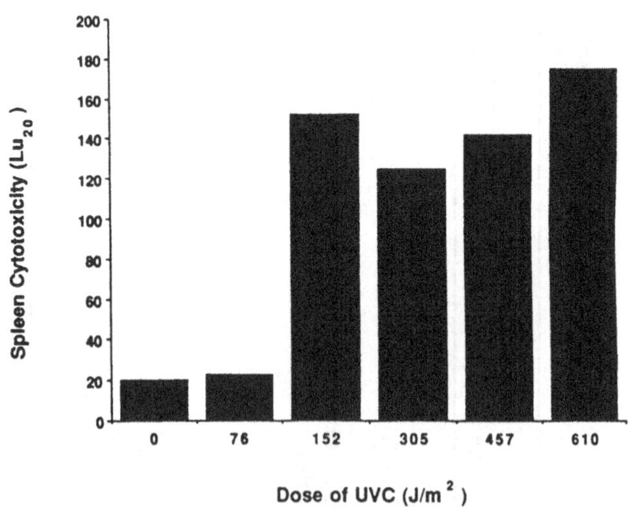

Fig. 2 Cytotoxic activity of normal spleen cells against MCA102 tumor cells irradiated with various doses of UVC light. MCA102 tumor cells were irradiated with 76-610 J/m^2 of UVC light. Cells were expanded and their sensitivity to cytotoxic activity of normal spleen cells was tested in an 18h ^{51}Cr release assay

III EFFECT OF UV LIGHT IRRADIATION OF TUMOR CELL SENSITIVITY TO TNF CYTOTOXICITY

NC activity of the effector cells appears to be mediated mainly by the release of TNF. Therefore, it was expected that the increase in NC sensitivity of the UV-treated tumor cells would be reflected in increased sensitivity to TNF. Indeed, MCA102UV and MCA105UV tumor cells showed significantly higher sensitivity to TNF cytotoxicity than the parental tumor cell line. The magnitude of TNF sensitivity of the MCA102UV and MCA105UV cells was comparable to that of WEHI-164 and L929 cells, which are considered to be among the most TNF-sensitive experimental tumor cells.

These experiments were performed with tumor cells irradiated with 2 doses of UV light (610 and 457 J/m^2). We studied whether lower doses of UV irradiation were also efficient in augmentation of tumor cell sensitivity to TNF lysis. MCA102 tumor cells that were irradiated with 76-610 J/m^2 were incubated with human recombinant TNF-α for 18h. The data obtained show that MCA102 tumor cells irradiated with 152-610 J/m^2 became sensitive to TNF lysis and doses of 76 J/m^2 did not increase tumor cell sensitivity to TNF. UV treatment was repeated and surviving cells were expanded. Two repeated courses of 76 J/m^2 did not increase tumor cell sensitivity, whereas other dose of repeated irradiation show further increase in TNF sensitivity in comparison to cells that received a single dose of UV treatment (Figure 3).

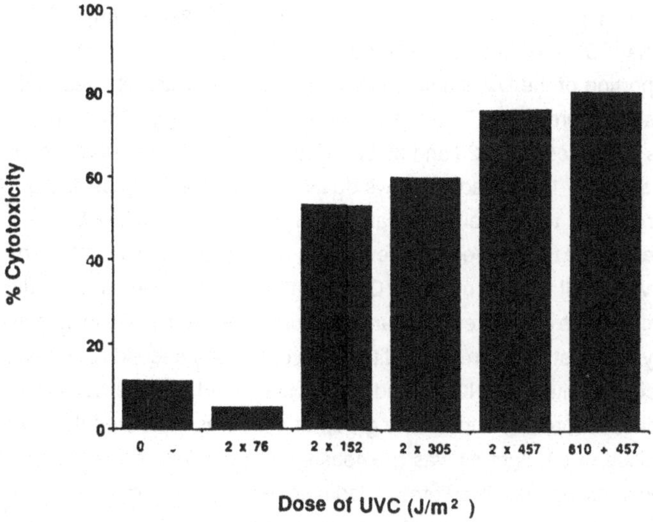

Fig. 3 Cytotoxic activity of human TNF against MCA102 tumor cells irradiated with 2 courses of UVC light. MCA102 tumor cells were irradiated twice with various doses of UVC light. Cells were expanded and their sensitivity to TNF lysis was tested after 18h of incubation with 1000 u/well of human recombinant TNF-alpha

Thus, these data indicate that treatment of MCA102 or MCA105 tumor cells with UVC light render them highly sensitive to TNF lysis. This effect of UV irradiation was rather stable since this TNF sensitivity was permanently induced in the treated tumor cell lines.

IV IMMEDIATE EFFECT OF UV LIGHT IRRADIATION ON TUMOR CELL SUSCEPTIBILITY TO NCMC AND RTNF

Tumor cell sensitivity to NCMC and TNF in the experiments mentioned above usually was tested 1-2 months after single or repeated doses of UVC irradiation in order to expend heavily irradiated cells. It remains unknown how fast UV irradiation could increase tumor cell sensitivity to TNF and NCMC and whether it is a result of selection of the UV resistant variants that happen to be more sensitive to TNF and NCMC or it is due to immediate UV light induction of this sensitivity that can be further maintained. In the present study the immediate effect of single and multiple UV treatments on susceptibility of MCA102 tumor cells to lysis by TNF and spleen cells was investigated.

To study this we irradiated ^{51}Cr-labeled MCA102 tumor cells with a range of doses of UVC light (38-304 J/m^2). Irradiated tumor cells were incubated for 18h with spleen cells of C57BL/6 mice or human rTNF and levels of their lysis was determined. UV irradiation significantly increased MCA102 tumor cell sensitivity to lysis by either spleen cells or TNF. The increase in tumor cell sensitivity to lysis by spleen cells was similar regardless of the dose of UV light applied. In contrast, increase in TNF sensitivity was higher with 152-304 J/m^2 UVC irradiation than with 38-76 J/m^2. No increase in the spontaneous ^{51}Cr release was observed.

A portion of the UV irradiated cells (5x10^5), that was not used in the cytotoxicity assay, was transferred into T-25 flasks and cultured in vitro. During the first 48h some tumor cells irradiated with 228 and 304 J/m^2 died and detached from the plastic, the rest continued to grow. It resulted in 3 days delay in saturation of these flasks in comparison to other groups of tumor cells irradiated with lower doses of UV light. The saturated cultured cells were transferred into new flasks and no obvious inhibition in tumor growth was observed in all groups of cells. Cells continue to culture during additional transfer generation. Twelve days after UV irradiation tumor cell sensitivity to spleen cell and TNF cytotoxicity was tested again. In addition, the effect of a second round of UV irradiation on tumor cell sensitivity to NCMC and TNF was studied. Tumor cells were labeled with ^{51}Cr and divided into 2 groups. One group of cells was irradiated the second time with the same dose of UV light as was previously utilized. The second portion of the tumor cell suspensions was not irradiated. Radiolabeled tumor cells were mixed with normal spleen cells or TNF and percent of cytotoxicity was determined after 18h of incubation.

Tumor cells that were irradiated 12 days before still showed a higher level of sensitivity to spleen cell cytotoxicity than the nonirradiated parental MCA102 cells (Figure 4A), to about the same extent as that observed just after UV treatment. Tumor cells exposed to a second course of UV irradiation showed a further increase in their susceptibility to lysis by spleen cells (Figure 4A). Applied doses of UV irradiation did not affect tumor cells immediate viability since it did not increase the level of the spontaneous ^{51}Cr release.

At twelve days after the first UV treatment, tumor cells also maintained higher sensitivity to TNF cytotoxicity (Figure 4B). The second UV treatment was found to be very effective in further augmentation of the tumor cells' sensitivity to TNF lysis. As seen initially, the repeated doses of 152-304 J/m^2 were more effective in augmentation of TNF sensitivity than lower doses (38-76 J/m^2) of UV light (Figure 4B).

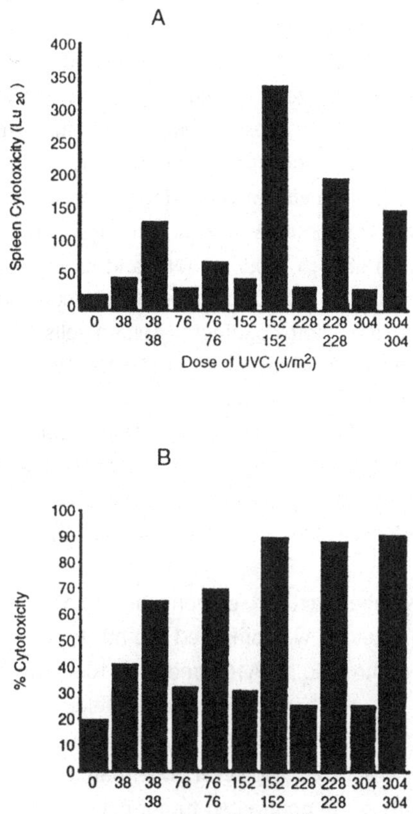

Fig. 4 Immediate effect of UVC irradiation on tumor cells sensitivity to lysis by spleen Fcells (A) or TNF-alpha (B). MCA102 tumor cells were irradiated with various doses of UVC light and 12 days later cells were labeled with ^{51}Cr and some cells were irradiated again. Immediately after irradiation radiolabeled cells were mixed with normal spleen cells of C57BL/6 mice (A) or human recombinant TNF-alpha (1000 u/well) (B). Percent tumor cell lysis was determined after 18h of incubation. Percent spontaneous ^{51}Cr release in UV-treated and nontreated tumor cells was around 30-35%

Thus, these data indicate that UVC irradiation render MCA102 tumor cells sensitive to NCMC and TNF. This effect of UV irradiation appeared shortly after irradiation. As we demonstrated above MCA102 tumor cells treated once or 2 times with UV irradiation (152-610 J/m^2) could be maintained in culture as permanent lines at least for 3 years without loosing their sensitivity to NCMC and TNF as well as their immunogenicity (14).

V INTERACTION OF NK AND NC CELLS IN LYSIS OF UV-TREATED TUMOR CELLS

Although spleen cells are capable of lysing tumor cells by NK and/or NC cell-mediated mechanisms, involvement of these mechanisms in tumor cell lysis depends on tumor cell sensitivity. Thus, YAC-1 cells are lysed by NK and WEHI-164 by NC-mediated cytotoxicity. As we showed above, UV-treated MCA102 and MCA105 tumor cells are lysed by both NK and NC cells. NK cell activity can be stimulated by various agents or lymphokines (IFN, IL-2), whereas no stimulation of NC activity can be achieved by these or any other approaches (3-6). It was of interest to investigate whether stimulation of NK cells could affect the NC contribution in tumor cell lysis. This prediction is based on findings that NK cells are able to produce TNF and its production is potentiated with stimulation of NK cells with IL-2 or biological response modifiers (17).

For this purpose the cytotoxic effect of spleen cells of mice stimulated with poly I:C (100 ug/mouse) against UV-treated and nontreated tumor cells was tested in the presence or absence of anti-TNF antibodies.

Poly I:C treatment significantly increased spleen cell cytotoxicity against YAC-1, MCA105UV and MCA102UV cells, but not against parental MCA105 and MCA102, confirming higher NK sensitivity of UV-treated tumor cells. Anti-TNF Abs reduced the cytotoxicity of NK-stimulated spleen cells but less efficiently than that of normal spleen cells.

IL-2 could substantially activate spleen cell cytotoxicity even after a short period (18 h) of incubation. Therefore, we compared the ability of IL-2 stimulated spleen cells to lyse UV-treated and nontreated MCA102 and MCA105 tumor cells. IL-2 (1000/ml) was added directly into the culture of spleen and radiolabeled tumor cells during the 18 h of their incubation. IL-2 stimulation resulted in increased lysis of both UV-treated and nontreated tumor cells (Table 2). However, UV-treated tumor cells were significantly more sensitive to lysis than the nontreated tumor cells. Anti-TNF antibodies were less efficient in blocking the cytotoxicity of IL-2-stimulated than normal spleen cells, indicating that the majority of tumor cells were killed by TNF-independent mechanisms (Table 2). When LAK cells were generated after 3 days of incubation of spleen cells with IL-2 (1000 u/ml), they showed high lytic activity against MCA102 and MCA105 tumor cells, although UV-treated tumor cells were still more sensitive to lysis by LAK cells. Anti-TNF antibody failed to substantially inhibit the cytotoxicity of LAK cells in an 18 h cytotoxicity assay, suggesting that the LAK cells only minimally utilized TNF-mediated mechanisms in their cytotoxicity. This is somewhat surprising in the light of findings that after IL-2 stimulation a substantial increase in TNF production was usually observed (17).

Table 2
Cytotoxic effects of IL-2-stimulated spleen cells on UV-treated tumor cells in the absence or presence of anti-TNF antibodies

Spleen cells[a]	Anti-TNF Abs	% Cytotoxicity, E:T ratio 100:1 MCA102		MCA102UV		MCA105		MCA105UV	
Normal	--	15.2		62.3		34.8		52.2	
	+	7.6	(50)[b]	37.3	(40)	5.4	(85)	40.7	(22)
Normal	--	51.2		85.8		40.8		74.4	
+ IL-2	+	38.2	(25)	72.5	(15)	25.1	(38)	69.5	(7)

[a]Spleen cells of C57BL/6 mice were mixed with the radiolabeled tumor cells and IL-2 (1000 u/ml) was added. In some groups, anti-TNF Abs (3000 nu/well) were also added. Differences between mean cytotoxicity above 7% were significant ($p < 0.05$).

[b]In parenthesis: percent reduction of spleen cytotoxicity in the presence of anti-TNF Abs.

One possible explanation for this finding was that the stimulated NK cells might be able to lyse tumor cells rapidly, before a major contribution by the NC cell/TNF mechanism that can be blocked by anti-TNF Abs. To analyze this possibility, we compared the cytotoxicity of normal and poly I:C stimulated spleen cells in 4 and 18 h ^{51}Cr-release assay in the presence or absence of anti-TNF antibodies. Although normal spleen cells showed low cytotoxic activity after 4h of incubation with the radiolabeled target cells, Poly I:C-stimulated spleen cells were rather efficient against MCA105UV cells even after short incubation (Figure 5). WEHI-164 cells that considered to be NK resistant also show some increase in lysis by Poly I:C-stimulated spleen cells after 4h of incubation. However, after 18h of incubation no differences in lysis of WEHI-164 by normal or Poly I:C-stimulated spleen cells was found. In the presence of anti-TNF antibodies cytotoxicity of normal spleen cells was substantially inhibited. These antibodies were less efficient in reduction of cytotoxic activity of Poly I:C-stimulated spleen cells after 18h of incubation (Figure 5). Thus, the failure of anti-TNF antibodies to effectively block the cytotoxicity of Poly I:C - stimulated spleen cells was due to their ability to rapidly lyse tumor cells by TNF-independent mechanisms prior to contribution of released TNF. The ability of anti-TNF antibodies to neutralize spleen cell-mediated lysis of various tumor cells seems to be influenced considerably by the level of NK cell activity and the NK or NC sensitivity of the tested targets.

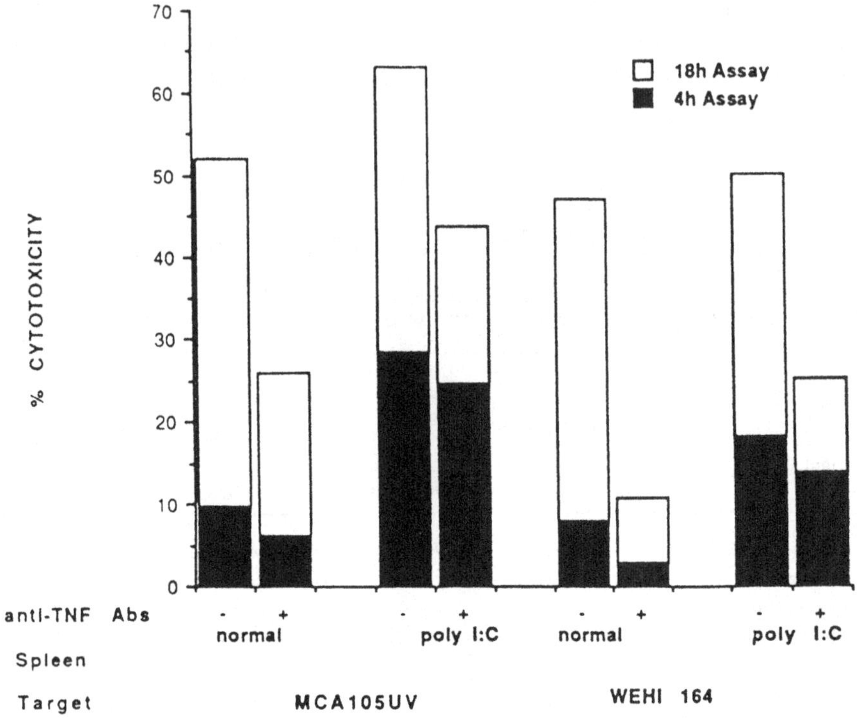

Fig. 5 Cytotoxic activity of normal and Poly I:C stimulated spleen cells after short (4h) or long (18h) incubation in the presence or absence of anti-TNF antibodies. C57BL/6 mice were treated with Poly I:C (100 ug/mouse). Spleen cells of normal or Poly I:C stimulated mice were mixed with the radiolabeled tumor cells at effector:target ratio 200:1. In some wells the anti-TNF antibodies (3000 nu/well) were added. Percent cytotoxicity was determined after 4h (close bars) and 18h (open bars) of incubation

If TNF sensitivity is an important factor in determining tumor cells sensitivity to lysis by NC cells, one could expect that the selection of tumor cell for resistance to TNF would result in resistance to NC-mediated cytotoxicity. It was also of interest to study whether resistance to TNF/NC activity would affect the sensitivity of UV-treated tumor cells to NK cell lysis. To test this, MCA102UV tumor cells were cultured with increasing concentrations of TNF (125-10000 u/ml) and maintained in the presence of 1000 u/ml of TNF.

MCA102UV tumor cells cultured in the presence of TNF became resistant to TNF cytotoxicity. In parallel, these cells ($MCA102UV_r$) became less sensitive to lysis by spleen cells. Lysis of $MCA102UV_r$ cells was not significantly affected by anti-TNF antibodies, suggesting that their lysis was mediated by NK cells.

Next we tested whether TNF-induced resistance of UV-treated tumor cells to spleen cell cytotoxicity could be overcome by their stimulation with IL-2, as well as whether the addition of exogenous IL-2 and TNF to the mixture of tumor and spleen cells could potentiate the destruction of the tumor cells.

In the presence of IL-2, the cytotoxic activity of spleen cells against TNF sensitive and TNF-resistant target cells substantially increased. However, differences in lysability of these tumor cells still remained. By adding exogenous TNF, an increase in lysis of only TNF-sensitive MCA102UV was observed and no significant addition in lysis of MCA102 and $MCA102UV_r$ lines were found. Potentiation of spleen cell lysis of MCA102UV targets by the exogenous TNF suggests that the production of the endogenous TNF by spleen cells was not optimal to provide maximal lysis of these tumor cells. However, when TNF and IL-2 were added to the spleen and tumor cell culture, TNF did not show further potentiation of the cytotoxic activity of spleen cells (Table 3).

Table 3
Effect of exogenous IL-2 or TNF on the cytotoxic activity of spleen cells against MCA102 and MCA102UV tumor cells

Spleen cells[a]	IL-2	TNF	% Cytotoxicity MCA102	% Cytotoxicity MCA102UV	% Cytotoxicity $MCA102UV_r$[b]
--	--	+	9.0	34.5	6.8
+	--	--	10.8	36.6	14.9
+	+	--	23.9	63.0	52.8
+	--	+	14.5	54.3	21.4
+	+	+	24.9	69.5	42.9

[a]Radiolabeled tumor cells were incubated with normal spleen cells of C57BL/6 mice (effector:target ratio 100:1) in the presence or absence of IL-2 (1000 u/ml) and/or TNF (100 u/well). After 18 h of incubation percent of cytotoxicity was determined.
Differences in mean cytotoxicity above 10% were significant ($p < 0.05$).
[b] $MCA102UV_r$ tumor cells were cultured in the presence of increasing concentrations of TNF (125-10000 u/ml).

NK and NC cells differ in the kinetic of tumor cell lysis and utilization of different cytolytic machinery and manifest different requirement for Ca^{++}. It was previously demonstrated that NK cell cytotoxicity is Ca^{++} dependent (2). In contrast, lysis of tumor cells by rTNF was found to be Ca^{++} independent (18). Although the tumor cells lysis by TNF could be Ca^{++} independent, it remains unknown whether NC activity of spleen cells requires Ca^{++}. NC activity of spleen cells includes several steps such as interaction of

NC and tumor cells, and the stimulation of the production and release of TNF by NC cells. To test whether these processes are Ca^{++} dependent, spleen cell cytotoxicity against NC and NK sensitive tumor cells was tested in the presence of the Ca^{++} channel blocker, Verapamil, or an inhibitor of intracellular Ca^{++} mobilization, 8-(diethylamino) octyl-3,4,5-trimethoxybenzoate hydrochloride (TMB-8). Culture of tumor and spleen cells with the calcium chelator EGTA during 18 h was highly toxic for these cells and, therefore, this chemical was not used in our experiments.

When freshly prepared spleen cells were incubated with the radiolabeled WEHI-164 and MCA105UV tumor cells in the presence of Verapamil or TMB-8, these chemicals at the tested concentrations did not affect significantly the viability of spleen and tumor cells. In confirmation of the previous findings, Verapamil and TMB-8 did not significantly inhibit the cytotoxic activity of TNF against WEHI-164 or MCA105UV tumor cells. However, Verapamil at concentrations 10, 5 and 2.5 x 10^{-5} M inhibited spleen cell lysis of MCA105UV tumor cells by 72, 45 and 26%, respectively. TMB-8 only at the highest tested concentrations (5 x 10^{-5} M) inhibited lysis of these cells, by 38%. In contrast, TMB-8 and Verapamil did not significantly affect lysis of WEHI-164 by spleen cells.

Thus, these data demonstrate that TNF production and release during tumor and NC cell interaction as well as TNF-mediated lysis of tumor cells are Ca^{++} independent processes. In contrast, lysis of MCA105UV tumor cells could be substantially reduced by the presence of the Ca^{++} channel blocker. This is probably due to the involvement of NK cells in lysis of these tumor cells.

To test the affect of Verapamil and TMB-8 on NK cell activity, spleen cells were incubated with YAC-1 in the presence of these chemicals. However, Verapamil and TMB-8 were rather toxic for YAC-1 cells and induced a high level of ^{51}Cr release. Therefore, in the next set of experiments, spleen cells were preincubated overnight with Verapamil or TMB-8, washed and their NC and NK activity was tested against WEHI-164 and YAC-1 cells in an 18 h ^{51}Cr release assay. The cytotoxic activity of spleen cells preincubated with Verapamil or TMB-8 against YAC-1 targets substantially reduced. In contrast, NC activity of spleen cells tested against WEHI-164 was not significantly impaired by this pretreatment. Therefore, lysis of MCA105 tumor cells that is mediated by both NK and NC cells can be reduced but not abolished in the absence of Ca^{++}, since residual cytotoxicity is mediated by NC cells. Similar data were obtained by Richards and Djeu (19) when IL-2 cultured NK line was tested on their ability to express NK and NC activities in the presence or absence of Ca^{++}.

VI POSSIBLE MECHANISMS FOR INCREASED SUSCEPTIBILITY OF UV-TREATED TUMOR CELLS TO MCMC

To analyze the mechanisms responsible for the increase in NK sensitivity of UV-treated MCA102 and MCA105 tumor cells, we studied whether UV irradiation affected the recognition or lytic phase of the effector-target cell interactions.

Using a cold target inhibition assay, we did not find a significant difference between UV-treated and nontreated tumor cells in their ability to compete with YAC-1 cells for the effector cells, suggesting that the increase in NK sensitivity after UV irradiation was not associated with changes in the expression of NK-recognizable determinants. As an alternative possibility, UV light increased tumor cell sensitivity to the NK cell-derived lytic molecules.

It is believed that the cytolytic granules of LGLs play a crucial role in NK cell-mediated lysis of tumor cells (2). Therefore, it was of interest to test whether the differences in NK sensitivity of UV-treated and untreated tumor cells were paralleled by differences in their lysability by LGL-derived cytolytic granules. Indeed, MCA102UV and MCA105UV tumor cells were significantly more sensitive to lysis by LGL granules than nontreated parental tumor cells (Table 4).

Table 4
Sensitivity of UV-treated tumor cells to the cytotoxic action of purified LGL granules

	Spleen cells, E:T				LGL granules, units			
Target cells	200:1	100:1	50:1	25:1	60	30	15	7.5
MCA102	23	18	18	15	26	9	0	0
MCA102UV	45	36	30	22	58	39	20	10
MCA105	17	20	18	13	21	14	9	5
MCA105UV	61	54	41	28	38	29	22	19

Radiolabeled UV-treated and nontreated tumor cells were mixed with granules derived from rat LGL lymphoma cells and their cytotoxicity was determined after 1 h of incubation. One unit of the granules is equivalent to the amount of the granules required to lyse 50% of sheep erythrocytes. In parallel the sensitivity of the tested tumor cells to lysis by the normal C57BL/6 spleen cells was assessed in an 18 h assay.

An increase in MCA102UV and MCA105UV tumor cell sensitivity to NC cell-mediated cytotoxicity as could be shown to be a result of a UV-induced increase in their sensitivity to lysis by TNF. The ability of UV light to permanently increase NK, NC as well as TNF sensitivity of tumor cells is rather unique and no other modalities were reported to have the same effects. It was reported that TNF sensitivity of tumor cells could be augmented by chemicals that are capable of inhibiting RNA, DNA, protein synthesis (21-23), topoisomerase II activity (24,25), and ADP-ribosylation (26), as well as disrupting cytoskeleton (27), or altering arachidonic acid or phospholipase metabolism (28). However, TNF sensitivity induced by these agents is rather transient and observed only in the presence of these chemicals.

Our data indicate that MCA102UV and MCA105UV tumor cells have a 3-4 times higher rate of proliferation than nontreated parental MCA102 and MCA105 cells. This seems to exclude the possibility that inhibition of RNA, DNA or protein synthesis could account for increased TNF sensitivity of UV-treated tumor cells.

In spite of intensive studies, the mechanisms of TNF-mediated cytotoxicity remain unclear (28). It is considered that TNF binds to specific receptors (p55 and p75) and activates phospholipase A_2 (PLA_2), with release of arachidonate. Directly or indirectly, it can lead to activation of protein kinases. By unidentified mechanisms, TNF can induce production of membrane- and DNA-damaging oxygen free radicals or other reactive oxygen species. Such putative mechanisms may lead to activation of an endogenous endonuclease, which cuts the DNA into nucleosome-sized fragments in multiples of 185 bp (28). DNA fragmentation appears to be a central component of the apoptotic type of cell death mediated by TNF (29).

Which of these mechanisms might be affected by UV light and, in turn, render tumor cells sensitive to TNF lysis? We analyzed whether TNF sensitivity of UV-treated tumor cells was associated with alterations in TNF binding, internalization and/or degradation. In multiple experiments, UV-treated tumor cells did not show an increase in TNF binding. On the contrary, a substantial reduction in the ability to bind TNF was observed, with no significant differences in TNF internalization and degradation by MCA105 and MCA105UV tumor cells. Thus, the observed differences in TNF sensitivity cannot be explained by their ability to bind, internalize or metabolize TNF.

It has been reported that TNF lyses tumor cells by induction of DNA fragmentation. It is quite possible that short wave-length UV irradiation of tumor cells can induce some changes in their DNA, that make it more vulnerable to TNF-induced DNA fragmentation. To test this, tumor cells were labeled with ^{3}H-thymidine and then incubated for 18 h with TNF or spleen cells. After exposure of tumor cells to TNF (1000 U/well) 73.9% of DNA of the UV-treated MCA105 tumor cells was fragmented in comparison to 46.8% of the DNA parental MCA105 tumor cells.

In parallel, the ability of spleen cells to induce DNA fragmentation of tumor cells was tested. As a result of incubation of MCA105 and MCA105UV tumor cells with spleen cells, 21.4 and 56.9%, respectively, of their DNA was found to be fragmented. Thus, higher levels of DNA fragmentation induced by TNF or spleen cells in MCA105UV vs. MCA105 tumor cells were paralleled by their higher sensitivity to the cytotoxic action of these modalities.

It was previously demonstrated that Zn^{++} could inhibit the activity of endogenous endonucleases and, in parallel, reduce CTL-mediated lysis of the tumor cells (20). It remains unclear whether TNF cytotoxicity could be inhibited by Zn^{++}. To test this possibility WEHI-164, MCA102UV and MCA105UV tumor cells were incubated with rTNF

(100 U/well) in the presence of $ZnSO_4$. Concentrations above 100 mM of $ZnSO_4$ were toxic for tumor cells during 18 h of incubation. Therefore, tumor cells were incubated with TNF in the presence of 25-100 mM of $ZnSO_4$. Cytotoxic effect of TNF against the tested targets was significantly inhibited in the presence of $ZnSO_4$ (100 mM). These data suggest that TNF and CTLs are using common pathways in activation of the endogenous endonucleases and fragmentation of tumor cell DNA. However, it remains unknown how and by which mechanisms UV irradiation render tumor cell sensitive to TNF-mediated cytotoxicity. It is assumable that the stable UV-induced increase in TNF sensitivity of tumor cells might be due to the induction of DNA damage, that makes it more vulnerable to TNF-induced fragmentation and affects some genes that are involved in tumor cell protection from TNF damage. It is well established that UVC light irradiation might have immediate and long lasting effects on cellular DNA, with induction of cyclobutyl pyrimidine dimmer and (6-4) photoproduct formation. In contrast, long wavelength UVA irradiation mostly induces single and double-strand DNA breakage (30,31). Our preliminary data indicate that increase of MCA102 tumor cell sensitivity to NCMC and TNF was found after their irradiation with UVC and UVB, but not UVA light. Since the majority of UV-induced DNA damage is quickly repaired this type of damage could make tumor cell DNA more vulnerable to TNF-induced DNA fragmentation shortly after UV irradiation. However, in the permanently established UV-treated line the cyclobutyl pyrimidine dimers and (6-4) photoproducts will be repaired or diluted out with DNA replication. Therefore, in these lines UV-irradiation could induce permanent genetic changes that are responsible for regulation of TNF sensitivity.

Further investigation of this phenomenon will be rather important for characterization of the biological effects of UV irradiation. This study might have relevance for understanding the experimental and clinical data that demonstrate rather efficient antitumor effects of photodynamic therapy. This therapy utilizes photosensitizers and UVA irradiation of growing tumors, or extracorporal irradiation of the blood of leukemic patients (32,33). The precise mechanisms of antitumor effects of phototherapy remain unknown and hardly can be attributed solely to direct tumoricidal effects (32,33). Based on our data, it is possible to suggest that photodynamic therapy may increase tumor cell immunogenicity and their sensitivity to TNF and natural cell mediated immunity, and thereby help the host to reject the tumor cells. Thus, it will be important to test whether treatment with a photosensitizer and UVA irradiation could render tumor cells immunogenic and sensitive to TNF and NCMC, as well as whether the therapeutic efficacy of the phototherapy could be potentiated by stimulation of the host's immune system.

References

1. Gorelik, E. and Herberman, R. Role of natural killer (NK) cells in the control of tumor growth and metastatic spread. In: Cancer Immunology: Innovative

Approaches to Therapy. Herberman, R. (Ed.) New York Martinus Nijhoff Pubs. 1986; pp. 151-176.

2. Herberman, R., Reynolds, C. and Ortaldo, J. Mechanism of cytotoxicity by natural killer (NK) cells. Ann. Rev. Immunol.1986; 4:651-680.
3. Lattime, E., Pecoraro, G. and Stutman, O. Natural cytotoxic cells against solid tumors in mice. III. A comparison of effector cell antigenic phenotype and target cell recognition structure with those of NK cells. J. Immunol. 1981; 126:2011-2014.
4. Lattime, E., Pecoraro, G., Cuttito, M. and Stutman, O. Murine non-lymphoid tumors are lysed by a combination of NK and NC cells. Int. J. Cancer 1983; 32:523-528.
5. Lattime, E., Pecoraro, G. and Stutman, O. Natural cytotoxic cells against solid tumors in mice. IV. Natural cytotoxic (NC) cells are not activated natural killer (NK) cells. Int. J. Cancer 1981; 30:471-477.
6. Bykowski, M. and Stutman, O. The cells responsible for murine natural cytotoxic (NC) activity; a multi-lineage system. J. Immunol. 1986; 137:1120-1126.
7. Ortaldo, J., Mason, L., Mathieson, B., Liang, S., Frick, D. and Herberman, R. Mediation of mouse natural cytotoxic activity by tumor necrosis factor. Nature 1986; 321:700-702.
8. Okuno, T., Takagaki, Y., Pluznik, D. and Djeu, J. Natural cytotoxic (NC) cell activity in basophilic cells: Release of NC-specific cytotoxic factor by IgE receptor triggering. J. Immunol. 1986; 136:4652-4658.
9. Djeu, J., Lanza, E., Pastore, S. and Hepel, A. Selective growth of natural cytotoxic (NC) but not natural killer (NK) effector cells in interleukin-3. Nature 1983; 306:788-790.
10. Williamson, B., Carswell, E., Rubin, J., Prendergast, B., and Old, L. Human tumor necrosis factor produced by human B-cell lines: Synergistic cytotoxic interaction with human interferon. Proc. Natl. Acad. Sci. USA 1983; 80:5397-5401.
11. Gorelik, E., Peppoloni, S., Overton, R. and Herberman, R. Increase in H-2 antigen expression and immunogeneity of BL6 melanoma cells treated with N-methyl-N'-nitronitrosoquanidine. Cancer Res. 1985; 45:5341-5347.
12. Gorelik, E., Gunji, Y. and Herberman, R. H-2 antigen expression and NK sensitivity of BL6 melanoma cells. J. Immunol. 1988; 140:2096-2102.
13. Peppoloni, S., Herberman, R. and Gorelik, E. Induction of highly immunogenic variants of Lewis lung carcinoma by ultraviolet irradiation. Cancer Res. 1985; 45:2560-2566.
14. Gorelik, E., Begovic, M., Duty, L. and Herberman, R. Effect of ultraviolet irradiation on MCA102 tumor cell immunogenicity and sensitivity to tumor necrosis factor. Cancer Res. 1991; 51:1521-1528.

15. Begovic, M., Herberman, R. and Gorelik, E. Increase in immunogenicity and sensitivity to natural cell-mediated cytotoxicity following in vitro exposure of MCA105 tumor cells to ultraviolet radiation. Cancer Res. 1991; 51:5153-5159.
16. Begovic, M., Herberman, R. and Gorelik, E. Ultraviolet light induced increase in tumor cell susceptibility to TNF-dependent and TNF-independent natural cell-mediated cytotoxicity. Cell. Immunol. 1991; 138:349-359.
17. Smyth, M. and Ortaldo, J. Comparison of the effect of IL-2 and IL-6 on the lytic activity of purified human peripheral blood large granular lymphocytes. J. Immunol. 1991; 146:1380-1384.
18. Hasegawa, T. and Bonavida, B. Calcium-independent pathway of tumor necrosis factor-mediated lysis of target cells. J. Immunol. 1989; 142:2670-2676.
19. Richards, A. and Djeu, J. Calcium-dependent natural killer and calcium-independent natural cytotoxic activities in an IL-2-dependent killer cell line. J.Immunol. 1990; 145:3144-3151.
20. Duke, R., Chervenak, R. and Cohen, J. Endogenous endonuclease-induced DNA fragmentation. An early even in cell-mediated cytolysis. Proc. Natl. Acad. Sci. USA 1983; 80:6361-6365.
21. Wallach, D. Cytotoxins (tumor necrosis factor, lymphotoxin and others): molecular and functional characteristics and interactions with interferons. Interferon 1986; 7:89-99.
22. Reid, T., Tortis, F. and Ringold G. Evidence of two mechanisms by which tumor necrosis factor kills cells. J. Biol. Chem. 1989; 264:4583-4590.
23. Flick, D. and Gifford, G. Comparison of in vitro cell cytotoxic assay for tumor necrosis factor. J. Immunol. Methods 1984; 68:167-175.
24. Alexander, R., Nelson, W. and Coffey, D. Synergistic enhancement by tumor necrosis factor of in vitro cytotoxicity from chemotherapeutic drugs targeted at DNA topoisoimerase II. Cancer Res. 1987; 47:2403-2410.
25. Utsugi, T., Mattern, M., Mirabelli, C. and Hanna, N. Protentiation of topoisomerase inhibitor-induced DNA strand breakage and cytotoxicity by tumor necrosis factor: enhancement of topoisomerase activity as a mechanism of potentiation. Cancer Res. 1990; 50:2636-2640.
26. Agarwal, S., Drysdale, B.E. and Shin, H. Tumor necrosis factor-mediated cytotoxicity involves ADP-ribosylation. J. Immunol. 1988; 140:4187-4192.
27. Scanlon, M., Laster, S., Wood, J. and Gooding, L. Cytolysis by tumor necrosis factor is preceded by a rapid and specific dissolution of microfilaments. Proc. Natl. Acad. Sci., USA 1989; 86:182-189.
28. Larrick, J. and Wright, S. Cytotoxic mechanism of tumor necrosis factor-α. FASEB J. 1990; 4:3215-3223.
29. Vilcek, J. and Lee, T. Tumor necrosis factor. New insights into the molecular mechanisms of its multiple actions. J.Biol.Chem. 1991; 266:7313-7319.

30. Moan, J. and Peak, M. Effects of UV radiation on cells. J. Photochem. Photobiol. 1989; 4:21-34.
31. Mitchell, D., Nguyen, T. and Cleaver, J. Nonrandom induction of pyrimidine-pyrimidone (6-4) photoproducts in ultraviolet-irradiated human chromatin. J. Biol. Chem. 1990; 205:5353-5357.
32. Dougherty, T. Photosensitization of malignant tumors. Sem. Surg. Oncol. 1986; 2:24-39.
33. Nelson, J., Liaw, L. and Berns, M. Tumor destruction in photodynamic therapy. Photochem. Photobiol. 1987; 46:829-836.

11

Studies on NK Cell Precursors in Mice

Carlo Riccardi
Emira Ayroldi
Lorenza Cannarile
Domenico Delfino
Francesca D'Adamio
Luciano D'Adamio
Graziella Migliorati
Institute of Pharmacology
University of Perugia
Perugia, Italy

I INTRODUCTION

One of the most important problems in tumor immunology studies is the in vivo role of NK cells and, as a consequence, the regulation of NK cell development and activity (1-3). It is known that NK cells can attack neoplastic cells as well as normal non-neoplastic cells and that a number of endogenous and environmental factors can influence reactivity levels (4-6). It has been demonstrated that cells and soluble factors, including cytokines (CKs) and hormones, are able to modulate NK activity by regulating both the reactivity of mature (lytic) effector cells and the growth and differentiation of their precursors (7-12).

Over the last 10-15 years our research was directed towards analyzing the possible influence of cells and soluble factors in modulating NK cell precursors (2,12). We have, in particular, studied the possible effect of CKs on NK activity and our final goal was to understand whether the in vivo effects of CKs could in part be due to the modulation of NK cell development from less mature precursors, rather than an effect on mature effectors.

Supported by Italian Association for Cancer Research (AIRC) and by PF ACRO, CNR, Italy

Our results are not only important for understanding some of the mechanisms involved in NK cell development, but might also be of value as regards an optimal therapeutic approach. In fact, in most studies only the effects of immunomodulating agents on mature effectors are considered, whereas those on precursors are not. Stimulating precursors to develop mature NK cells may be an important aspect and should be considered when studying the mechanisms of NK activity regulation.

II IN VIVO STUDIES ON NK CELL PRECURSORS

In an attempt to study the possible effect of CKs on NK cell precursors in vivo, we carried out experiments using lethally irradiated and bone marrow (BM) reconstituted mice (2,13). It has been reported that NK cells are radioresistant (6) and in fact we have previously shown that a few hours after lethal irradiation spleen NK activity can still be measured and is comparable to that in untreated controls (14). However, 24 hours after lethal irradiation, both in vitro and in vivo NK activity are very low and remain at low or undetectable levels for days, suggesting that NK cells are radiosensitive and though not affected soon after irradiation, their reactivity is weakened after some hours (14).

When mice are lethally irradiated and reconstituted by syngeneic BM transplant, NK activity decreases to undetectable levels and remains low until day 4 after irradiation and BM graft when endogenous (recipient's) NK cells are not active and donor transplanted precursors have not yet developed to mature cytolytic effectors (13-15). On day 7-9, after transplant, regeneration of NK activity starts and returns to control levels by day 14 (13,15). These data confirm the radiosensitivity of mature NK cells and also suggest that two weeks are necessary for transplantable precursors to completely restore the NK activity in vivo (13,15,16).

Using this system we have shown that in vivo treatment with CKs, including IL-2 and interferons (IFNs), soon after BM graft or in vitro pretreatment of donor BM cells before transplant, results in a significant increase of NK cell activity reconstitution (13,17). These data suggest that IL-2 and IFNs can influence the development of NK cell precursors participating in the in vivo reconstitution of reactivity. Similar results have also been obtained with IL-1, TNFα and LT (2). Combination of two or more CKs induces a synergic effect and the increased reconstitution of NK activity correlates well with increased tumor resistance (2).

III STUDIES ON NK CELL PRECURSORS IN VITRO

We performed experiments to analyze the role of CKs on development of NK cell precursors as well as in vitro experiments. We used two different in vitro systems: 1) 7-day cultures of BM from 5-fluorouracyl-treated mice (FUBM). 2) Long term BM cultures (LTBMC).

A. FUBM Cultures

In an attempt to selectively analyze the possible effect of CKs on BM precursors, we cultured BM cells from donor mice, treated with 5-fluorouracil (5-FU), a treatment which has been shown to eliminate the more differentiated cells while sparing the less differentiated precursors (18).

Our results show that NK1-1-negative, MAC-1-negative cells are present in the FUBM, and in the presence of IL-2 give mature effector cells which lyse only NK-sensitive but not NK-resistant targets, express NK1-1, MAC-1, LFA-1, Ly5 and asialoGM1 antigens but do not express CD3, CD4, CD8, MAC-2, and MAC-3 antigens (19). NK cells are not generated in culture controls without IL-2. In vitro generated NK cells, like fresh NK cells, express the truncated form of the ß-chain (1.0 Kb) of the TCR but not the productive 1.3 or the gamma-chain, and also constitutively produce mRNA of TNFα (20). Furthermore these in vitro generated NK cells, like fresh NK cells (21,22), can be boosted by IFNs and IL-2 (20).

Studies to analyze the characteristics of less mature non-lytic precursors revealed that 5-FU-resistant non-lytic precursors are present in the BM. These precursor cells are MAC-1 and NK1.1-negative and can generate CD3-negative MAC-1 and NK1.1-positive NK cells when cultured with recombinant IL-2. The NK1.1-negative precursors proliferate three days from the beginning of the culture, indicating that differentiation rather than proliferation is the first requirement after which cells proliferate and generate mature NK cells acquiring the lytic activity as well as the expression of MAC-1 and NK-1-1 specificities (20).

Taken together the above data clearly indicate that NK precursors, unlike mature effectors, are present in the BM which generates NK cells when cultured in vitro with IL-2. Moreover, IL-2 is necessary and Abs against IL-2-receptor (IL-2/R) completely inhibit NK cell development (12,23).

One of the most interesting aspects of the in vitro generation of NK cells was the study of the role of different soluble factors, possibly involved in this process. In fact, while IL-2 was necessary for NK cell generation (24-26) it is clear that other factors are involved in the first phase of differentiation and acquisition of IL-2-responsiveness (24,27). Studies to analyze factors other than IL-2, show that IL-1α (28), TNFα (23), LT (29), and IFNs (30,31), when used alone are unable to induce NK cells but have synergic activity when used in combination with relatively high (50-100U/ml) concentrations of IL-2 and are essential when used in combination with low IL-2 concentrations (1-10U/ml). This synergic effect is due to an increased IL-2-responsiveness to NK precursors.

When high IL-2 concentrations were used (200U/ml), the above synergic effect was not seen and IL-2 alone was sufficient to induce NK cell activity. In an attempt to analyze the possible role of other endogenously produced soluble factors we did experiments in which we demonstrated that in the first three days, when proliferation is not detected, IL-2 induces TNFα, IFN-γ and IL-1α production (31,32). Production of TNF,

IFN-γ and IL-1 is important since addition of Abs against these cytokines can completely inhibit the IL-2-induced NK cell development (31,32).

The above data suggest that while IL-2 is necessary other factors are involved in this complex process and IL-1, TNF and IFNs could induce IL-2-unresponsive precursors to become IL-2-responsive so that IL-2 can induce development of fully differentiated NK cells (27,31-33).

Other soluble factors such as TGFß and CSF1 can inhibit the IL-2-induced NK cell development (29). Other factors, such as IL-3 and IL-4 (20,24,34) can inhibit or stimulate depending on the IL-2 concentration used. The effects of IL-4 and TNFα agree well with the expression of IL-2/Rα (p55) (20,23) suggesting that one of the mechanisms by which cytokines influence the IL-2-induced generation of NK cells is via the up or down modulation of IL-2/R.

B. Long Term Bone Marrow Cultures (LTBMC)

In an attempt to further analyze the nature of NK precursors and the possible role of CKs in NK cell development we cultured mouse BM cells in a long term bone marrow culture (LTBMC) system recently described by M.V. Den Brink, et al. in experiments in a rat system (35). Rat bone marrow cells cultured for four weeks in complete medium without growth factors, gave rise to an enriched population of NK precursor cells. As these cells were recultured with IL-2 and conditioned medium (CM) obtained at the end of the four-wk culture, lytic NK cells were generated within four days. Replacing CM with fresh medium before adding IL-2, markedly decreased NK cell generation, suggesting that endogenous factor(s) present in CM were necessary for IL-2 induction of NK cells and required a two-wk culturing of BM which then acquired the capacity to generate NK cells. This suggests that, as in the rat, differentiation of mouse NK cells in addition to IL-2 needs other factors which are present in the CM. Addition of anti-IFN-γ or anti-TNFα Abs inhibited generation of cytotoxic cells. Supplementing CM deprived LTBMC cultures, with IL-2 and TNFα or IL-2 and IFN-γ we obtained a significant increase in the generation of mature NK cells. These results demonstrate that endogenous TNFα and IFN-γ can replace CM and also confirm previous observations from our laboratory which show that these CKs play an essential role in the development of NK cells.

IV CONCLUSION

The above results suggest that mature NK cells can develop from less mature BM precursors in vitro and that this generation process requires a number of CKs. IL-2 is necessary to induce non-lytic precursors to become cytotoxic and acquire some NK markers such as NK1-1 and MAC-1 which while not restricted to LGL are expressed on most NK cells. IN LTBMC, IL-2 sensitive precursors were induced and/or enriched and a CM medium was produced containing TNFα and IFN-γ which by themselves are unable to induce NK cells but are necessary to make precursors IL-2-sensitive. Most of the observations suggest that NK cell development is a complex and multistep process requiring the interaction of a number of CKs including IL-2, IL-1, TNFα and IFNs.

References

1. Trinchieri, G. Biology of natural killer cells. In Advances in Immunology, Academic Press, New York, 1989; 87:187.
2. Riccardi. C,, Migliorati, G., Cannarile, L., D'Adamio, F., Frati, L., Herberman, R.B. In vivo effects of cytokines on development of natural killer cells and antitumor activity in lethally irradiated bone marrow transplanted recipients. J. Biol. Res. Modif. 1990; 9:15.
3. Cudkowicz, G. and Hochman, P.S. Do natural killer cells engaged in regulated reactions against self to ensure homeostasis? Immunol. Rev. 1979; 44:13.
4. Herberman, R.B., Nunn, M.E., Lavrin, D.H. Natural cytotoxic reactivity of mouse lymphoid cells against syngeneic and allogeneic tumors. I. Distribution of reactivity and specificity. Int. J. Cancer. 1975; 16:216.
5. Nunn, M.E., Herberman, R.B., Holden, H.T. Natural cell-mediated cytotoxicity in mice against non-lymphoid tumors cells and some normal cells. Int. J. Cancer. 1977; 20:381.
6. Haller, O., Kiessing, R., Orn, A. and Wigzell, H. Generation of natural killer cells: an autonomous function of the bone marrow. J. Exp. Med. 1977; 145:1411.
7. Ostensen, M.E., Thiele, D.L. and Lipsky, P.E. Tumor necrosis factor alpha enhances cytolytic activity of human natural killer cells. J. Immunol. 1987; 138:4185.
8. Herberman, R.B. (ed.) NK Cells and Other Natural Effector Cells. Academic Press, New York, 1982.
9. Kalland, T. Generation of natural killer cells from bone marrow precursors in vitro. J. Immunol. 1986; 57:493.
10. Koo, G.C. and Manyak, C.L. Generation of cytotoxic cells from murine bone marrow by human recombinant IL-2. J. Immunol. 1986; 37:1751.
11. Hackett, J., Bennet, M. and Kumar, V. Origin and differentiation of natural killer cells. I. Characteristics of a transplantable NK cell precursor. J. Immunol. 1985; 134:3731.
12. Migliorati, G., Cannarile, L., Herberman, R.B., Bartocci, A., Stanley, E.R. and Riccardi, C. Role of interleukin-2 (IL-2) and hemopoietin-1 (H-1) in the generation of mouse natural killer (NK) cells from primitive bone marrow precursors. J. Immunol. 1987; 138:3612.
13. Riccardi, C., Giampietri, A., Migliorati, G., Cannarile, L., D'Adamio, L. and Herberman, R.B. Generation of mouse natural killer (NK) cell activity: effect of

interleukin-2 (IL-2) and interferon (IFN) on the in vivo development of natural killer cells from bone marrow (BM) progenitor cells. Int. J. Cancer 1986; 38:553.

14. Riccardi, C., Santoni, A., Barlozzari, T. and Herberman, R.B. Role of NK cells in rapid in vivo clearance of radiolabeled tumor cells. In: Natural Cell-Mediated Immunity Against Tumors. Academic Press, New York, 1980; p.1121.

15. Riccardi, C., Migliorati, G., Giampietri, A., Ayroldi, E. and Herberman, R.B. Regulation of mouse NK activity. In Mechanisms of Cytotoxicity by NK Cells. R.B. Herberman and D.M. Callewaert eds. Academic Press, New York, 1985; p.421.

16. Riccardi, C., Barlozzari, T., Santoni, A., Cesarini, C. and Herberman, R.B. Regulation of in vivo reactivity of natural killer (NK) cells. In: NK cells and other natural effector cells. Academic Press, Inc. 1982; p.549.

17. Riccardi, C., Vose, M.B. and Herberman, R.B. Modulation of IL-2-dependent growth of mouse NK cells by interferon and T lymphocytes. J. Immunol. 1983; 130:228.

18. Van Zant, G. Studies of hemopoietic stem cells spared by 5-fluorouracil. J. Exp. Med. 1984; 159:679.

19. Migliorati, G., Moraca, R., Nicolett, I. and Riccardi, C. IL-2-dependent generation of natural killer cells from bone marrow: role of MAC-1, NK1.1 precursors. Cell. Immunol. In Press.

20. Migliorati, G., Cardinali, L. and Riccardi, C. Effect of Interleukin-4 on Interleukin-2-dependent generation of natural killer cells. Cell. Immunol. 1991; 136:194.

21. Ortaldo, J.R. and Herberman, R.B. Augmentation of natural killer activity. In Immunolb. of Natural Killer Cells. eds. E. Lotzova and R.B. Herberman, CRC Press, Boca Raton, FL, 1986; p.145.

22. Sayers, T.J., Mason, A.T. and Ortaldo, J.R. Regulation of human natural killer cell activity by interferon-gamma: lack of a role in interleukin-2-mediated augmentation. J. Immunol. 1986; 136:2176.

23. Ayroldi, E., Sorci, G., Cannarile, L. and Riccardi, C. Effect of recombinant murine Tumor Necrosis Factor on the generation of NK cells in bone marrow cultures. Nat Immun Cell Growth Regul, In Press.

24. Saikawa, Y., Hasui, M., Miura, M., Tachinami, T., Katayama, K., Takano, N., Miyawaki, T., Koizumi, S. and Taniguchi, N. Interleukin-3 enhanced interleukin-2-dependent maturation of NK progenitor cells in bone marrow from mice with severe combined immunodeficiency. Cell. Immunol. 1990; 136:2176.

25. Suzuki, R., Handa, H.I. and Kumagai, K. Natural killer (NK) cells as a responder to Interleukin-2 (IL-2). I. Proliferative response and establishment of cloned cells. J. Immunol. 1983; 130:981.

26. Domzig, W., Stadler, B.M. and Herberman, R.B. Interleukin-2 dependence of human natural killer (NK) cell activity. J. Immunol. 1983; 130:1970.

27. Kuribayashi, K., Gillis, S., Kern, D.E. and Henney, C.S. Murine NK cell cultures: effects of interleukin-2 and interferon on cell growth and cytotoxic reactivity. J. Immunol. 1981; 126:2321.

28. Migliorati, G., Cannarile, L., D'Adamio, L., Herberman, R.B. and Riccardi, C. Interleukin-1 augments the interleukin-2-dependent generation of natural killer cells from bone marrow precursors. Nat. Immun. Cell Growth. Regul. 1987; 6:306.

29. Migliorati, G., Cannarile, L., Herberman, R.B. and Riccardi, C. Effect of various cytokines and growth factors on the interleukin-2-dependent in vitro differentiation on natural killer cells from bone marrow. Nat. Immun. Cell Growth Regul. 1989; 8:48.

30. Migliorati, G., Cannarile, L., Herberman, R.B. and Riccardi, C. Role of interferon in natural killer cell generation from primitive bone marrow precursors. Int. J. Immunophar. 1988; 10:665.

31. Delfino, D., D'Adamio, F., Migliorati, G. and Riccardi, C. Growth of murine natural killer from bone marrow in vitro: role of TNF and IFN. Int. J. Immunophar. 1991; 13:943.

32. Ayroldi, E., Cannarile, L. and Riccardi, C. Natural killer (NK) cell generation in bone marrow cultures: role of IL-1. Immunophar. Immunotox. 1991, In Press.

33. Itoh, K., Shiba, K., Shimizu, Y., Suzuki, R. and Kumagai, K. Generation of activated killer (AK) cells by recombinant interleukin-2 (rIL-2) in collaboration with interferon-γ (IFNγ) J. Immunol. 1985; 134:3124.

34. Keever, C.A., Pekle, K., Gazzola, M.V., Collins, N.H., Bourhis, J.H. and Gillio, A. Natural killer and lymphokine-activated killer cell activities from human marrow precursors. II. The effects of IL-3 and IL-4. J. Immunol. 1989; 143:3241.

35. VanDenBrink, M.R.M., Boggs, S.S., Herberman, R.B. and Hiserodt, J.C. The generation of natural killer (NK) cells from NK precursors in rat long-term bone marrow cultures. J. Exp. Med. 1990; 172:303.

PART IV

Introduction

Tumor Microenvironment and Immune Effector Cells

Theresa L. Whiteside
Pittsburgh Cancer Institute and
University of Pittsburgh School of Medicine
Pittsburgh, Pennsylvania

For many years, tumor immunology has been concerned with the development and measurement of systemic immune responses to tumor-associated antigens (TAA). The presence of both humoral and cellular antitumor immunity in experimental animals bearing immunogenic tumors has been demonstrated (1-3). Furthermore, animals immunized with tumor cells or TAA were shown to develop systemic antitumor immunity and subsequently reject implants of the tumor expressing these TAA (4,5). These early experiments suggested the possibility that the immune system may be utilized to control or even arrest tumor metastases.

More recently, systemic administration of biologic response modifiers (BRMs) to tumor-bearing animals or patients with cancer has led to a significant augmentation of antitumor responses in many instances (6,7). Nevertheless, the presence of demonstrable systemic antitumor immunity in the tumor-bearing host has not always translated into a clinically beneficial outcome, and it has become necessary to explain these seemingly contradictory observations. The lack or low frequency of clinical responses in hosts demonstrating systemic antitumor immunity has focused attention on the complex and relatively poorly understood local interactions between the tumor and immune effector cells.

Solid tumors arise in tissues and from the very beginning their progress is dependent on the tissue microenvironment. The role of the tissue microenvironment in tumor growth and metastasis has been emphasized recently. As new information about interactions between tissue and immune cells becomes available, including insights into the processes of cell extravasation, adhesion, activation and production of cytokines, tumor progression is increasingly frequently viewed as a result of multiple complex events, which involve immune effector cells, vascular endothelium, extracellular matrix (ECM) components and tumor cells themselves. Pathologists have long ago observed that cells of the immune system are part of the tumor microenvironment (8,9), although some tissue sites (e.g., mucosa of the aerodigestive tract or upper airways), clearly contain a greater mononuclear cell component than other sites. As the primary tumor emerges and establishes its own vascular supply, immune cells acquire a better access to the site and accumulate in response to chemotactic stimuli generated at the site of tumor growth. The presence of immune cells, e.g., T lymphocytes in or around the tumor or its metastases has been generally considered as evidence of local antitumor immune response. The intense infiltration of tumors with mononuclear cells has been even considered by some to be a good prognostic factor (10,11). More recently however, when functional attributes of these tumor-infiltrating mononuclear cells were examined, it became apparent that they were inhibited in their proliferative and probably other activities (reviewed in 12). These observations led to formulating of a hypothesis that in the tumor microenvironment, immune cells lose their effectiveness, become partially or completely paralyzed, so that the balance between the immune system and tumor shifts in favor of the latter. Clearly, such a shift would create a microenvironment which is not only unfavorable to immune effector cells but also unique, depending on the ability of a given tumor to produce immunoinhibitory factors. A number of such factors produced by a variety of human as well as animal tumors have been described (13,14) and purified (15) recently, and these results lend credence to the hypothesis. In addition, evidence from experimental animal tumor models indicates that tumors which abundantly produce immunoinhibitory substances metastasize better and kill the host faster than those producing low levels or no immunosuppressive factors (16).

With the advent of adoptive immunotherapy (AIT) of cancer more recently, several issues concerning tumor-infiltrating mononuclear cells have emerged. First, their usefulness as effector cells in AIT has been questioned. However, it soon became apparent that these cells incubated ex vivo in the presence of interleukin 2 (IL2) or a combination of IL2 and other cytokines rapidly regained the ability to proliferate and mediate antitumor activities. In fact, tumor-infiltrating lymphocytes cultured in IL2 were shown by Rosenberg and his colleagues to be far more effective than peripheral blood LAK cells in eliminating established pulmonary or liver metastases in experimental animals (17). TILs have been considered a favored source of allegedly tumor-specific T cells, although in many cases their specificity for autologous tumor has not been rigorously tested. Today, large-scale selection and production of $CD8^+$ T lymphocytes from solid tumors obtained during surgery for AIT is feasible and autotumor reactivity of these effector cells is being extensively explored.

The second issue has been that of successful delivery of adoptively-transferred TIL to the tumor and of sustaining their antitumor functions in vivo. Systemic administration of effector cells and IL2 appears to result in the delivery of only small number of effector cells to the tumor (18). Therefore, recent interest has turned to locoregional adoptive therapy. In experimental models of tumor metastasis, locoregional delivery of activated effector cells appears to localize larger numbers of these cells to the established tumor metastases and more effective elimination of these metastases. However, these preliminary successes underscore the need for a better understanding of cellular interactions at the sites of tumor growth and metastasis. These interactions appear to be crucial for modifying the tumor microenvironment and tipping the balance in favor of adoptively transferred effector cells. In addition, the mechanisms of antitumor effects mediated by these effector cells are not understood. Studies are in progress to define these mechanisms, and this aspect of AIT appears of utmost importance, as it is likely that in the near future a choice of effector cells for cancer therapy may well depend on a particular functional attribute of a T cell or NK cell subset selected for such therapy.

The realization that local immune responses taking place in the tumor microenvironment are an important part of host defense against malignancy has altered our recent thinking about tumor progression. In this section of the book, we explore recent approaches to studies of immune cells localization to tumors or their metastases and the mechanisms of their interactions with tumor and tissue cells. Additionally, newer developments in therapeutic utilization of these immune cells are described.

References

1. Prehn, R.T. and Main, J.N. Immunity to methylcholanthrene-induced sarcomas. J. Natl. Cancer Inst. 1957; 18:769-778.
2. Klein, G. Tumor-specific transplantation antigens. Cancer Res 1968; 28:625-635.
3. Old LJ, Boyse EA, Clarke DA, Carswell EA. Antigenic properties of chemically-induced tumors. Annals NY Acad. Sci. 1962; 101:80-106.
4. Old, L.J. Cancer immunology: The search for specificity. Cancer Res. 1981; 41:361-375.
5. Klein, G., Sjogren, H.O., Klein, E. and Hellstrom, K.E. Demonstration of resistance against methylcholanthrene-induced sarcomas in the primary autochthonous host. Cancer Res. 1960; 20:1561-1572.
6. Greenberg, P.D., Chever, M.A. and Fefer, A. Irradiation of the disseminated murine leukemia by chemo/immunotherapy with cyclophosphamide and adoptively-transferred immune syngeneic lyt $1^{+}2^{-}$ lymphocytes. J. Exp. Med. 1981; 154:952-963.

7. Rosenstein, M., Eberline, T.J. and Rosenberg, S.A. Adoptive immunotherapy of established syngeneic solid tumors: Role of T lymphoid subpopulations. J. Immunol. 1984; 132:2117-2122.
8. Ioachim, H.L. The stroma reaction of tumors: An expression of immune surveillance. J. Natl. Cancer Inst. 1976; 57:465-475.
9. Lauder, I. and Ahern, E.W. The significance of lymphocytic infiltration in neuroblastoma. Br. J. Cancer 1972; 26:321-330.
10. Kreider, J.W., Bartlett, G.L. and Butkiewicz, B.L. Relationship of tumor leukocyte infiltration to host defense mechanisms and prognosis. Cancer Met. Rev. 1984; 3:53-74.
11. Brocker, E.B., Kolde, G., Steinhausen, D., Peters, A. and Macher, E. The pattern of the mononuclear infiltrate as a prognostic parameter in flat superficial spreading melanomas. J. Cancer Res. Clin. Oncol. 1984; 107:48-52.
12. Whiteside, T.L., Jost, L.M. and Herberman, R.B. Tumor-infiltrating lymphocytes: Potential and limitations to their use for cancer therapy. Crit. Rev. Oncol. Hematol. 1992; 12:25-47.
13. Roth, J.A., Osborne, B.A. and Ames, R.S. Immunoregulatory factors derived from human tumors. J. Immunol. 1983; 130:303-308.
14. Cianciolo, G.J., Copeland, T.D., Oroszlan, S. and Snyderman, R. Inhibition of lymphocyte proliferation by synthetic peptide homologous to retroviral envelope protein. Science. 1985; 230:453-455.
15. Bodmer, S., Strommer, K., Frei, K., Siepl, C., DeTribolet, N., Heid, I. and Fontana, A. Immunosuppression and transforming growth factor beta in glioblastoma. J. Immunol. 1989; 143:3222-3229.
16. Tada, T., Ohzeki, S., Utsumi, K., Takiuchi, H., et al. Transforming growth factor beta-induced inhibition of T cell function. Susceptibility difference in T cells of various phenotypes and functions and its relevance to immunosuppression in the tumor-bearing state. J. Immunol. 1991; 146:1077-1082.
17. Rosenberg, S.A., Speiss, P. and Lafreniere, R. A new approach to the adoptive immunotherapy of cancer with tumor-infiltrating lymphocytes. Science. 1986; 223:1318-1321.
18. Griffith, H., Read, E.J., Carrasquillo, J.A., et al. In vivo distribution of adoptively-transferred ^{111}indium-labeled tumor-infiltrating lymphocytes and peripheral blood lymphocytes in patients with metastatic melanoma. J. Natl. Cancer Inst. 1989; 81:1709-1717.

12

Tumor-Infiltrating Lymphocytes in Human Solid Tumors

Theresa L. Whiteside
Pittsburgh Cancer Institute and
University of Pittsburgh School of Medicine
Pittsburgh, Pennsylvania

I PHENOTYPIC CHARACTERISTICS OF TCL

Many human solid tumors are infiltrated by mononuclear cells which are mostly localized in the tumor stroma or in tissues immediately surrounding the tumor (1,2). The proportion of lymphocytes to monocytes in the infiltrate varies in different human tumors as does its intensity (3). The lymphoid component of these infiltrates consists almost entirely of T and B lymphocytes, while natural killer (NK) cells are infrequently found among tumor-infiltrating lymphocytes (TIL) (4,5). By immunoperoxidase staining in tissue sections or by flow cytometry of freshly-isolated TIL using monoclonal antibodies (mAbs) to lymphocyte surface antigens, it has been determined that TIL-T in most human solid tumors express the T cell receptor (TCR) α/β (6) and that a considerable proportion of TIL-T has the phenotype associated with activated T cells (7). Figure 1 shows phenotypic data obtained with fresh TIL isolated from several types of human solid tumors. The presence of considerable but variable proportions of $CD3^+HLA\text{-}DR^+$ or $CD3^+CD25^+$ T cells among TIL obtained from these tumors indicates that activation antigens are commonly expressed by TIL in situ, and in some tumors, e.g., ovarian carcinomas, up to 70% of TIL may be $HLA\text{-}DR^+$ (8). When compared with autologous peripheral blood lymphocytes, TIL contain significantly higher proportions of activated T cells (9). Also, TIL-T are mainly primed or memory T cells, as determined phenotypically on the basis of CD29 or CD45RO expression, while the proportion of naive T cells ($CD45RA^+$) is significantly decreased in fresh TIL, in comparison to autologous PBL (10). Additional two-color flow cytometry analyses showed that the CD4/CD8 ratio among fresh TIL is variable but often is greater than 1, suggesting a predominance of $CD8^+$ over $CD4^+$ T lymphocytes in many, but clearly not all, human solid tumors (10). This enrichment in $CD8^+$ T cells among TIL contrasts with non-malignant inflammatory infiltrates, which contain mainly $CD4^+$ T cells

(11). It has been suggested that the CD8$^+$ T cell content of TIL may have prognostic significance, as tumors infiltrated predominantly with CD8$^+$ T cells seem to have a better prognosis (12,13). However, only very limited number of careful correlative studies linking the presence of CD8$^+$ T cell infiltrates at the time of surgery with prognosis or survival are available to support this hypothesis, and additional prospectively-collected data are needed to confirm it. Overall, recent phenotypic studies of freshly isolated TIL-T have emphasized the presence of activated, primed T lymphocytes in most human solid tumors.

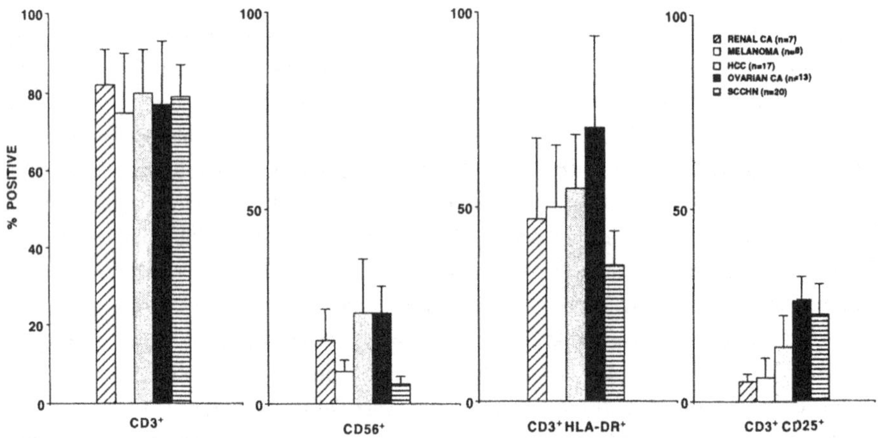

Fig. 1 Phenotypic characteristics of tumor-infiltrating lymphocytes (TIL) obtained from a variety of human primary or metastatic tumors and studied by two-color flow cytometry. The data are mean percentages of positive cells ± SD

Both phenotypic and functional studies of TIL-T have been limited by difficulties in obtaining sufficient numbers of these cells for complete and systematic analyses. One of the more controversial issues has been the use of enzymes for tissue disaggregation necessary for TIL isolation from solid tumors, because of a possibility that even mild enzymatic digestion might alter TIL function. Recently adopted protocols for TIL isolation have been optimized for recovery and preservation of TIL functions by combining a mild, short-term collagenase treatment with mechanical disaggregation (14). Another frequently encountered problem concerns the separation of TIL from tumor cells prior to functional assays. The most effective and practical separation of TIL on differential density gradients often yields only partially-purified TIL-T preparations, and additional steps are necessary for further purification (15).

II FUNCTIONAL RESPONSES OF TIL

Functional characteristics of fresh TIL-T obtained from a variety of solid tumors have been studied in several laboratories with consistently disappointing results: both in bulk preparations and in limiting dilution assays (LDA), fresh TIL-T were functionally not as responsive as autologous or normal PBL-T (4,16,17). In many cases, under in vitro conditions optimized for functional measurements (i.e., in the presence of antigen-presenting cells and exogenous interleukin 2; IL2), TIL-T have been found to be functionally unresponsive (17). Not only antitumor cytotoxicity of TIL-T against autologous or allogeneic tumor cell targets, but also their profoundly reduced or absent proliferative responses to mitogens, alloantigens and low doses of IL2 have been well documented in the literature (18). Table 1 shows, for example, that TIL-T isolated from human melanomas proliferated poorly in response to phytohemagglutinin (PHA), in comparison to normal PBL-T tested under the same experimental conditions (6,17). Similarly, we have shown repeatedly that cloning of TIL-T obtained from different types of solid tumors by LDA, performed under conditions optimized for T cell growth, was significantly reduced compared to autologous or normal PBL-T (16). In Figure 2, results of a representative LDA of TIL-T obtained from a patient with renal cell carcinoma are shown to illustrate a reduced frequency of proliferating T-cell precursors (PTL-p) among these TIL. Another function of lymphocytes, locomotion in collagen gels, has been recently described to be impaired in TIL isolated from human melanomas or renal cell carcinomas (18). We have also examined functional responses of lymph node lymphocytes (LNL) from patients with head and neck cancer (19), and found that these cells were unable to mediate cytotoxicity or proliferate in culture in response to exogenous stimuli (Table 2).

The apparent functional impairment of freshly-isolated TIL-T is not in agreement with the phenotypic data reviewed above. It could be expected, based on the presence of numerous activated memory T cells at the tumor site, that TIL-T would mediate high levels of antitumor cytotoxicity and proliferate readily in response to non-specific or specific stimuli. The fact that these TIL-T are immunoincompetent requires an explanation, and to find a plausible reason for this behavior of TIL, several alternative hypotheses have been advanced as follows:

(a) intrinsic lack of ability of TIL to respond normally;
(b) presence of activated suppressor cells among TIL;
(c) influence of the tumor microenvironment on TIL functions.

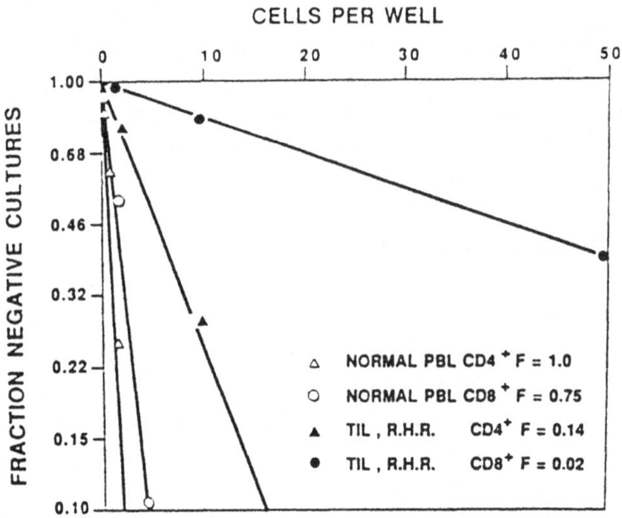

Fig. 2 Limiting dilution analysis (LDA) performed with $CD4^+$ or $CD8^+$ T cells obtained by sorting of tumor-infiltrating lymphocytes (TIL) from the patient with a renal cell carcinoma or of normal peripheral blood lymphocytes (PBL). F = frequency as determined by LDA performed under conditions described previously (20). Reproduced with permission from ref. 20

Table 1.
Proliferation of fresh TIL obtained from human melanomas in response to phytohemagglutinin[a]

TIL	-PHA	+PHA	S.I.
1.	0.9±0.2	1.9±0.3	2.1
2.	1.6±0.1	2.1±0.1	1.3
3.	0.4±0.1	0.6±0.1	1.5
4.	3.8±0.3	2.7±0.2	<1.0
5.	0.6±0.1	0.9±0.1	1.5
PBL & enz[b]	0.5±0.1	26.0±2	52
Tonsil & enz	0.3±0.1	8.5±0.7	28

[a]The data are mean cpm x 10^{-3} ±SD from triplicate wells. 5×10^5 TIL/well were cultured in the presence of irradiated feeder cells (allogeneic PBL) and PHA (1%) and pulsed with ^{3}[H]thymidine for 18h before harvest.

[b]As positive controls, normal PBL or tonsil lymphocytes were pretreated with collagenase and DNA-ase prior to culture for the same period of time as that used for TIL isolation.

Table 2
Functional responses of fresh lymph node lymphocytes obtained from patients with head and neck cancer[a]

	Patients			Normal controls	
	PBL	I-LNL	NI-LNL	PBL	p value[d]
Proliferation[b]:					
PHA	44±5	14±5	13±3	34±2	<0.05
rIL2	7±1	4±1	4±1	NT	
AuTu	1.5±0.4	1±2	1±1	NT	
Cytotoxicity[c]:					
K562	93±1	6±10	7±8	99±43	<0.001
Daudi	4±3	0	0	3±1	
Allo Tu	6±6	2±4	3±6	7±6	
P815	112±51	3±2	4±3	35±27	<0.05

[a]Cells were recovered from freshly-harvested lymph nodes (n=10) without enzymatic treatment. As controls, peripheral blood lymphocytes (PBL) from 10 normal individuals were studied. I-LNL = lymphocytes from tumor-involved lymph nodes; NI-LNL = lymphocytes from tumor-uninvolved lymph nodes. NT = not tested.

[b]Proliferation was determined by ^{3}[H]thymidine incorporation (18h pulse) into cells (2×10^4/well, in triplicate) cultured for 5 days in the presence of 5 μg/well of phytohemagglutinin (PHA), recombinant interleukin 2 (rIL2) at a concentration of 100 IU/well or irradiated autologous tumor cells (AuTu; 2×10^3/well). The data are mean cpm $\times 10^{-3} \pm$ SD.

[c]Cytotoxicity was measured in 4h ^{51}Cr-release assays at the E:T ratios ranging from 50:1 to 6:1. The data are mean LU/$10^7 \pm$ SD. Allogeneic (Allo-Tu) squamous cell carcinoma of the head and neck cell line (PCI-1) was used as a target. To measure antibody dependent cellular cytotoxicity (ADCC), P815 targets coated with anti-P815 antibodies (Ab) were used.

[d]The p values are for differences between normal PBL and patients' LNL.

In previously reported studies with TIL, it has been demonstrated that these cells have intrinsically impaired functional responses, and that the lack of or decreased ability to respond is not due to the absence of antigen presenting cells, cytokines or other extrinsic factors (reviewed in 10,14). Furthermore, fresh CD8$^+$ TIL-T were more unresponsive than CD4$^+$ TIL-T in proliferative assays (20). Evidence for the presence of activated suppressor cells in TIL-T has been inconclusive. In some studies, activated CD8$^+$CD11b$^+$ cells (21) or CD3$^+$CD4$^+$CD8$^-$ cells (22) have been identified as responsible for suppression of TIL-T functions. However, we have been unable to confirm the presence of lymphoid cell-mediated suppression in a series of mixing experiments utilizing LNL from tumor-involved lymph nodes of patients with head and neck cancer (Table 3). For example, purified CD8$^+$ LNL populations before or after activation with IL2 were unable to suppress generation of LAK activity (Table 3) or proliferative responses of allogeneic normal PBL (19). The ability of human tumors to produce factors which suppress functions of immune cells has been documented in the literature (23,24). It can be seen in Figure 3 that proliferative responses of normal PBL incubated in the presence of tumor cell supernatants or on monolayers of tumor cells were profoundly inhibited (17). Various tumor-derived immunoinhibitory substances have been identified including transforming growth factor-ß (TGF-ß), prostaglandin E_2 (PGE_2), the p-15E retrovirus-related protein or tumor-associated gangliosides (23,24,25). These observations strongly support the hypothesis that reduced in situ functional responses of TIL-T are related to the ability of the tumor to secrete one or more of these immunosuppressive factors. Fresh TIL-T isolated from the immunosuppressive tumor microenvironment are in the state of functional paralysis which may not be complete and which appears to be reversible in vitro upon culture in the presence of exogenous IL2 (26). Incidently, studies of tumor-derived macrophages have also demonstrated impairments in their functions, indicating that effector cells other than lymphocytes found in the tumor microenvironment may not be as competent as those found in normal tissues (27,28).

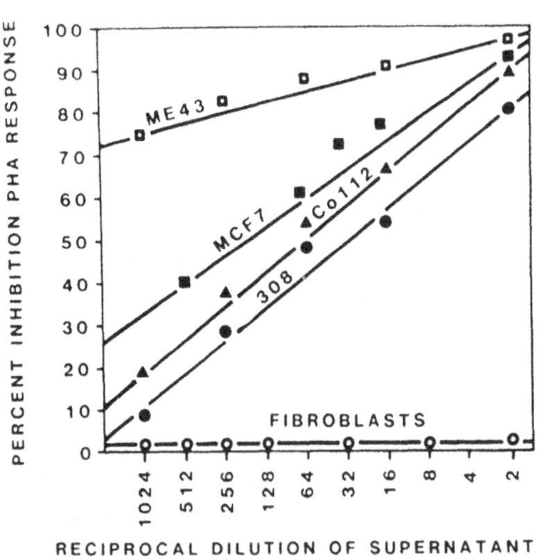

Fig. 3 Inhibition of responses to PHA of normal PBL incubated with varying dilutions of supernatants of tumor cell lines or normal adult dermal fibroblasts. PBL were incubated with diluted supernatants for 3h at 37°C, washed once and stimulated with PHA. The cells were incubated for 3 days and ^{3}[H]thymidine incorporation was measured. The % inhibition of PHA responses was calculated as previously described (17). Reproduced with permission from ref. 17

Table 3
Lymphokine-activated killer (LAK) activity of normal PBL incubated in the presence of interleukin 2 (IL2) and CD8$^+$ LNL obtained from patients with head and neck cancer[a]

PBL: CD8$^+$ LNL[b]	K562	Daudi	PCI-1
1:0	3537	2016	1216
0:1	701	256	192
1:10	2414	638	628
1:1	2207	709	658
1:0.1	2431	636	776

[a]PBL obtained from normal individuals were incubated in the presence of IL2 (6,000 IU/ml) and irradiated CD8$^+$ LNL isolated from lymph nodes of patients with head and neck cancer and added to the cultures on day 0. On day 5 of culture, cells were tested for cytotoxicity in 4h ^{51}Cr-release assays. PCI-1 is an NK-resistant squamous cell carcinoma cell line. The data are mean LU/10^7 cells from 7 experiments.

[b]CD8$^+$ LNL were obtained by panning LNL on flasks coated with anti-CD8 monoclonal antibody (AIS, Menlo Park, CA). Different ratios of PBL and irradiated CD8$^+$ LNL were used to determine effects of CD8$^+$ cells on generation of LAK activity by normal PBL.

III STUDIES OF TIL-TUMOR INTERACTIONS IN SITU

As a result of conflicting observations regarding the activated phenotype expressed by TIL-T in situ and their apparent failure to function normally, interactions between the tumor and TIL-T have been more carefully scrutinized recently. Since activated lymphocytes are likely to produce cytokines, we examined the ability of fresh TIL obtained from head and neck tumors to produce IL1-ß, IL2, IFN-γ or TNF-α spontaneously as well as in response to exogenous activators. The results of these experiments indicated that TIL, compared to autologous PBL, produced lower levels or no IL1-ß or TNF-α in vitro. These TIL produced normal levels of IFN-γ or IL2 in vitro (Table 4). Again, these data have suggested that fresh TIL are not able to produce a full range of cytokines, as would be expected of normal activated T cells (29). To extend the analysis of the cytokine profile of TIL, we next performed in situ hybridization with radiolabeled cDNA probes for cytokine gene expression in several types of human solid tumors (30). This approach offers a unique opportunity to assess a functional potential of TIL in tumor tissues without TIL isolation, which could result in a selective loss of subsets responsible for cytokine production. In addition, in situ hybridization allows for the determination of tissue localization of cells which express mRNA for cytokines. In human tumors containing abundant lymphoid infiltrates (e.g., head and neck tumors, mucin-producing breast carcinomas), we detected numerous cells positive for cytokine mRNA (30). On the other hand, in other human tumors (e.g., ovarian carcinomas or non-mucinous invasive ductal breast carcinomas) few infiltrating cells expressed genes for cytokines, as measured by in situ hybridization (30). It appears from these results that tumors differ in the ability to modulate gene expression for cytokines, and that either the tumor immunogenicity or ability to induce and sustain the influx of inflammatory cells or both may determine the state of TIL activation, as evidenced by the expression of genes for cytokines.

Table 4
Cytokine production by fresh tumor-infiltrating lymphocytes (TIL) obtained from patients with squamous cell carcinoma of the head and neck[a]

		TNF-α (pg/ml)		IL1-β (pg/ml)		IL2 (U/ml)		IFN-γ (U/ml)	
		S	IN	S	IN	S	IN	S	IN
1.	TIL	0	0	74	68	ND	ND	ND	ND
	PBL	140	880	1,150	1,400	0	8	0	34
2.	TIL	0	0	190	120	7	48	0	9
	PBL	0	0	0	0	12	20	0	6
3.	TIL	90	120	30	50	0	13	0	158
	PBL	810	810	4,600	4,980	3	39	4	247
4.	TIL	0	0	0	0	0	50	0	13
5.	TIL	0	10	0	0	0	7	0	36
6.	TIL	40	230	36	32	0	4	0	45

[a] Tumor-infiltrating lymphocytes (TIL) were recovered from 6 squamous cell carcinomas of the head and neck, using mild enzymatic and mechanical disaggregation (4) and differential density gradients (15). In three cases, patients' PBL were also studied. Spontaneous (S) and induced (IN) production (with lipopolysaccharide for IL1-β and TNF-α or PHA for IL2 and IFN-γ) of cytokines by TIL or PBL was determined. The cells were incubated in medium alone or in the presence of activators for 24h. Culture supernatants were harvested and assayed by immunoassays (ELISA) for each cytokine. The data are cytokine levels measured in individual supernatants.

By in situ hybridization, strong support for local activation of TIL found in the immediate proximity to the tumor cells has been obtained (see Figure 4a). However, in addition to expressing mRNA for IFN-γ, IL2, IL1-ß, TNF-α or IL2R, TIL which were localized next to the tumor cells contained abundant message for TGF-ß (Fig.4b). Thus, as a result of interactions between tumor cells and TIL, activated lymphocytes were induced to express mRNA for, and presumably produce, an immunoinhibitory cytokine, TGF-ß. This would suggest that in the tumor microenvironment, activated TIL themselves may be a source of immunosuppressive cytokines, which are known to inhibit proliferation of lymphoid cells (31). The ability of at least some human tumors to induce activation of TIL-T, which, in turn, leads to TIL-mediated release of immunosuppressive factors provides an example of how tumors might escape or disarm the immune system.

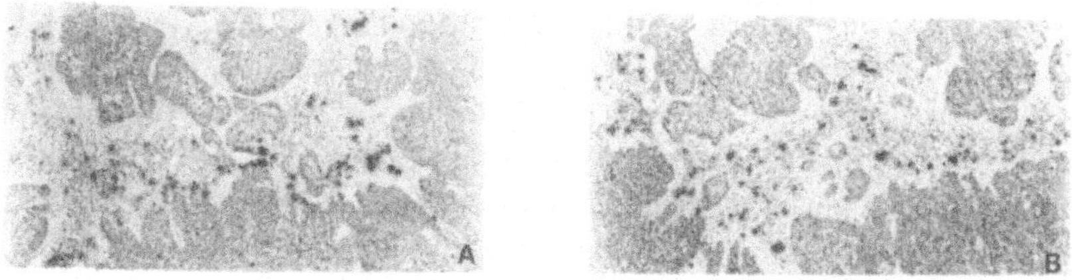

Fig. 4 In situ hybridization for mRNA for INF-γ (A) and for TGF-ß (B) on cryostat sections of a human squamous cell carcinoma of the head and neck. ^{35}[S]-labeled cDNA probes were used under conditions described by us earlier (30). Note numerous positive cells located in the proximity to tumor cells in the tumor stroma. Mag x 100

The molecular basis of tumor-induced immunosuppression appears to involve several distinct mechanisms. In addition to tumor-induced production of cytokines by TIL, as indicated above, the production by tumors themselves of immunosuppressive cytokines: IL10 by B cell lymphomas (32); TGF-ß by glioblastomas (33); GM-CSF by mammary adenocarcinomas (34) has been recently documented. A variety of other soluble factors (e.g., chemotactic or angiogenic factors) produced in the tumor microenvironment are likely to profoundly affect tumor growth and its interactions with immune effector cells. The ability of the tumor to downregulate or cause deletion of T cells positive for the TCR Vß2 genes from the T cell repertoire (35) is another example of tumor-mediated modulation of effector cells.

Interactions between TIL-T and tumor cells in the tumor microenvironment are being actively investigated at present. In the last several years, we have focused on these local interactions to be able to explain the lack of effective antitumor functions of lymphocytes accumulating in the tumor. Insights into the mechanisms responsible for immunosuppression at the tumor site are important for future therapeutic efforts, which might be directed at elimination of inhibitory signals and selective activation of TIL in situ. These efforts are also necessary to optimize antitumor effects of TIL delivered to the tumor as a part of adoptive immunotherapy.

References

1. Ioachim, H.L. The stroma reaction of tumors: An expression of immune surveyance. J. Natl. Cancer Inst. 1976; 57:465-475.
2. Vose, B.M. and Moore, M. Human tumor-infiltrating lymphocytes: A marker of host response. Semin. Hematol. 1985; 22:27-40.
3. Svennevig, J-L. and Svaar, H. Content and distribution of macrophages and lymphocytes in solid malignant human tumors. Int. J. Cancer 1979; 24:754-758.
4. Whiteside, T.L., Miescher, S., Hurlimann, J., Moretta, L. and VonFliedner, V. Separation, phenotyping and limiting dilution analysis of lymphocytes infiltrating human solid tumors. Int. J. Cancer 1986; 37:806-811.
5. Finke, J.H., Tubbs, R., Connely, B., Pontes, E. and Monti, E.J. Tumor infiltrating lymphocytes in patients with renal cell carcinoma. Ann. NY. Acad. Sci. 1988; 532:387-394.
6. Miescher, S., Stoeck, M., Qiao, L., Barras, C., Barrelet, L. and VonFliedner, V. Proliferative and cytolytic potentials of purified human tumor-infiltrating T lymphocytes. Impaired response to mitogen-driven stimulation despite T-cell receptor expression. Int. J. Cancer 1988; 42:659-666.
7. Shimizu, Y., Iwatsuki, S., Herberman, R.B. and Whiteside, T.L. Effect of cytokines on in vitro outgrowth of tumor-infiltrating lymphocytes obtained from human primary and metastatic liver tumors. Cancer Immunol. Immunother. 1991; 32:280-288.
8. Vaccarello, L., Wang, Y.L. and Whiteside, T.L. Sustained outgrowth of autotumor-reactive T lymphocytes from human solid tumors in the presence of tumor necrosis factor-alpha and interleukin 2. Human Immunol. 1990; 28:216-227.
9. Snyderman, C.H., Heo, D.S., Johnson, J.T., D'Amico, F., Barnes, L. and Whiteside, T.L. Functional and phenotypic analysis of lymphocytes isolated from the tumor and lymph nodes in head and neck cancer. Arch. Otolaryngol. Head Neck Surg. 1991; 117:899-905.
10. Whiteside, T.L., Jost, L.M. and Herberman, R.B. Tumor-infiltrating lymphocytes. Potential and limitations to their use for cancer therapy. Crit. Rev. Oncol./Hematol. 1991; 12:25-47.

11. Pitzalis, C., Kingsley, G., Haskard, D. and Panay, I.G. The preferential accumulation of helper-inducer T lymphocytes in inflammatory lesions: Evidence for regulation by selective endothelial and homotypic adhesion. Eur. J. Immunol. 1988; 18:3097-1404.
12. Snyderman, C.H., Heo, D.S., Chen, K., Whiteside, T.L. and Johnson, J.T. T cells in tumor-infiltrating lymphocytes of head and neck cancer. Head Neck 1989; 11:331-336.
13. Wolf, G.T., Hudson, J.L., Peterson, K.A., Miller, H.L. and McClatchey, K.D. Lymphocyte subpopulations infiltrating squamous carcinomas of the head and neck: Correlations with extent of tumor and prognosis. Otolaryngol. Head Neck Surg. 1986; 95:142-151.
14. Elder, E.M. and Whiteside, T.L. Processing of tumors for vaccine and/or tumor infiltrating lymphocytes. In: Rose, N.R., deMacario. E.C., Fahey, J.L., Friedman, H., Penn, G.M. (eds). Manual of Clinical Laboratory Immunology, Fourth edition. Amer Soc Microbiol, Washington, DC. 1992; 123:817-819.
15. Whiteside, T.L., Miescher, S., MacDonald, R.H. and VonFliedner, V. Separation of tumor-infiltrating lymphocytes from human solid tumors. A comparison of velocity sedimentation and discontinuous density gradients. J. Immunol. Methods 1986; 90:221-233.
16. Miescher, S., Whiteside, T.L., Moretta, L. and VonFliedner, V. Clonal and frequency analysis of tumor-infiltrating T lymphocytes from human solid tumors. J. Immunol. 1987; 138:4004-4011.
17. Miescher, S., Whiteside, T.L., Carrel, S. and VonFliedner, V. Functional properties of tumor-infiltrating and blood lymphocytes in patients with solid tumors: Effects of tumor cells and their supernatants on proliferative responses of lymphocytes. J. Immunol. 986; 136:1899-1907.
18. Applegate, K.G., Balch, C.M. and Pellis, N. In vitro migration of lymphocytes through collagen matrices: Arrested locomotion in tumor-infiltrating lymphocytes. Cancer Res. 1990; 50:7153-7158.
19. Letessier, E., Sacchi, M., Johnson, J.T., Herberman, R.B. and Whiteside, T.L. The absence of lymphoid suppressor cells in tumor-involved lymph nodes of patients with head and neck cancer. Cell. Immunol. 1990; 130:446-458.
20. Miescher, S., Stoeck, M., Qiao, L., Barras, C., Barrelet, L. and VonFliedner, V. Preferential clonogenic deficit of CD8-positive T lymphocytes infiltrating human solid tumors. Cancer Res. 1988; 48:6992-6998.
21. Cozzolino, F., Torcia, M., Carossino, A.M., et al. Characterization of cells from invaded lymph nodes in patients with solid tumors. Lymphokine requirement for tumor-specific lymphoproliferative response. J. Exp. Med. 1987; 166:303-318.
22. Mukherji, B., Ergin, M.T., Juha, A., et al. Clonal analysis and regulatory T cell responses against human melanoma. J. Exp. Med. 1989; 169:1961-1976.

23. Roth, J.A., Osborne, B.A. and Ames, R.S. Immunoregulatory factors derived from human tumors. J. Immunol. 1983; 130:303-308.

24. Cianciolo, G.J., Copland, T.D., Oroszlan, S. and Snyderman, R. Inhibition of lymphocyte proliferation by synthetic peptide homologous to retroviral envelope protein. Science. 1985; 230:453-455.

25. Robb, R.J. The suppressive effect of gangliosides upon IL2-dependent proliferation as a function of inhibition of IL2 receptor association. J. Immunol. 1986; 136:971-976.

26. Whiteside, T.L., Heo, D.S., Takagi, S., Johnson, J.T., Iwatsuki, S. and Herberman, R.B. Cytolytic antitumor cells in long-term cultures of human liver infiltrating lymphocytes in recombinant interleukin 2. Cancer Immunol. Immunother. 1988; 26:1-10.

27. Economou, J.S., Colquhoun, S.D., Anderson, T.M., McBride, W.W., Golub, S., Holmes, E.C. and Morton, D.L. Interleukin 1 and tumor necrosis factor production by tumor-associated mononuclear leukocytes and peripheral mononuclear leukocytes in cancer patients. Int. J. Cancer 1988; 42:712-714.

28. Erroi, A., Sironi, M., Chiaffarino, F., Zhen-Guo, C., Mengozzi, M. and Mantovani, A. IL1 and IL6 release by tumor-associated macrophages from human ovarian carcinoma. Int. J. Cancer 1989; 44:795-801.

29. Vitolo, D., Letessier, E.M., Johnson, J.T. and Whiteside, T.L. Immunologic effector cells in head and neck cancer. J. Natl. Cancer Inst. Monograph #13: Biology of and Novel Therapeutic Approaches for Epithelial Cancers of the Aerodigestive Tract. In press, 1992.

30. Vitolo, D., Zerbe, T., Kanbour, A., Dahl, C., Herberman, R.B. and Whiteside, T.L. Expression of mRNA for cytokines in tumor-infiltrating mononuclear cells in ovarian adenocarcinoma and invasive breast cancer. Int. J. Cancer 1992; In press.

31. Inge, T.H., Hoover, S.K., Susskind, B.M., Barrett, S.K. and Bear, H.D. Inhibition of tumor-specific cytotoxic T lymphocyte responses by transforming growth factor ß1. Cancer Res. 1992; 52:1386-1392.

32. Howard, M. and O'Garra, A. Biological properties of interleukin 10. Immunol. Today 1992; 13:198-200.

33. Bodmer, S., Strommer, K., Frei, K., Siepl, C., deTribolet, N., Heid, I. and Fontana, A. Immunosuppression and transforming growth factor ß in glioblastoma. J. Immunol. 1989; 143:3222-3229.

34. Fu, Y., Watson, G., Jimenez, J.J., Wang, Y. and Lopez, D.M. Expansion of immunoregulatory macrophages by granulocyte-macrophage colony-stimulating factor derived from a murine mammary tumor. Cancer Res. 1990; 50:227-234.

35. Wang, H. and Wei, W-Z. Characterization of Vß2 depleting activity associated with mouse mammary preneoplastic hyperplastic alveolar nodule. The FASEB J. 1992; 6:A1689.

13

Localization of Immune Effector Cells to Tumor Metastases

Per Basse
Ronald H. Goldfarb
Pittsburgh Cancer Institute and
University of Pittsburgh School of Medicine
Pittsburgh, Pennsylvania

I INTRODUCTION

Adoptive immunotherapy with interleukin 2 (IL-2) and lymphokine activated killer (LAK) cells or plastic adherence-enriched LAK (A-LAK) cells (1,2) has produced dramatic reductions in the number of metastatic lesions in several animal systems (3-7). In clinical settings, LAK cell therapy has also been successful, with complete or partial responses in 20-30% of the patients with advanced cancer, especially malignant melanoma and renal cell carcinoma (8-12). Nevertheless, frequent long-term survival and complete eradication of metastases, has yet to be achieved, indicating the need for improvement of this therapeutic modality. Therefore, the mechanism(s) behind the in vivo anti-cancer effect of LAK and A-LAK cells must be further elucidated in order to optimize the use of these cells in therapy of cancer.

At present, very little is known about how LAK cells exert their cytotoxic activity in vivo, but it seems logical that the degree of accumulation of effector cells within or near the microenvironment of malignant lesions might be decisive for the efficacy of adoptive immunotherapy. It is therefore important to analyse the migratory potential of these immune effector cells to clarify to what extent they localize to tumor metastases and whether direct contact between effector and target cell, which is a prerequisite for LAK cell-mediated killing in vitro, also takes place in vivo. The results presented herein describe our attempts to answer these questions.

II IN VIVO IDENTIFICATION OF ADOPTIVELY TRANSFERRED EFFECTOR CELLS

In several studies, the biodistribution of injected effector cells has been assessed by using various radiolabels. However, radionuclides may not provide sufficient sensitivity for precise detection of small proportions of the injected cells accumulating in malignant lesions located in organs such as lung, liver and spleen, where considerable accumulation of effector cells is reported to occur even in the absence of tumors (13). Furthermore, the usual assessment levels of radiolabel in an organ or tissue do not permit discrimination between infiltration of malignant and normal tissue within the specimen. In order to more sensitively and precisely determine the in vivo distribution of LAK cells, we therefore have utilized fluorescent dyes, rhodamine or the lipophilic dye PKH26, for identification of the injected cells ((14,15). Rhodamine, previously used by Butcher et al (16,17) in studies of lymphocyte traffic, and the lipophilic, membrane-bound PKH26 (18,19) allows for the identification of individual LAK cells in frozen sections of various organs and makes it possible to determine their exact relation to malignant lesions. Used in low concentrations, these dyes alter neither the proliferative properties, nor the cytotoxic capacity of the labeled cells. However, due to leak of dye from labeled cells, the use of rhodamine is limited to 16-24 hours after labeling. Apparently, PKH26 does not leak from the labeled cells, but within 24-48 hours most of the dye redistributes from the surface membrane of the cell into small membrane bound vesicles in the cytosol, making it increasingly difficult to identify the labeled cells.

III INFILTRATION OF PULMONARY METASTASES

To evaluate the localization of adoptively transferred effector cells into metastatic lesions, A-LAK cells labeled with rhodamine were injected i.v. into mice bearing lung metastases of B16 melanoma origin. One hour after injection, the number of labeled cells was higher in the normal lung tissue than in the malignant lesions (Figure 1). However, during the following hours, A-LAK cells were gradually cleared from the normal tissue, whereas more and more labeled cells accumulated in the metastatic lesions. As early as 6 hours after injection, significantly more cells ($p < 0.001$) were seen in the metastatic tissue compared to the surrounding normal lung tissue. Identical results were obtained using PKH26 as cell label, showing that the observed accumulation was not restricted to rhodamine-labeled cells. Injection of fluorescent cells, killed prior to the injection, revealed that non-specific accumulation of rhodamine (or PKH26) in the metastases did not occur. Furthermore, the localization of A-LAK cells into malignant tissues was not restricted to B16 melanoma metastases since accumulation of A-LAK cells was seen also in metastases of Lewis lung carcinoma and MCA102 sarcoma origin.

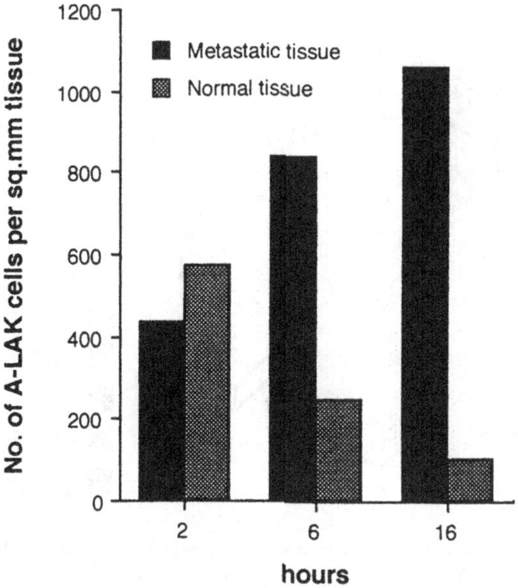

Fig. 1 Time course of A-LAK cell localization into pulmonary metastases. 2.5-3 x 10^6 rhodamine-labeled A-LAK cells were injected i.v. into animals bearing 12-day-old pulmonary metastases of the B16 melanoma. Mice received i.p. injections of 50,000 U IL-2 at 0, 4 and 8 h. Two to 16 hours after injection, mice were killed, and the lungs were removed and processed for fluorescence microscopy. Values represents mean numbers of fluorescent cells/mm^2 of metastatic (black columns) or normal (gray columns) lung tissue. Modified from Journal of the National Cancer Institute (15)

IV EFFICACY OF THE INFILTRATION

Thus, adoptively transferred A-LAK cells indeed have the ability to find and infiltrate malignant lesions in a time dependent fashion. Despite the the fact that infiltration of pulmonary metastases is highly specific (a 10-fold higher number of effector cells in the metastatic lesions compared to the normal tissue at 16 h), it is nonetheless a very inefficient process. After injection of 15 x 10^6 A-LAK cells, an E:T ratio better than 1:5 was seen in only one third of the lesions (Figure 2) and the E:T ratio was even lower in the other two thirds. The total number of infiltrating A-LAK cells at 16 h was estimated to be less than 5% of the injected dose, i.e. 95% of the cells probably never made it to the malignant lesions. This means that in order to get more cells into the tumors, more cells have to be injected, until we learn how to direct the injected cells to specific locations. Preliminary studies seem to indicate that infiltration is dose dependent, in that i.v. injection of 50 x 10^6 A-LAK cells raised the number of tumor infiltrating cells to approximately 2000 cells per mm^2. This resulted in an average E:T ratio better than 1:1 in most metastases. Nevertheless, the total number of infiltrating A-LAK cells at 16 h was still very low, only ~2.5 millions cells or 5.5% of the total number of injected cells.

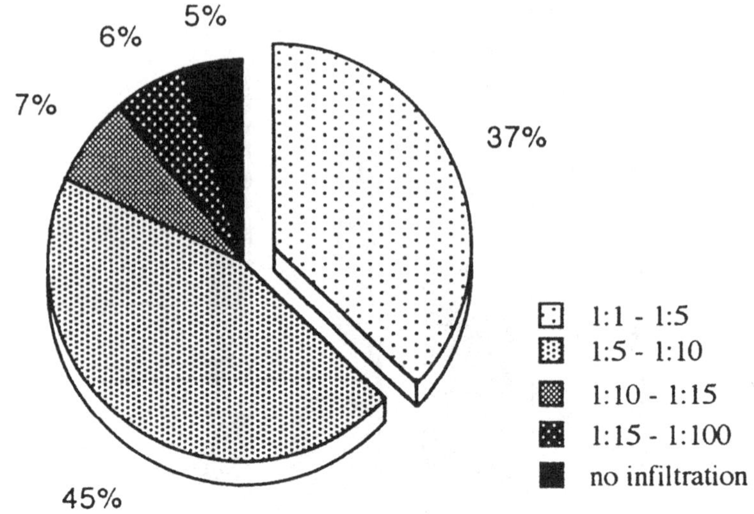

Fig. 2 Variation in A-LAK cell:tumor cell ratio among individual lung metastases 16 hours after injection of 15 x 10^6 A-LAK cells into animals bearing 9-day-old pulmonary metastases of the B16 melanoma. Mice received i.p. injections of 50,000 U IL-2 at 0, 4 and 8 h. The A-LAK cell:tumor cell ratio was better than 1:5 in only 37% of the metastatic lesions. Modified from Journal of Experimental Medicine (14)

Importantly, after i.v. injection of high numbers of A-LAK cells (~50 x 10^6), where the vast majority of lung metastases was highly infiltrated (~2000 cells/sq.mm), approximately 5% of the lesions remained totally devoid of infiltrating A-LAK cells. These non-infiltrated lesions, often located very close to highly infiltrated lesions, were almost exclusively growing around larger vessels and injection of FITC-labeled microbeads immediately before removal of the lungs, revealed that these non-infiltrated lesions were mainly avascular. The existence of such lesions, perhaps due to heterogeneity in vascular access (20), might in part explain why LAK or A-LAK cell therapy, although capable of an 80-95% numerical reduction of metastases, seldom results in complete eradication of tumor burden (4-7).

When comparing infiltration of smaller (3 day old) and larger (12 day old) metastases, no difference in the number of infiltrating A-LAK cells per mm^2 tumor tissue was seen. We can not exclude, however, that the infiltration of larger tumors might be less sufficient due to increasing areas of necrosis and impaired vascular access.

V PHENOTYPE OF TUMOR INFILTRATING LAK CELLS

If the anti-neoplastic effect of activated killer cells is dependent on direct contact with the tumor cells, the optimal therapeutic environment would exist when high numbers of the most cytotoxic LAK cell type are present in the in the tumor. Judged from in vitro studies, the cytotoxicity of A-LAK cells is superior to that of standard (st.-) LAK cells, but A-LAK cells, rich in activated NK cells, might not be as infiltrative as the st.-LAK cells, rich in activated T-cells. We therefore analyzed the phenotype of different effector cell subpopulations and compared their ability to infiltrate pulmonary B16 metastases. The results are summarized in Figure 3. Apparently, non-activated spleen cells are not able to infiltrate the tumors at all. In contrast, st.-LAK cells grown for 3 days in IL-2 had some ability to infiltrate the metastatic lesions, but less efficiently than the infiltration by A-LAK and st.-LAK cells, both grown for 5 days in IL-2. It is worth noting that when grown for 5 days in IL-2, A-LAK cells are composed almost exclusively of activated NK cells and contain less than 7% activated CD8+ T-cells, whereas st.-LAK cells contain more than 50% CD8+ T-cells and only half the number of NK cells as A-LAK cells. Since both subpopulations of effector cells infiltrated the metastastic lesions equally well, we conclude that both activated NK and T-cells have this ability. Of the surface markers examined so far, the only one that correlated with infiltration, was asialo-GM1. This might suggest that this surface structure is directly involved in the migration process (21,22), but it is more likely that expression of asialo-GM1 reflects a stage of activation required before effective infiltration of malignant tissue can occur by NK or NK-like cells.

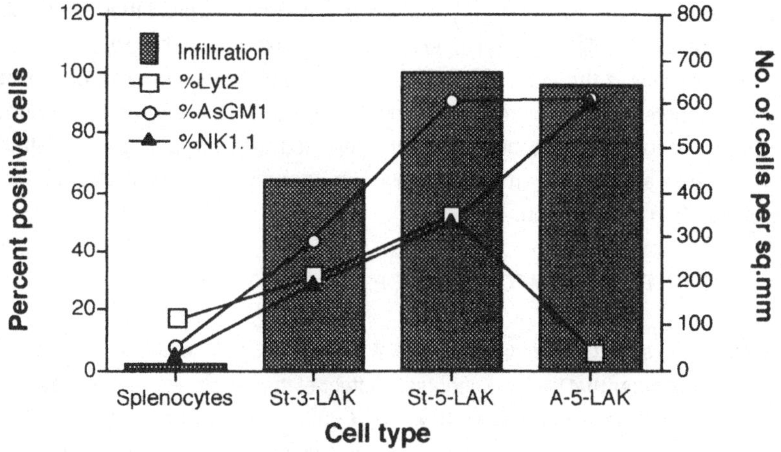

Fig. 3 Phenotype of tumor-infiltrating effector cells cells. 15 x 10^6 splenocytes, standard LAK cells grown for 3 days (St-3-LAK) or 5 days (St-5-LAK) in IL-2 and A-LAK cells grown for 5 days in IL-2 (A-5-LAK) were injected i.v. into mice bearing lung metastases of the B16 melanoma. Mice received i.p. injections of 50,000 U IL-2 at 0, 4 and 8 h. Two to 16 hours after injection, mice were killed, and the lungs were removed and processed for fluorescence microscopy. The mean numbers of fluorescent cells/mm^2 of metastatic tissue (gray columns) were estimated and compared to the expression of Lyt-2 (triangles), NK1.1 (squares) and Asialo GM1 (circles) by the effector cell subpopulations

VI INFILTRATION OF EXTRA-PULMONARY METASTASES BY LAK CELLS

Injection of LAK cells by the intravenous (i.v.) route results in direct delivery of these cells to the lungs and lung metastases since all i.v. injected cells initially reach the lungs. In order to infiltrate liver metastases, LAK cells must first traverse the lung vasculature and, for those cells entering the liver via the portal vein, the capillary bed of the intestinal tract. We have recently reported that intraarterial (left ventricular) inoculation of tumor cells (23) and A-LAK cells (manuscript submitted) resulted in increased distribution to the liver when compared to i.v. injection. Furthermore, A-LAK cells are large, rigid and less deformable than other lymphoid cells (24). The capillary beds of the lungs and the intestinal tract might therefore impede the distribution of i.v. injected cells into the liver. Indeed, after i.v. injection very few LAK cells were seen in normal and metastatic liver tissue of the recipients (Figure 4). To determine if the use of a more direct route of injection could enhance the degree of infiltration of liver metastases, A-LAK cells were inoculated into the portal system via a branch of the inferior mesenteric vein. By this route of administration, a five-fold increase in the number of injected A-LAK cells per mm^2 normal liver tissue one hour after injection was achieved (Figure 4). By 16 hours, the number of A-LAK cells in the normal liver tissue decreased slightly, whereas the number of A-LAK cells per mm^2 tumor tissue increased dramatically. These data, together with our recent analysis of the tissue distribution of ^{125}IUdR-labeled LAK and A-LAK cells (25) indicate that these cells circulate very inefficiently and that improved therapeutic efficacy might be obtained by injecting them directly into the vessels nourishing the tumor bearing organ. In the case of the liver, this means injection into the portal vein or the hepatic artery (for tumors nourished mainly by arterial blood). This assertion is supported by findings of Lafreniere & Rosenberg (3), who reported that intraportal administration of LAK cells was significantly more effective for therapy of liver MCA 105 sarcoma metastases, than i.v. injection.

VII THE ANTITUMOR MECHANISM OF LAK CELLS

Given the highly selective localization of LAK and A-LAK cells into malignant tissue, it is intriguing to consider that the anti-neoplastic effect of these cells is mediated by a contact dependent lysis. It was not possible, however, from microscopy of frozen tissue sections to judge whether the infiltrating LAK cells were located intra- or extra-vascularly, i.e. whether they established physical contact with the tumor cells or not. To answer this question, metastatic tissue samples were analyzed by electron microscopy in collaboration with Drs. Ulf Nannmark and Bengt Johansson (University of Göteborg, Sweden (15)). In most pulmonary melanoma nodules, adoptively transferred LAK cells were found inside capillaries as well as in various stages of exit from the microvessels into the interstitium of the metastases; some LAK cells were completely surrounded by endothelium, some were partly in contact with endothelium and partly with tumor cells and yet others were completely surrounded by melanin-producing B16 tumor cells.

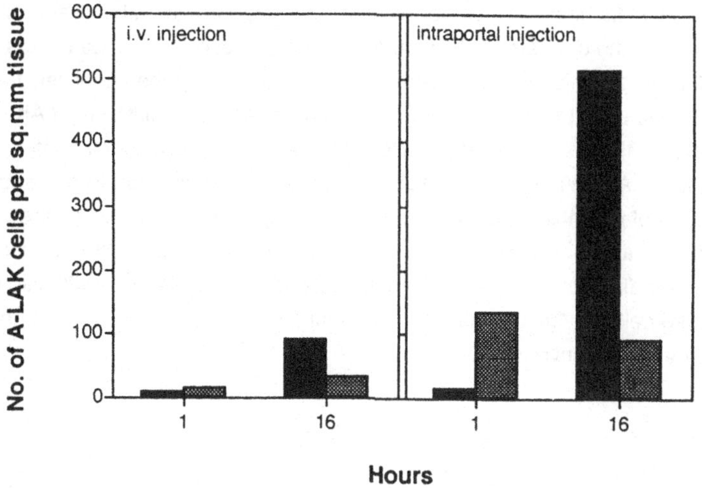

Fig. 4 Number of A-LAK cells per square mm normal (gray columns) or day-9 metastatic liver tissue (black columns) one and 16 hour after i.v. or intraportal injection of 15 x 10^6 rhodamine labeled A-LAK cells. Statistical analysis (by Student's double tailed t-test): Number of A-LAK cells in tumor tissue one hour vs. 16 hour after i.v. injection: $p<0.01$; after intraportal injection: $p<0.001$. Number of A-LAK cells in normal tissue one hour vs. 16 hour after i.v. injection: $p<0.01$; after intraportal injection: $p<0.001$. Modified from Journal of Experimental Medicine (14)

These observations unambiguously demonstrate the ability of the adoptively transferred effector cells to establish contact with extravascular metastatic tumor cells. They do not, however, allow us to draw any conclusion as to the mechanism behind the in vivo anti-neoplastic effects of LAK cells. Even though the LAK cells are able to bring themselves into intimate contact with the tumor cells, the relatively low effector-to-target ratio in most lesions (often less than 1:1) makes it unlikely that the in vivo therapeutic results is due solely to contact mediated cytotoxicity. Alternatively, local production of cytokines by the LAK cells, such as TNF alpha, interferon gamma and possible other factors, could amplify their effects and might recruit and activate other (endogenous) effector cells. Yet another possibility is that relatively few LAK cells, attached to the tumor endothelium, might damage or activate the endothelial cells, leading to the observed antitumor effects. However, the successful treatment of 3 day old, avascular micrometastases by use of LAK cells (4,6,26) excludes the tumor endothelium as being the only target tissue.

VIII CONCLUSIONS AND PERSPECTIVES

Overall, our data clearly demonstrate that immune effector cells of both NK and T-cell origin are able to localize selectively into tumor metastases in a time- and dose-dependent fashion and that they establish physical contact with extravascular tumor cells. However, the data also indicate that homing of these cells to the tumors is a very inefficient process which is only executed by less than 5% of the adoptively transferred cells. It remains to be seen whether these relatively few infiltrating LAK cells are responsible for the observed antitumor effect. Further, the very low circulatory potential of LAK and A-LAK cells strongly indicates a local route of injection to be used, at least until analysis of the molecular basis for LAK cell migration enables us to direct the cells to specified locations. An improved homing to the tumor areas might not only enhance the direct or indirect cytotoxic efficacy of the injected LAK cells, but it will also make the use of these cells as "guided missiles", carrying toxic drugs selectively into metastatic deposits, more attractive.

References

1. Vujanovic, N.L., Herberman, R.B., Al Maghazachi, A. and Hiserodt, J.C. Lymphokine-activated killer cells in rats. III. A simple method for the purification of large granular lymphocytes and their rapid expansion and conversion into lymphokine-activated killer cells. J. Exp. Med. 1988; 167:15.
2. Gunji, Y., Vujanovic, N.L., Hiserodt, J.C., Herberman, R.B. and Gorelik, E. Generation and characterization of purified adherent lymphokine-activated killer cells in mice. J. Immunol. 1989; 142:1748.
3. Lafreniere, R. and Rosenberg, S.A. Successful immunotherapy of murine experimental hepatic metastases with lymphokine-activated killer cells and recombinant interleukin 2. Cancer Res. 1985; 45:3735.
4. Mazumder, A. and Rosenberg, S.A. Successful immunotherapy of natural killer-resistant established pulmonary melanoma metastases by the intravenous adoptive transfer of syngeneic lymphocytes activated in vitro by interleukin 2. J. Exp. Med. 1984; 159:495.
5. Mule, J.J., Shu, S., Schwarz, S.L. and Rosenberg, S.A. Adoptive immunotherapy of established pulmonary metastases with LAK cells and recombinant interleukin-2. Science 1984; 225:1487.
6. Mule, J.J., Shu, S. and Rosenberg, S.A. The anti-tumor efficacy of lymphokine-activated killer cells and recombinant interleukin 2 in vivo. J. Immunol. 1985; 135:646.

7. Schwarz, R.E., Vujanovic, N.L. and Hiserodt, J.C. Enhanced antimetastatic activity of lymphokine-activated killer cells purified and expanded by their adherence to plastic. Cancer Res. 1989; 49:1441.
8. Rosenberg, S.A., Lotze, M.T., Muul, L.M., Leitman, S., Chang, A.E., Ettinghausen, S.E., Matory, Y.L., Skibber, J.M., Shiloni, E., Vetto, J.T., et al . Observations on the systemic administration of autologous lymphokine-activated killer cells and recombinant interleukin-2 to patients with metastatic cancer. N. Engl. J. Med. 1985; 313:1485.
9. Rosenberg, S.A., Lotze, M.T., Muul, L.M., Chang, A.E., Avis, F.P., Leitman, S., Linehan, W.M., Robertson, C.N., Lee, R.E., Rubin, J.T., et al . A progress report on the treatment of 157 patients with advanced cancer using lymphokine-activated killer cells and interleukin-2 or high-dose interleukin-2 alone. N. Engl. J. Med. 1987; 316:889.
10. Rosenberg, S.A., Packard, B.S., Aebersold, P.M., Solomon, D., Topalian, S.L., Toy, S.T., Simon, P., Lotze, M.T., Yang, J.C., Seipp, C.A., et al . Use of tumor-infiltrating lymphocytes and interleukin-2 in the immunotherapy of patients with metastatic melanoma. A preliminary report. N. Engl. J. Med 1988; 319:1676.
11. Schoof, D.D., Gramolini, B.A., Davidson, D.L., Massaro, A.F., Wilson, R.E. and Eberlein, T.J. Adoptive immunotherapy of human cancer using low-dose recombinant interleukin 2 and lymphokine-activated killer cells. Cancer Res. 1988; 48:5007.
12. West, W.H., Tauer, K.W., Yannelli, J.R., Marshall, G.D., Orr, D.W., Thurman, G.B. and Oldham, R.K. Constant-infusion recombinant interleukin-2 in adoptive immunotherapy of advanced cancer. N. Engl. J. Med 1987; 316:898.
13. Maghazachi, A.A.., Goldfarb, R.H., Kitson, R.P., Hiserodt, J.A. Giffen, C.C, and Herberman, R.B. 1990. In vivo distribution of interleukin-2 activated cells. In Interleukin-2 and killer cells in cancer. E. Lotzova and R. B. Herberman, editors. CRC Reviews, CRC Press, Boca Raton, Florida. 260.
14. Basse, P., Herberman, R.B., Nannmark, U., Johansson, B.R., Hokland, M., Wasserman, K. and Goldfarb, R.H. Accumulation of adoptively transferred adherent, lymphokine-activated killer cells in murine metastases. J. Exp. Med 1991; 174:479.
15. Basse, P.H., Nannmark, U., Johansson, B.R., Herberman, R.B. and Goldfarb, R.H. Establishment of cell-to-cell contact by adoptively transferred adherent lymphokine-activated killer cells with metastatic murine melanoma cells. J. Natl. Cancer Inst. 1991; 83:944.
16. Butcher, E.C. and Weissman, E.L. Direct fluorescent labeling of cells with fluorescein or rhodamine isothiocyanate. 1. Technical aspects. J. Immunol. Methods 1980; 37:97.

17. Butcher, E.C. and Weissman, E.L. Direct fluorescent labeling of cells with fluorescein or rhodamine isothiocyanate. II. Potential application to studies of lymphocyte migration and maturation. J. Immunol. Methods 1980; 37:109.
18. Horan, P.K. and Slezak, S.E. Stable cell membrane labelling. Nature 1989; 340:167.
19. Slezak, S.E. and Horan, P.K. Fluorescent in vivo tracking of hematopoietic cells. Part I. Technical considerations. Blood. 1989; 74:2172.
20. Jain, R.K. Delivery of novel therapeutic agents in tumors: physiological barriers and strategies. J. Natl. Cancer Inst. 1989; 81:570.
21. Kleinman, H.K., Martin, G.R. and Fishman, P.H. Ganglioside inhibition of fibronectin-mediated cell adhesion to collagen. Proc. Natl. Acad. Sci. U. S. A. 1979; 76:3367.
22. Perkins, R.M., Kellie, S., Patel, B. and Critchley, D.R. Gangliosides as receptors for fibronectin? exp. cell. res 1982; 141:231.
23. Basse, P., Hokland, P., Heron, I. and Hokland, M. Fate of tumor cells injected into left ventricle of heart in BALB/c mice: role of natural killer cells. J. Natl. Cancer Inst. 1988; 80:657.
24. Sasaki, A., Jain, R.K., Maghazachi, A.A., Goldfarb, R.H. and Herberman, R.B. Low deformability of lymphokine-activated killer cells as a possible determinant of in vivo distribution. Cancer Res. 1989; 49:3742.
25. Basse, P.H., Herberman, R.B., Hokland, M.E. and Goldfarb, R.H. Tissue distribution of adoptively transferred adherent lymphokine-activated killer cells assessed by different cell labels. Cancer Immunol. Immunother. 1992; 34:221.
26. Mule, J.J., Ettinghausen, S.E., Spiess, P.J., Shu, S. and Rosenberg, S.A. Antitumor efficacy of lymphokine-activated killer cells and recombinant interleukin-2 in vivo: survival benefit and mechanisms of tumor escape in mice undergoing immunotherapy. Cancer Res. 1986; 46:676.

14

Antitumor Effector Cells: Extravasation and Control of Metastasis

Theresa L. Whiteside
Ronald H. Goldfarb
Pittsburgh Cancer Institute
and University of Pittsburgh School of Medicine
Pittsburgh, Pennsylvania

I INTRODUCTION

Mononuclear cells often accumulate at the sites of tumor growth or metastasis, and these tissue-infiltrating cells are considered to be important in the control of tumor spread. Development of effective antitumor local immune responses depends on the ability of lymphocytes to extravasate and migrate into nonlymphoid solid tissues. At the same time, the antimetastatic therapeutic potential of tissue-infiltrating lymphocytes depends on their effectiveness in inhibition or elimination of tumor cells within established metastases. In this chapter, we focus on this potential of lymphocytes as antitumor effector cells. Considerable progress has been made recently in studies of the mechanisms involved in lymphocyte movement through tissues, including interactions with extracellular matrix (ECM) components, vascular elements and other tissue cells. Furthermore, not only surface molecules, which may be involved in lymphocyte extravasation (1), but also products of these cells, which may facilitate extravasation and antimetastatic effects including degradative proteases of effector cells (2), have been identified and at least in part, characterized. The conceptual framework for lymphocyte extravasation that has evolved in recent years on the basis these new insights emphasizes the role of specific receptors on the migrating cells and ECM as well as of cellular products, both enzymes and cytokines, in regulation of these important functions of immune cells.

II ADHESION MOLECULES ON LYMPHOCYTES

Lymphocytes express a variety of surface receptors, some of which have been identified as signal transducing structures important for cell-cell or cell-substratum interactions. These adhesion receptors belong to several distinct families of membrane glycoproteins, and their structure and expression on lymphoid as well as nonlymphoid cells have been a subject of several recent and comprehensive reviews (3,4). One family of cell surface glycoproteins that mediate cell adhesion of lymphocytes are integrins, heterodimeric molecules similar in structure to immunoglobulins and consisting of noncovalently associated α and β subunits. Various combinations of α and β subunits derived from at least 16 integrins are responsible for great diversity of integrins among cell types (5). The integrins expressed on lymphocytes belong to the β_1 family, also referred to as the very late antigen (VLA) family, and the β_2 family, which includes lymphocyte function-associated antigen-1 (LFA-1), Mac-1, and p150/95 (6,7).

Adhesion molecules expressed on human lymphocytes and their ligands are listed in Table 1. This selected listing focuses on the β_1 and β_2 families of integrins (8,9) and is not meant to be all inclusive. Lymphocytes often express both an adhesion molecule and receptor for it ("counter receptor"), so that homotypic lymphocyte-lymphocyte interactions are common. Also, integrins play a crucial role in lymphocyte-ECM interactions. For example, some adhesion proteins recognize and bind to the tripeptide RGD (arginine-glycine-aspartic acid) which is commonly found in ECM macromolecules (30). With the availability of monoclonal antibodies to many adhesion molecules, it is now possible to perform binding inhibition studies and elucidate the functional importance of adhesion molecules expressed on different lymphocyte subsets (31).

III LYMPHOCYTE-ENDOTHELIAL CELL INTERACTIONS

The vascular endothelium, long considered to be a passive lining of blood vessels has been recently identified as an active participant in a variety of inflammatory and immune reactions, including lymphocyte extravasation into tissues (32). Lymphocytes in the circulation begin the process of extravasation by attachment to vascular endothelial cells (EC). The adhesion molecules mediate the attachment of lymphocytes to EC, and the increased expression of these molecules on the lymphocyte surface due to an activating signal delivered by a suitable ligand initiates the process of extravasation. The luminal surface of EC, which receives vascular responses, contains receptors able to transmit environmental signals. It has been suggested that EC might themselves provide activating signals to lymphocytes (32). In the areas of the vasculature, where EC become activated as a result of environmental signals, changes occur in the cell surface that allow leukocytes to attach and migrate through the endothelium (32). The process of EC activation involves changes in the level of expression of endothelial-leukocyte adhesion molecule 1 (ELAM-1), which appears to be endothelium specific (33), and a variety of other surface proteins including adhesion and activation molecules. Activated EC produce and release a variety of cytokines, which facilitate interactions between the endothelium and leukocytes, act as chemoattractants and modulate the coagulation pathway (34,35). As a result of interactions with EC, nonadherent lymphocytes become adherent, polarized, and poised for the next step of the extravasation process, namely, migration from the vessel lumen into tissue.

The mechanisms involved in the lymphocyte adherence to and extravasation through the endothelial monolayer are not fully understood and are under careful investigation. Certain adhesion molecules appear to play an important role in the endothelial migration of lymphoid cells. For example, expression of LFA-1 appears to be important for the transendothelial passage of T lymphocytes (36). During the transendothelial passage, lymphocytes appear to migrate between EC, squeezing through intercellular spaces to reach the basement membrane and move on into the interstitial matrix. Cytokines, both those released by activated lymphocytes and those produced by EC, not only modulate interactions between the endothelium and leukocytes (35), but also facilitate the transendothelial passage of lymphocytes. More recently, it has been shown that the vascular endothelium in various organs might be specialized and express organ-specific antigens (37). Such organ-related differences in EC may be important for the binding and transendothelial migration of lymphocytes.

IV LYMPHOCYTE-ECM INTERACTIONS

Lymphocytes which cross the vascular endothelial monolayer and the basement membrane, find themselves in a microenvironment of ECM (Figure 1). Contrary to earlier beliefs, components of ECM such as fibronectin, laminin, hyaluronic acid serve not only to provide structural support for lymphoid cells in tissue but are thought to be actively involved in several stages of lymphocyte function and development. Consequently, ECM-lymphocyte interactions are currently of great interest. It is not a surprise that lymphocytes express several distinct ECM receptors (Table 1), although the β_1 family of the VLA integrins represents the predominant class of receptors involved in interactions with ECM. ECM is thought to influence several distinct lymphocyte functions including their migrations, activations, signal transduction, proliferation, and differentiation (38,39). Various components of ECM appear to be involved in costimulating signal delivery to lymphocytes, and evidence, largely derived from in vitro studies with purified ECM components, suggests that specific receptor/ECM ligand interactions are involved in modulation lymphocyte functions (40).

Table 1
A selected list of adhesion molecules on lymphocytes

CD Designation	Alternate Name	mol. wt. (kd)	Name of receptor or ligand	Reference
CD11a/CD18	Integrin LFA-1	α180	ICAM-1 (CD54)	10,11
	αLß2	ß95	ICAM-2	
CD11b/CD18	Integrin Mac-1	α165	ICAM-1	12,13
	αMß2	ß95	C3bi, FN, others	
CD11c/CD18	Integrin p150/95	α150	C3bi?, others	14
	αXß2	ß95		
CD2	SRBC receptor	50	LFA-3 (CD58)	15,16
	LFA-2			
	T11, Leu5			
CD4	T4, Leu3	55	MHC class II	17
CD8	T8, Leu2	30-38	MHC class I	18
CD44	ECMR-III, Pgp-1	90 &	HA, COLL?	19,20
	Hermes	200	MECA-367	
CD49a/CD29	Integrin VLA-1	$\alpha_1$200	LN, COLL	6
	α1ß1	ß110		
CD49b/CD29	Integrin VLA-2	$\alpha_2$160	LN, COLL	6,21
	α2ß1	ß110		
CD49c/CD29	Integrin VLA-3	$\alpha_3$150	FN, LN, COLL	6,22
	α3ß1	ß110		
CD49d/CD29	Integrin VLA-4	$\alpha_4$150	FN, VCAM-1	6,23,24
	α4ß1, LPAM-1	ß110		
CD49e/CD29	Integrin VLA-5	$\alpha_5$155	FN	6,25
	α5ß1, FNR	ß110		
CD49f/CD29	Integrin VLA-6	α140	LN	6,23
	α6ß1	ß110		
CD49d/CD?	Integrin, LPAM-2	α150	?HEV	26,27
	α4ßp	ß110		
Mel-14	Selectin, gp90,	90	HEV, ELAM-1	28,29
	Leu8, LAM-1, lec-CAM			

Abbreviations: LFA = leukocyte function associated antigen; ICAM = intercellular adhesion molecule; C3bi = breakdown product of the third component of complement; FN - fibronectin; SRBC = sheep red blood cell; MHC = major histocompatibility complex; HA = hyaluronic acid; COLL = collagen; LN = laminin; ECMR = endothelial cell membrane receptor; MECA = mucosal endothelial cell addressin; Pgp = platelet glycoprotein; VCAM = vascular cell adhesion molecule; VLA = very late activation antigen; LPAM = lymphocyte - Payer's patch adhesion molecule; LAM = leukocyte adhesion molecule; lec-CAM = lectin-binding cellular adhesion molecule; ELAM = endothelial leukocyte adhesion molecule.

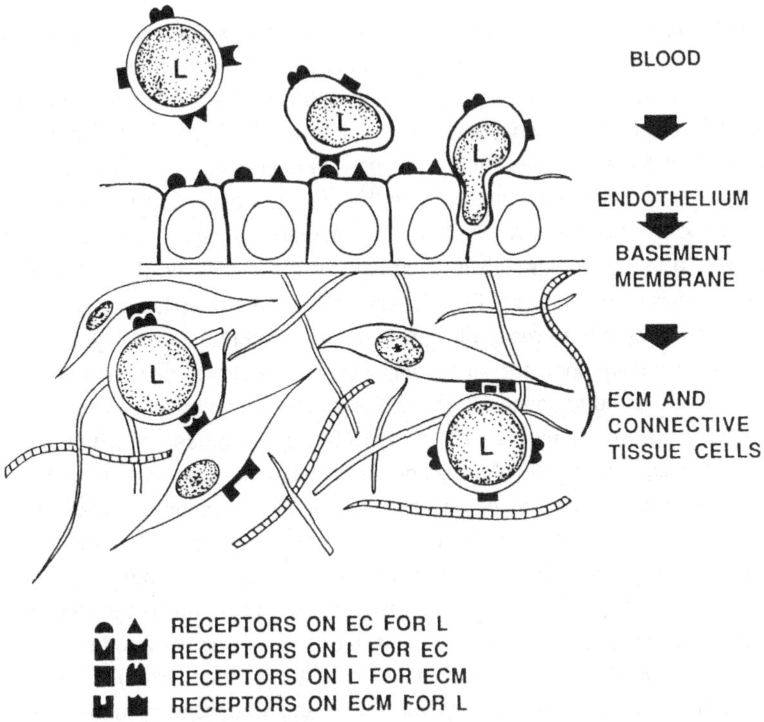

Fig. 1 A schematic representation of lymphocyte extravasation from a vessel lumen through the vascular endothelium and basement membrane to the ECM. The arrows indicate the direction of lymphocyte migration. Extravasation through the vascular endothelium is accomplished by squeezing in between individual endothelial cells and follows receptor-dependent interactions between lymphocytes and endothelial cells. The movement of lymphocytes through the ECM is also dependent on the presence of receptors for the ECM components on lymphocytes and counter-receptors for lymphocyte antigens on connective tissue cells. The receptor- counter-receptor interactions between lymphocytes and tissue cells seem to be short-term interactions, so that lymphocytes move through the tissue by engaging and, in turn, disengaging their receptors. The abbreviations used are as follows: L=lymphocytes; EC=endothelial cells; ECM=extracellular matrix. Reproduced with permission from Whiteside and Herberman (59)

V MIGRATION CHARACTERISTICS OF DIFFERENT LYMPHOID CELL POPULATIONS

Although all mature lymphocytes are circulating cells and have the potential to reach almost all locations in the body, it is thought that lymphocytes follow preferential migration pathways and are not randomly distributed in the various tissues. Thus, while B cells migrate preferentially to gut-associated lymphoid tissue, T cells like to migrate to peripheral lymph nodes (41). The ability of a lymphoid cell to extravasate from the blood into tissues depends on various cellular attributes as follows: a) the state of lymphocyte activation, which is related to the upregulation of adhesion molecule on its surface and to cytokine production, both important for lymphocyte interaction with the vascular endothelium and ECM; b) lymphocyte size and shape; it is well known that large, i.e., blastic, and small lymphocytes show differences in motility and migration; c) distribution and organization of adhesion molecules in the presence of specialized structures such as pseudopodia or podosomes, which facilitate migration and/or adhesion; d) the rigidity of lymphocyte membrane, which is important for the capability of the lymphocyte to squeeze in between the vascular EC. For example, IL2-activated lymphocytes which are larger and more rigid than resting lymphocytes (42) might be less likely to extravasate than non-activated lymphocytes; e) the ability of lymphocytes to respond to chemotactic stimuli and/or respond to specific signals. For example, primed lymphocytes appear to extravasate more efficiently than naive cells (43); g) the presence in the lymphocyte microenvironment of migration-promoting or-inhibiting factors; and h) the ability of lymphocytes to produce enzymes capable of degrading components of the ECM (see below). This rather extensive list of cellular characteristics implies that various types of lymphoid cells might differ in terms of their migration properties. Indeed, differences in migration between T lymphocytes, B lymphocytes and NK cells have been described (44), and it is well known that resting lymphocytes and lymphocytes activated by cytokines or other agents show striking differences in their migratory properties (45). Since migration characteristics of lymphocytes are dependent on and influenced by numerous factors, including cellular morphology and cell surface properties, the states of activation and differentiation, tissue localization and ability to interact with tissue cells, it is not surprising that in pathologic conditions, and especially in diseases associated with changes in activation, differentiation or distribution of selected lymphocyte subpopulations, migration of lymphocytes into tissue may be altered or may be even responsible for a pathologic condition. For example, one mechanism by which a solid tumor might escape the host immune system is through the inhibition of lymphocyte migration, thus preventing access of effector cells to the tumor. If human solid tumors are capable of inhibiting migration of lymphocytes through ECM and into the tumor, as recent data seem to indicate, then one approach to therapy might be to deliver competent, ex vivo activated effector cells to the tumor site in the hope of enhancing antitumor responses in the tumor microenvironment.

VI POTENTIAL ROLE FOR EFFECTOR CELL PROTEASES IN DEGRADATION OF ECM

Substantial evidence exists indicating that cellular proteases contribute to degradation of subendothelial extracellular matrices, including basement membranes. Such degradative proteolytic enzymes include various matrix metalloproteinases, e.g., type IV collagenase, (46-48) plasminogen activator/plasmin, cathepsin B, heparanase, etc.

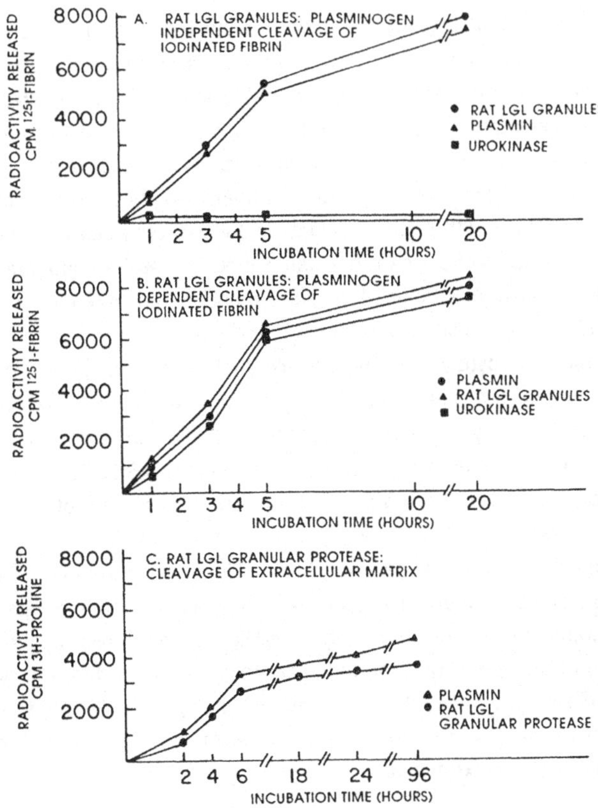

Fig. 2 Proteolytic activity of rat LGL granular proteases. Purified reference enzymes and granular protease-containing samples were assayed for their capacity to degrade iodinated fibrin in the presence or absence of plasminogen. Degradation of extracellular matrices utilized a basement membrane-like extracellular matrix containing incorporated, tritiated proline

Moreover, proteases have been found to directly exhibit both chemotactic properties as well as adhesive properties. For example, plasminogen activator and alpha thrombin have been found to be chemotactic and to play a role in cellular migration. In addition, proteolytic fragments of extracellular matrix molecules such as laminin and fibronectin are known to be chemotactic and/or haptotactic for various cell types (reviewed in (46)). Indeed, some proteolytic enzymes have been implicated in regulating aspects of adhesion and have been colocalized with their substrates at sites of adhesion (49).

Recent studies have shown that CTL produce a trypsin-like protease that degrades sulfated proteoglycans in the subendothelial matrix (50,51). In addition, we have recently shown that proteases enriched from highly purified rat large granular lymphocytes (LGL), i.e., NK cells, cleave various fluorogenic substrates as well as macromolecular substrates. Macromolecular substrates include iodinated fibrin, casein, various extracellular matrices and isolated matrix components (Figure 2). Our results show that rat LGL granular proteolytic activities are able to cleave a tritiated proline-radiolabeled smooth muscle-derived basement membrane-like extracellular matrix. Our studies have also revealed that this enzymatic activity directly cleaves laminin and other glycoprotein components of the extracellular matrix. It therefore appears quite possible or likely that degradative proteases of NK cells might contribute to extravasation through degradation of host extracellular matrices.

In addition to cytolytic granule-associated proteases of NK cells, we have described additional NK cell proteolytic enzymes including: plasma-membrane-associated, extracellularly released proteases, and high Mr cytosolic, non-granzyme proteases related to the multicatalytic proteinase complex (46,52,53). The discovery and biochemical purification of non-granular proteolytic enzymes of NK cells identify candidates for potential roles in NK cell functions including degradation of host extracellular matrices and NK cell migration. In view of our recent findings that adoptively transferred interleukin-2 activated NK cells accumulate within established metastases (54,55), it is possible that proteases of NK cells might contribute to detachment from microvascular endothelial cells, invasion through subendothelial basement membranes, and cellular migration. NK cell proteases might therefore play a role analogous to that exhibited by proteases of invasive tumor cells in matrix degradation and cellular extravasation.

VII THERAPEUTIC TRANSFERS OF LYMPHOCYTES

Although cellular adoptive immunotherapy has been used in recent years for treatment of selected metastatic malignancies, many questions remain unanswered as to the mechanism by which adoptively transferred effector cells localize to the tumor or tumor metastases and mediate their antitumor effects <u>in vivo</u>. In particular, the movement of these effector cells through vessels and migration into tumors are poorly understood. In another chapter in this volume, mechanisms of antitumor effects of adoptively transferred activated natural killer (NK) cells are discussed in greater detail (see Chapter 13). To additionally address these issues, a series of experiments was performed by our group using IL2-activated, fluorescently-labeled human NK (A-NK) cells (56). These cells were injected intraarterially with low dose IL2 into normal (mature granulation) tissue or an implant of VX-2 carcinoma growing in the rabbit ear chamber (56). Normal granulation tissue or VX-2 tumor were grown in transparent Sandison-type chambers implanted in the ears of white New Zealand rabbits. Highly enriched human IL2-activated NK (92% $CD3^-$ $CD56^+$) were labeled with acridine orange and injected with or without 6,000 IU/ml of IL2 (10^5-10^6 cells) into the auricular artery supplying the ear chamber. The arrival, mechanical entrapment, adhesion to EC, detachment, extravasation and departure of these cells were observed using a modified intravital microscope, video taped, and analyzed using computer-assisted image analysis (56). Observations of activated NK (A-NK) cell movement through both normal and tumor vasculature demonstrated that interactions with the endothelium could be described as short-term (<1 sec), long-term (≥1 sec) or extended (≥60sec). Movement through the vasculature of A-NK cells was visualized as a series of rapid motions, interrupted by brief adhesions to the endothelium. The duration of short-term adhesive events for A-NK cells was not significantly different between normal and tumor tissues. However, the duration of each long-term adhesive event was significantly ($p<0.05$) longer for A-NK cells binding to tumor vasculature than to normal vasculature. A-NK cells injected without IL2 were retained for significantly shorter times than A-NK cells with IL2 in the tumor tissue. The majority of the interactions between the adoptively transferred xenogeneic A-NK cells and tumor vasculature were relatively brief, and only small fractions (<0.01%) of the injected cells remained in the tumor in excess of an hour. However, this was sufficient to produce widespread tumor-associated stasis in the tumor vasculature, which was observed within 1-2 days after injection. Subsequent necrosis of the tumors was observed along with diffuse infiltrates of lymphocytes, monocytes, and granulocytes in the interstitial spaces within the tumor. These findings suggest that the number of A-NK cells required to produce antitumor effects is relatively small and that the vasculature changes which occur are rapid. Also, the development of necrosis despite a very low effector to target cell ratio indicates that in addition to direct cytotoxicity, the therapeutic effects of adoptively transferred A-NK cells might be mediated via the tumor vasculature. Attachment of A-NK cells to the tumor vasculature might produce damage to the endothelial cell at that site. Such damage in the tumor neovasculature might reduce or shut down the blood supply to the tumor. This may lead to subsequent ischemia through the production of small thrombi at the site of vascular damage. Indeed, such effects were observed in the tumor chambers after A-NK administration in our experiments. It is also possible that cytokines produced by the A-NK cells, such as IFNγ, TNF-α, IL2, or granulocyte-monocyte colony-stimulating factor (57) may contribute to tumor necrosis either directly or by recruitment of other cytotoxic cells to the area of the tumor. In addition to functional attributes of A-NK cells (antitumor cytotoxicity and cytokine production) or their morphologic characteristics (high expression of CAM and their structural rigidity), the structure of the tumor vasculature may also contribute to the selective adhesion of A-NK cells (58). The tumor vasculature contains large numbers of turns and irregularities which cause resistance to blood flow and increase the likelihood of successful adhesive events. Also, it is possible that EC in the tumor vasculature, which usually have a high rate of proliferation, may have increased adhesiveness for A-NK cells. This series of experiments showed that both the characteristics of the tumor vasculature and of systemically delivered A-NK cells contribute to the therapeutic efficacy of adoptively transferred effector cells.

Cascade for Accumulation of Activated NK (A-NK)
Cells in Tumor Metastasis

Adoptive transfer of A-NK cells (i.v. or
locoregional e.g. portal circulatory system)

↓

Adhesion to microvascular endothelial cells of
target organ (e.g. internal lumen of blood vessels)

↓

Penetration through target organ subendothelial matrices

↓

Binding to microvascular endothelial cells and
metastatic tumor cells within microenvironment
of metastases: accumulation within metastases

↓

Further activation of A-NK cells:
- Cellular enlargement?
- Cytokine induction?
- Protease induction?
- Enhanced cytolytic function?
- Effects on microvasculature?
- Recruitment of additional effector cells?
- Activation of endothelial cells?

Fig. 3 A schematic outline of events that might take place in the vasulature and/or tumor site when immune effector cells, e.g., A-NK cells, are used for therapy of cancer

VIII SUMMARY

To develop effective antimetastic response, effector cells have to reach the tissue site of tumor metastasis. Different lymphocyte subsets vary in their ability to extravasate and reach sites of tissue injury or tumor growth. The ability of lymphoid cells to extravasate is a funciton of several different parameters, including cellular activation state and ability to release cytokines, proteolytic enzymes or to respond to chemotactic stimuli present in the tumor microenvironment. Adoptively transferred effector cells can only be therapeutically effective if they extravasate, reach the tumor or its microenvironment and interfere with tumor growth or metastasis. A large number of events that might be involved in accumulation of effector cells such as, e.g., A-NK cells in tumor metastases, are illustrated in the schema in Figure 3. The future challenge will be to gain an in-depth understanding of these events, including effector cell migration and extravasation, to assure optimal transfers of these cells to patients with metastatic disease.

References

1. Butcher, E.C. Cellular and molecular mechanisms that direct lymphocyte traffic. Am. J. Pathol. 1990; 136:3-11.
2. Goldfarb, R.H., Wasserman, K., Herberman, R.B. and Kitson, R.P. Nongranular proteolytic enzymes of rat IL2-activated natural killer cells. J. Immunol. 1992; In press, Vol. 149.
3. Shaw, S. Lymphocyte interactions with extracellular matrix. FASEB J. 1991; 5:2292-2299.
4. Dustin, M.L. and Springer, T.A. Role of lymphocyte adhesion receptors in transient interactions and cell locomotion. Ann. Rev. Immunol. 1991; 9:27-66.
5. Ruoslahti, E. Integrins. J. Clin. Invest. 1991; 87:1-5.
6. Hemler, M.E. VLA proteins in the integrin family: Structures, functions and role on leukocytes. Ann. Rev. Immunol. 1990; 8:365-400.
7. Patarroyo, M., Prieto, J., Rancon, J., Timonen, T., et al. Leukocyte-cell adhesion: A molecular process fundamental in leukocyte physiology. Immunol. Rev. 1990; 114:67-108.
8. Albelda, S.M. and Buck, C.A: Integrins and other cell adhesion molecules. FASEB J. 1991; 4:2868-2880.
9. Duijvestijn, A. and Hamann, A. Mechanisms and regulation of lymphocyte migration. Immunol. Today 1989; 10:23-28.
10. Patarroyo, M. and Makgoba, M.W. Luekocyte adhesion to cells. Scand J Immunol 1989; 30:129-164.
11. Marlin, S.D. and Springer, T.A. Purified intercellular adhesion molecule-1 (ICAM-1) is a ligand for lymphocyte function-associated antigen 1 (LFA-1). Cell 1987; 51:813-819.
12. Pardi, R., Bender, J.R., Dettori, C., Giannazza, E. and Engleman, E.G. Heterogeneous distributionand transmembering signalling properties of leukocyte function-associated antigen (LFA-1) in human lymphocyte subsets. J. Immunol. 1989; 143:3157-3166.

13. Altieri, D.C., Bader, R., Mannucci, P.M. and Edgington, T.S. Oligospecificity of the cellular adhesion receptor MAC-1 encompasses an inducible recognition specificity for fibrinogen. J. Cell. Biol. 1988; 107:1893-1900.
14. Sanchez-Madrid, F., Nagy, J.A., Robbins, E., Simon, P. and Springer, T.A. A human leukocyte differentiation antigen family with distinct alpha subunits and a common beta subunit: The lymphocyte function-associated antigen (LFA-1), the C3bi complement receptor (OKM-1, MAC-1), and the p150, 95 molecule. J. Exp. Med. 1983; 158:1785-1803.
15. Takai, Y., Reed, M.L., Burakoff, S.J. and Herrmann, S.H. Direct evidence for a receptor-ligand interaction between the T cell surface antigen CD2 and lymphocyte-function-associated antigen 3. PNAS USA 1987; 84:6864-6868.
16. Plunket, E.M.L., Sanders, M.E., Selvaraj, P., Dustin, M.L. and Springer, T.A. Rosetting of activated human T lymphocytes with autologous erythrocytes: Definition of receptor and ligand molecules as CD2 and lymphocyte function-associated antigen 3 (LFA-3). J. Exp. Med. 1987; 165:664-676.
17. Doyle, C. and Strominger, J.L. Interaction between CD4 and class II MHC molecule mediates cell adhesion. Nature 1987; 330:256-259.
18. Salter, R.D., Benjamin, R.J., Wesley, P.K., Buxton, S.E., Garrett, T.P.J., Clayberger, C., Krensky, A.M., Normant, A.M., Littman, D.R. and Parham, P. A binding site for the T-cell call-receptor CD8 on the alpha 3 domain of HLA-A2. Nature 1990; 345:41-46.
19. Aruffo, A., Stamenkovic, I., Melnick, M., Underhill, C.B. and Seed, B. CD44 is the principle cell surface receptor for hyaluronate. Cell 1990; 61:1303-1313.
20. Shimizu, Y., van Seventer, G.A., Siraganian, R., Wahl, L. and Shaw, S. Dual role of the CD44 molecule in T cell adhesion and activation. J. Immunol. 1989; 143:2457-2463.
21. Elices, M.J. and Hemler, M.E.: The human integrin VLA-2 is a collagen receptor on some cells and a collage laminin receptor on others. PNAS USA 1989; 86:9906-9910.
22. Dang, N.H., Torimoto, Y., Schlossman, S.F. and Morimoto, C. Human CD4 helper T cell activation: Functional involvement of two distinct collagen receptors, 1F7 and VLA integrin family. J. Exp. Med. 1990; 172:649-652.
23. Hemler, M., Huang, C. and Schwarz, L. The VLA protein family. J. Biol. Chem. 1987; 3300-3309.
24. Elices, M.J., Osborne, L., Takada, Y., Crouse, C., Luhowjsky, J.S., Hemler, M.E. and Lobb, R.R. VCAM-1 on activated endothelium interacts with the leukocyte integrin VLA-4 at a site distinct from the VLA-4/fibronectin binding site. Cell 1990; 60:577-584.
25. Yamada, A., Nikaido, T., Nojima, Y., Schlossman, S.F. and Morimoto, C. Activation of human CD4 T lymphocytes. Interaction of fibronectin with VLA-5

receptor on CD4 cells induces the AP-1 transcription factor. J. Immunol. 1991; 146:53-56.

26. Holzmann, B. and Weissman, I.L. Integrin molecules involved in lymphocyte homing to Peyer's patches. Immunol. Rev. 1989; 108:45-60.

27. Holzmann, B., McIntyre, B.W. and Weissman, I.L. Identification of a murine Peyer's patch-specific lymphocyte homing receptor as an integrin molecule with an alpha chain homologous to human VLA-4 alpha. Cell 1989; 56:37-46.

28. Tedder, T.F., Penta, A.C., Levine, H.B. and Friedman. A.S. Expression of the human leukocyte adhesion homing molecule, LAM-1: Identity with the TQ-1 and Leu-8 differentiation antigen. J. Immunol. 1990; 144:532-540.

29. Camerini, D., James, S.P., Stamenkovic, I. and Seed, D. Leu-8/TQ-1 is the human equivalent of the mel-14 lymph node homing receptor. Nature. 1989; 342:78-82.

30. Ruoslahti, E. and Pierschbacher, M.D. Arg-Gly-Asp: A versatile cell recognition signal. Cell 1986; 44:517-518.

31. Pigott, R., Needham, L.A., Edwards, R.M., Walker, C. and Power, C. Structural and functional studies of the endothelial activation antigen endothelial leukocyte adhesion molecule 1 using a panel of monoclonal antibodies. J. Immunol. 1991; 147:130-135.

32. Pober, J.S.: A cytokine mediated activation of vascular endothelium: Physiology and pathology. Am. J. Pathol. 1988; 133:426-433.

33. Bevilacqua, M.P., Pober, J.S., Mendrick, D.L., Cotran, R.S. and Gimbrone, M.A., Jr. Identification of an inducible endothelial-leukocyte adhesion molecule. PNAS 1987; 84:9238-9242.

34. Carlos, T.M., Schwartz, B.R., Kovach, N.L., Yee, E., Rosa, M., Osborne, L., Chi-Rosso, G., Newman, B., Lobb, R. and Harlan, J.M. Vascular cell mediated lymphocyte adherence to cytokine-activated cultured human endothelial cells. Blood 1990; 76:965-970.

35. Mantovani, A. and Dejana, E. Cytokines as communication signals between leukocytes and endothelial cells. Immunol. Today 1989; 10:370-375.

36. Kavanaugh, A.F., Lightfoot, E., Lipsky, P.E. and Oppenheimer-Marx, N. Role of CD11-CD18 in adhesion and transendothelial migration of T cells. Analysis utilizing CD18-deficient T cell clones. J. Immunol. 1991; 146:4149-4156.

37. Picker, L.J., Kishimoto, T.K., Smith, C.W., Warnock, R.A. and Butcher, E.C. ELAM-1 is an adhesion molecule for skin-homing T cells. Nature 1991; 349:796-799.

38. Levesque, J.P., Hatzfeld, A. and Hatzfeld, J. Mitogenic properties of major extracellular proteins. Immunol. Today 1991; 12:258-262.

39. Davis, L,S,, Oppenheimer-Marx, N., Bednarczyk, J.L., McIntyre, B.W. and Lipsky, P.E. Fibronectin promotes proliferation of naive and memory T cells by signalling

through both VLA-4 and VLA-5 integrin molecules. J. Immunol. 1990; 145:785-793.

40. Springer, T.A. Adhesion receptors of the immune system. Nature 1990; 346:425-434.
41. Woodruff, J.L., Clarke, L.M. and Chin, Y.H. Specific cell-adhesion mechanism determining migration pathways of recirculating lymphocytes. Ann. Rev. Immunol. 1987; 5:201-222.
42. Sasaki, A., Jain, R.K., Al Maghazachi, A., Goldfarb, R.H. and Herberman, R.B. Low deformability of lymphokine-activated killer cells as a possible determinant of in vivo distribution. Cancer Res. 1989; 49:3742-3746.
43. Shimizu, Y., van Seventer, G.A., Horgan, K.J. and Shaw, S. Regulated expression and function of three VLA (ß1) integrin receptors and T cells. Nature 1990; 345:250-253.
44. Issekutz, T. Effects of six different cytokines on lymphocyte adherence to microvascular endothelium and in vivo lymphocyte migration in the rat. J. Immunol. 1990; 144:2140-2146.
45. Ratner, S. and Heppner, G.H. Motility and tumoricidal activity of interleukin-2 stimulated lymphocytes. Cancer Res. 1988; 48:3374-3380.
46. Goldfarb, R.H. and Liotta, L.A. Proteolytic enzymes in cancer invasion and metastasis. Sem. Thrombosis Hemostasis 1986; 12:25694-307.
47. Goldfarb, R.H. Proteolytic enzymes in tumor invasion and degradation of host extracellular matrices. In: Mechanisms of Cancer Metastasis: Potential Therapeutic Implications. (edit. Honn, K.V., Powers, W.E. and Sloane B.F.) Martinus Nijhoff, Boston, 1986; pp 341-375.
48. Goldfarb, R.H., Timonen, T. and Herberman, R.B. Productgion of plasminogen activator by human natural killer cells (large granular lymphocytes) J Exp Med 1984; 2159:935-951.
49. Beckerlie, M.C., Burridege, K., De Martino, G.M. and Croall, D. Colocalization of calcium-dependent protease II and one of its substrates at sites of cell adhesion. Cell 1987; 51:569-577.
50. Kramer, M.D. and Simon, M.M. Are proteinases functional molecules of T lymphocytes? Immunol. Today 1987; 8:140-142.
51. Simon, M.M., Simon, H.G., Fruth, U., Epplen, J., Muller-Hermelilnk, H.K. and Kramer, M.D. Cloned cytolytic T-effector cells and their malignant variants produce an extracellular matrix degrading trypsin-like serine protease. Immunol. 1987; 60:219-230.
52. Goldfarb, R.H., Wasserman, K.W., Herberman, R.B. and Kitson, R.P. Nongranular proteolytic enzymes of rat IL-2 activated natural killer cells. I. Subcellular localization and functional role. J. Immunol., In Press.

53. Wasserman, K., Kitson, R.P., GGabauer, M.K., Miller, C., Herberman, R.B. and Goldfarb, R.H. Zymographic analysis of cell-associated and released proteases of rat interleukin 2 activated natural killer (A-NK) cells. 1992; Submitted.

54. Basse, P., Herberman, R.B., Nannmark, U., Johansson, B.R., Hokland, M., Wasserman, K. and Goldfarb, R.H. Accumulation of adoptively transferred adherent, lymphokine-activated killer cells in murine metastasis. J. Exp. Med. 1991; 174:479-488.

55. Basse, P.H., Nannmark, U., Johansson, B.R., Herberman, R.B. and Goldfarb, R.H. Establishment of cell-to-cell contact by adoptively transferred adherent lymphokine-activated killer cells with metastatic murine melanoma cells. J. Natl. Cancer Inst. 1991; 83:944-950.

56. Sasaki, A., Melder, R.J., Whiteside, T.L., Herberman, R.B. and Jain, R.K. Preferential localization of human adherent lymphokine-activated killer cells in tumor microcirculation. J. Natl. Cancer Ins. 1991; 83:433-437.

57. Vujanovic, N.L., Vitolo, D., Rabinowich, H., Lee, Y.J., Herberman, R.B. and Whiteside, T.L. IL2-induced activation of human NK cells: In situ hybridization of IL2 receptor and cytokine genes. Nat. Immunity Cell Growth Reg. 1991; 10:154.

58. Jain, R.K. Determinants of tumor blood flow: A review. Cancer Res. 48:2641-2658, 1988.

59. Whiteside, T.L. and Herberman, R.B. Extravasation of antitumor effector cells. Invasion Metastasis. 1992; In Press.

15

Tumor Microenvironment and Immune Effector Cells: Isolation, Large Scale Propagation and Characterization of CD8$^+$ Tumor Infiltrating Lymphocytes From Renal Cell Carcinomas

T. Juhani Linna[1]
Dewey J. Moody
Lee Ann Feeney
Thomas B. Okarma
Applied Immune Sciences, Inc.
Santa Clara, California

Cho Lea Tso
Arie Belldegrun
UCLA School of Medicine
Los Angeles, California

I INTRODUCTION

It has been well documented that murine and human tumors can harbor highly cytolytic effector cells, which have protective anti-tumor activity against intravenous tumor challenge and against established tumors (1,2,3). Metastatic renal cell cancer has been one of the malignancies in focus for therapeutic attempts with such in vitro expanded tumor infiltrating lymphocytes (TIL) (4,5). These studies have been done with IL-2 activated "bulk" effector T cells after expansion in vitro. We briefly present here studies, in which a monoclonal antibody and device-based technology, developed by Applied Immune Sciences, has been used to capture CD8$^+$ subpopulations present in renal cell

[1] Present address: Syntex Research, Division of Syntex (U.S.A.) Inc., Palo Alto, California

carcinoma-derived TIL. The captured $CD8^+$ TIL have been successfully expanded in vitro on a large scale, applicable to clinical therapeutic studies. Such studies are in progress.

II ISOLATION AND CULTURE OF $CD8^+$ TIL

Renal cell carcinoma (RCC) samples were obtained from resected kidneys post-surgery after obtaining patient informed consent. The tumor samples were digested as described previously (4), separated over Histopaque 1077, washed and resuspended in a serum-free medium (AIM-V, GIBCO, Chagrin Falls, OH) containing recombinant interleukin-2 (rIL-2, courtesy of M. Gately, Hoffmann-LaRoche, Nutley, NJ). Aliquots of bulk cultured TIL were transferred into standard tissue culture flasks and maintained at 37°C (5% CO_2). Separations of lymphocyte subsets were performed using CD8 AIS-CELLectors™ (Applied Immune Sciences, Menlo Park, CA). The capture flasks are sterile tissue culture flasks to which monoclonal antibodies against CD8 are covalently bound (7). Cells from bulk-cultured TIL were concentrated, and each device was loaded with the concentrated bulk cultured TIL. The devices were then incubated at room temperature. The non-adherent fraction was removed and the capture surface rinsed leaving the desired cells on the surface. Some TIL samples were cultivated in RPMI 1640 containing 10% human AB serum and 400 U/ml rIL-2. Most cultures were propagated with serum free AIM-V medium containing rIL-2. The captured cells generally were incubated for 3 days (37°C), at which time the $CD8^+$ cells were recovered. Large scale propagation of the $CD8^+$ TIL in rIL-2 containing medium was carried out using SteriCell bags (Terumo, Elkton, MD). Cultures were monitored for rates of proliferation, phenotype and lytic activity. Effector cell lytic activity was measured in vitro using a 4h ^{51}Cr-release assay against labeled target cells K562, DAUDI, M14, and cryopreserved autologous and allogeneic RCC. Percent specific lysis was determined and lytic activity against each target cell expressed in lytic units (LU_{30}).

The first phase of TIL production involved lymphocyte cultivation from whole tumor digest. Initial lymphocyte composition was highly variable, ranging from 27% to 70% (49 ± 15, mean ± 1 standard deviation) immediately post-digestion. There were no significant correlations between the sample mass (grams) and either the yield of cells or percentage of lymphocytes recovered. The bulk cultures were propagated ex vivo for 13 to 28 days prior to the $CD8^+$ selection process in order to permit sufficient lymphocyte expansion. The pre-separation phenotypic composition of the bulk cultures was heterogeneous. T cells, as determined by CD3 surface expression, ranged from 9% to 96% (57 ± 34). The proportion of $CD8^+$ T cells ranged from 24% to 88% (39 ± 21) prior to separation. $CD8^+$ TIL cultures demonstrated sustained ex vivo proliferation for periods ranging from 19 to 41 days (post-separation) in the absence of appreciable numbers of autologous $CD4^+CD8^-$ cells. This was observed with $CD8^+$ TIL cultures isolated as early as day 13 and as late as day 28 ex vivo. Six of seven $CD8^+$ TIL cultures isolated from the bulk tumor cell suspensions proliferated well. The fold expansions for all cultures ranged from a low of 2-fold to a high of 1877 fold. The isolated TIL retained the highly enriched $CD8^+$ phenotype up to the time of culture termination. The $CD8^+$ TIL cultures always contained less than 8% contaminating $CD4^+$ cells. Dual color flow cytometry analysis of these contaminating $CD4^+$ cells showed that the majority were also co-expressing CD8. The proportion of TIL expressing CD16, a surface marker associated with NK cells, was less than 4% in these cultures just after separation and less than 1% following culture expansion. Thus, there was no evidence to suggest that $CD8^+$ NK cells were a significant component of these cell cultures. Three of 7 $CD8^+$ cultures presented with CD56 expression in excess of 20% of the population. The levels of $\alpha\beta$TCR expression (90% to 99% positive, 97 ± 3) in the $CD8^+$ cultures establish that the cells were mature T-cells. Cytolytic activities against NK (K562) and LAK (DAUDI, M14) susceptible target started relatively high but decayed quickly. There was also an initial early decay in lytic capacity against RCC targets followed by a sustained plateau of lytic activity against RCC autologous targets.

There was a significant ($p<0.005$) correlation between the fold expansion of a $CD8^+$ TIL culture and the magnitude (at time of culture termination) of the autologous anti-tumor lytic activity. A significant ($p<0.05$) correlation was also found between the anti-tumor cytotoxicity and the number of days that the $CD8^+$ TIL culture was propagated ex vivo. Consistent with expectations, a significant correlation was observed between the CD56 expression and cytotoxic activity against NK ($p<0.01$) and LAK ($p<0.01$) sensitive target cells.

III CONCLUSIONS

These studies illustrate that $CD8^+$ TIL can be successfully isolated from TIL preparations and propagated on a therapeutic scale. These cells are capable of rapid ex vivo expansion even in the virtual absence of helper $CD4^+$ TIL. Highly enriched $CD8^+$ TIL, under the influence of rIL-2, proliferate well without detectable loss of phenotypic homogeneity. Residual in vitro anti-tumor activity is detectable even after several weeks of propagation. These highly enriched and expanded $CD8^+$ TIL should be used, with cytokine support, to assess possible efficacy in metastatic renal cell cancer, thus expanding the laboratory and clinical findings with bulk TIL in renal cell cancer and other malignancies (4,5,6,7,8). Such a clinical study is in progress in metastatic renal cell cancer. This approach may be of value in the experimental treatment of other disseminated malignancies as well.

References

1. Wong, R.A., Alexander, R.B., Puri, R.K.and Rosenberg, S.A. In vivo proliferation of adoptively transferred tumor-infiltrating lymphocytes in mice. J. Immunol. 1991; 10:120-130.
2. Yamasaki, T., Handa, H., Yamashita, J., Watanabe, Y., Namba, Y. and Hamaoka, M. Specific adoptive immunotherapy with tumor specific CTL clone for murine malignant gliomas. Cancer Res. 1984; 44:1776-1782.
3. Rosenberg, S.A., Packard, B.S., Abersold, P.M., Solomon, M.D., Topalian, J.L., Toy, S.T., Simon, P., Lotze, M.T., Yang, J.C., Seipp, C.A., Simpson, C., Carter, C., Bock, S., Schwartzentruber, D., Wei, J.P. and White, D.E. Use of tumor-infiltrating lymphocytes and interleukin-2 in the immunotherapy of patients with metastatic melanoma. A preliminary report. New Engl. J. Med. 1988; 319:1676-1680.
4. Belldegrun, A.B., Muul, L.M. and Rosenberg, S.A. Interleukin 2 expanded tumor-infiltrating lymphocytes in human renal cell cancer: Isolation, characterization, and antitumor activity. Cancer Res. 1988; 48:206-214.
5. Finke, J.N., Rayman, P., Alexander, J., Edinger, M., Tubbs, R.R., Connelly, R., Pontes, E. and Bukowski, R. Characterization of the cytolytic activity of $CD4^+$ and $CD8^+$ tumor-infiltrating lymphocytes in human renal cell carcinoma. Cancer Res. 1990; 50:2363-2371.
6. Morecki, S., Topalian, S.L., Myers, W.W., Okrongly, D., Okarma, T.B. and Rosenberg, S.A. Separation and growth of human $CD4^+$ and $CD8^+$ tumor-infiltrating lymphocytes and peripheral blood mononuclear cells by direct positive panning on covalently attached monoclonal antibody-coated flasks. J. Biol. Resp. Mod. 1990; 9:463-472.
7. Heo, D.S., Whiteside, T.L., Johnson, J.T., Chen, K., Barnes, E.L. and Herberman, R.B. Long-term interleukin-2 dependent growth and cytotoxic activity of tumor-infiltrating lymphocytes from human squamous cell carcinomas of the head and neck. Cancer Res. 1989; 47:6353-6362.
8. Topalian, S.L., Muul, L.M., Solomon, D. and Rosenberg, S.A. Expansion of human tumor infiltrating lymphocytes for use in immunotherapy trials. J. Immunol. Methods 1991; 102:127-141.

PART V

Introduction

Immunomodulation and Antitumor Mechanisms

Theresa L. Whiteside
Ronald H. Goldfarb
Pittsburgh Cancer Institute and
University of Pittsburgh Medical Center
Pittsburgh, Pennsylvania

In tumor immunology, immunomodulation represents an approach to increasing antitumor activity of the immune system. By definition, this approach is based on a hypothesis that a vigorous immune response against autologous tumor can be induced and sustained in a patient with cancer and that this response will be effective in reducing or eliminating tumor growth and metastasis. Experience accumulated over many years has indicated that this may be a somewhat naive and perhaps unrealistic expectation. While there is considerable evidence in the literature indicating that a variety of biologic agents are active in modulating the antitumor functions of immune cells both in vitro and in vivo, few reliable studies can be quoted in support of a notion that such functionally augmented immune cells are effective in protecting the host from developing cancer or ridding the host of established tumor metastases. The issue concerning the involvement of immune cells in the control of tumor spread or destruction of micrometastases has been just about as controversial as the existence of tumor-specific antigens (TAA) or their capacity to induce an effective immune response against progressive tumor. Both issues remain seemingly unresolved today.

Nevertheless, there is something intuitively appealing and biologically logical in the idea that protective immunity against autologous tumor can be induced and maintained in vivo. Furthermore, newer insights into the mechanisms of antitumor responses suggest that both tumor-specific and tumor-nonspecific immune responses operate in tumor-bearing hosts, that TAA are able under certain circumstances to elicit demonstrable systemic and/or local immune responses, that these responses can be modified by means of exogenous agents, and that, under certain well-defined conditions, protective immune responses, including generation of specific memory cells, can be elicited. It remains to be determined to what extent specific MHC-restricted antitumor responses are more effective in vivo than non-MHC-restricted, natural immunity in control of tumor growth. The role(s) of these two antitumor cellular mechanisms play in, e.g., different tumor types or various stages of tumor development has been much debated but remains unclear. Only incomplete data are available regarding correlations between the presence or magnitude of these immune responses and prognosis or survival of patients with cancer. To date, no meaningful prognostic in vitro indicator has been identified which provides evidence for immunologic control of tumor growth and metastasis. A need for such an indicator obviously exists, and it is possible that the judicious use of immunomodulatory agents may lead to its definition.

Regulation of the antitumor immune response by biologic response modifiers (BRM) to benefit the host and disadvantage the tumor has been, and remains, a central goal of immunotherapy. For example, while studies have indicated that adoptively transferred A-NK cells can accumulate within established metastases, it may be that accumulation is necessary but not sufficient for the functional reactivity of these effector cells at the site of tumor metastases and that a cascade of subsequent activation steps of cells must take place within the microenvironment of the metastases. At the same time, it has been acknowledged that in order to rationally modify the immune response, basic knowledge of the mechanisms involved in tumor-effector cell interactions is essential. The ability to modulate antitumor immune responses by immopharmacologic intervention remains an important direction in antitumor drug discovery employing natural products and synthetic analogs, synthetic heterocyclic compounds that induce or mimic cytokine function, as well as the use of novel biologicals and/or enriched or specialized effector cell populations for adoptive immunotherapy. The many components of such interactions, including putative tumor antigens, antigen-presenting cells, cytokines involved in antigen-specific as well as non-specific amplification of immune responses, and the ability of effector cells to localize at the tumor site, all represent potential targets for immunomodulation. The challenge is to precisely and correctly select the target which is not only highly responsive to intervention with a given BRM but which is critically important for antitumor effects and thus, hopefully, useful for therapeutic modeling. A future immunotherapist must also bear in mind the fact that tumors have become remarkably adept at avoiding the immune system. For example, tumor-induced immunosuppression appears to be a common feature of most human tumors. Perhaps one of the most desirable therapeutic goals is to learn how to counteract such immunosuppression. In addition to a variety of cytokines whose activities are being explored in this regard, genetic modifications of tumor cells or other cells present in the tumor microenvironment (e.g. fibroblasts or mononuclear cells) are being explored as means of making immunotherapy more effective. These and other approaches focused on inducing changes in the tumor microenvironment as well as modulating functions of immune effector cells seem promising antitumor strategies of the near future.

The chapters in this section deal with these issues of immunomodulation and antitumor mechanisms. The chapter by Dr. Guido Forni reviews the role of cytokines in tumor rejection and that of Dr. Uchida discusses the biological significance of autologous tumor killing. The chapter by Dr. West is an overview of biological response modifiers and chemotherapeutic agents that alter interleukin-2 activities. These chapters therefore represent important insights into immunomodulation and antitumor mechanisms.

16

Interleukin-Induced Tumor Immunogenicity

Federica Cavallo
Mirella Giovarelli
Institute of Microbiology
University of Turin
Turin, Italy

Alberto Gulino
Alesandra Vacca
University of L'Aquila
L'Aquila, Italy

Giuseppe Scala
University of Naples
Naples, Italy

Guido Forni
CNR-Immunogenetics and Histocompatibility Center
Turin, Italy

I INTERLEUKINS AND THE REGULATION OF THE IMMUNE RESPONSE

Interleukins (ILs) are a communication code naturally regulating the induction, progression and effector phases of the immune response (1). They interact in a network by inducing each other, and display synergistic, additive or antagonistic effects. A few have been made unlimitedly available by recombinant DNA technology (2) and the current challenge is to find ways of using these ILs to compose meaningful immune messages.

This is a difficult task as ILs do not behave like immune hormones, but rather as very local cell-cell transmitters quickly cleared from the circulation (3). Moreover, the highly interactive nature of the distinct mechanisms forming an immune response makes it difficult to tease apart individual IL activities in vivo. Perhaps the presence of an unaccompanied IL is even a nonsense in in vivo immune communication. It certainly never occurs in natural conditions. A meaningful message is solely formed by the association of distinct ILs. Indeed, the in vivo pharmacologic efficacy of an exogenous IL may simply rest on its ability to induce the production of other ILs and the expression of their receptors.

In recent years, we have studied the possibility of using recombinant (r) ILs to interfere with the control mechanisms of the immune system during the growth of several poorly or non-immunogenic murine tumors, and human tumors and make them immunogenic (4). If immunogenicity is regarded as the ability to generate an immune response, the overwhelming majority of spontaneous tumors are non-immunogenic. This inability rests mainly on several functional features of the immune response, rather than on the lack of tumor associated antigens (TAA) on the tumor cell surface (5,6); poor presentation of TAA antigens by the very few glycoproteins of the major histocompatibility complex expressed by tumor cells; inefficient indirect TAA presentation by professional antigen presenting cells; absence of the appropriate IL repertoire; the rise of dominant suppressor activities (7). The addition of pharmacologic amounts of rILs could thus reverse a state of unresponsiveness. Many data from non-tumor systems indicate that this is often true (review in 8).

II THE ABILITY OF EXOGENOUS ILs TO ENHANCE TUMOR IMMUNOGENICITY

A. Methods

Normal BALB/c mice (Charles River Lab., Calco, Italy) were challenged in the left inguinal flank with the minimal 100% tumor inducing dose of: methylcholanthrene induced fibrosarcomas CA-2 and CE-2 (9); a lung carcinoma that arose spontaneously, Madison 109 (10); a Moloney virus induced sarcoma, MS-2 (11); and a spontaneous mammary adenocarcinoma, TS/A (12). Starting 1-48 hours after challenge, some mice received ten daily injections of 0.4 ml of Hanks balanced salt solution supplemented with 2% fetal bovine serum (HBSS-FBS, both from Flow Laboratories, Opera, Italy) alone (controls) or with various amounts of rILs. Neoplastic masses were measured with calipers in the two perpendicular diameters for 60 days. Mice tumor-free at this time were classed as survivors.

B. LATI efficacy

A lymphokine activated tumor inhibition (LATI) was induced by this treatment. While its efficacy depends on the tumor and the rIL injected, it nonetheless displays many hallmark features. The dose-response curve of each rIL follows a bell shape. Each rIL has its own range of efficacy. Evaluation of dose-response relationship over a large dose range produced several successive bell-shaped curves. As little as 10 U IL-2 (13). 12-3,000 IU IFN-γ (14), 10 pg IL-4 (15), 1-10 pg IL-1ß (16) and 10 pg IL-6 (F. Pericle et al., submitted for publication), gave the best results. Efficacy is directly related to the number of tumor infiltrating lymphocytes (13). LATI requires recipient mice reactivity, as it is abolished when the mice are sublethally irradiated or treated with cyclosporin A (15, 17).

C. LATI mechanisms

LATI is consistently marked by infiltration of many granulocytes and mononuclear cells, and a few lymphocytes. In the presence of rILs, T-lymphocytes and NK cells become able to directly kill tumor cells. However, experimental data suggest that their major role is to attract, trigger and guide the lytic action of inflammatory cells both through the release of ILs and other factors, and by physical interaction. These factors act as chemoattractants, enhance cell-cell interactions and trigger many distinct effector functions. The non-specific immune cells, thus guided and boosted by multiple T-lymphocyte messages, efficiently perform their lytic activity. Moreover, the cytokines sustaining the inflammatory phases apparently build a favorable milieu for tumor recognition and induction of a tumor-specific, systemic and persistent immunity. $CD8^+$ T-lymphocytes are responsible for this immune memory (4,8,15,17).

D. Synergism between rILs

Since ILs are never alone in natural conditions, the possibility that the pharmacologic efficacy of a given rIL is enhanced by its association with other ILs was investigated as the ability to both elicit LATI and induce tumor immunogenicity. No additive effect in LATI of TS/A adenocarcinoma was obtained through the association of rIL-1ß and rIL-4, whereas antagonisms stemmed from that of rIL-2 and rIL-4. Similar antagonisms in LATI of CE-2 followed by rIL-2 plus rIFN-γ, and rIL-2 plus rIL-4, whereas rIL-4 plus rIL-1ß provided only a small additive effect (15,18). However, in both tumor systems, a striking protection against a second lethal challenge of the same tumor was evident after rIL-4 plus rIL-1ß LATI (15). These findings are surprising, as CE-2 fibrosarcoma is poorly immunogenic and TS/A adenocarcinoma is apparently non-immunogenic in BALB/c mice (9,12).

Local administration of rILs thus reverses the natural immune unresponsiveness to tumors and induces immunogenicity.

III INTERLEUKIN-INDUCED TUMOR IMMUNOGENICITY (IITI)

A. Rationale

IITI is probably the most seminal finding of LATI. The data obtained with local rIL administration have been confirmed with several tumors producing ILs as a result of gene insertion. Engineered neoplastic cells secreting relatively low amounts of IL-4 (19), IL-2 (20,21), IFN-γ (22) and G-CSF (23) induce a protective immune response against a subsequent lethal dose of parental, untreated tumor cells. These data emphasize the concept that the local presence of small amounts of ILs is enough to induce immunogenicity to an otherwise non-immunogenic tumor. IITI often elicits an immune response that is strong and prompt enough to suppress tumor growth.

B. TS/A cell transfection with human IL-1ß gene

Expression vectors were used to introduce the cDNA coding for human IL-1ß or murine IL-2 into TS/A adenocarcinoma cells. To obtain the eukaryotic expression vector p220.2-IL-1ß, the SV40 promoter-IL-1ß cDNA-SV40 poly A transcriptional box was isolated from DspIL-1ß vector and ligated to the Sa1I site of p220.2 plasmid by standard techniques (G. Scala et al., submitted for publication). Cells were transfected by electroporation, washed and resuspended in 0.8 ml of Ca^{2+} and Mg^{2+} -free PBS at a concentration of $1.5x10^7$/ml in the presence of 10 μg of intact p220.2 or p220.2-IL-1ß plasmids. They were subjected to a single electrical pulse (0.2 kV, 960 μFD) using a Bio Rad apparatus, recovered, and cultured for 48 hours in Dulbecco's modified minimal essential medium (DMEM, Flow) supplemented with 10% FBS before selection in 450 μg/ml of hygromycin B. After 3 weeks, stable bulk cultures were subjected to limiting dilution cultures in RPMI 1640 medium (Flow) supplemented with 10% FBS (RPMI-FBS) in the presence of hygromycin B. At present, a single clone (A1) of IL-1ß transfected cells has been isolated and stabilized. $1x10^5$ A1 cells cultured for 48 hours in 2 ml of RPMI-FBS medium release 170-220 pg/ml of IL-1ß, as shown by Elisa assay (Eurogenetics, Turin, Italy). The <u>in vitro</u> growth rate of the IL-1ß transfected clone (hereafter denominated A1.200), TS/A cells transfected with the p220.2 plasmid containing the hygromycin B resistance gene only (A0.0), and that of untreated parental TS/A cells are comparable. BALB/c mice were then challenged in the left flank with $1x10^5$ cells to compare their oncogenicity. TS/A cells grew and killed all the mice while A0.0 cells killed 80% of the mice. Mitomycin-C treated TS/A cells (Mit-C) and A1.200 cells did not grow at all.

Comparison between survival after challenge with A1.200 cells or TS/A cells and repeated daily injection of rIL-1ß is made questionable by the kinetics of growth and survival of the transfected cells and differences in the IL-1ß delivery route. Nevertheless, the overall picture is similar in the two systems (Figure 1).

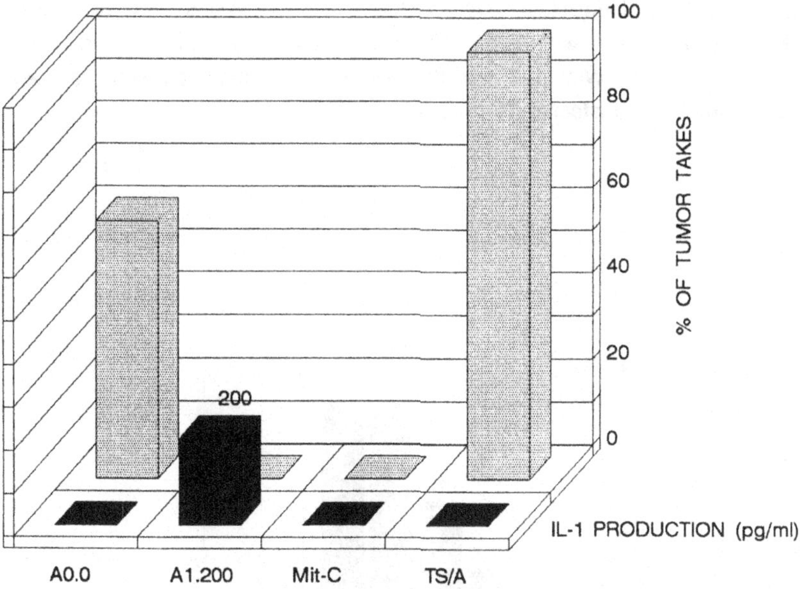

Fig. 1 In vivo growth of IL-1ß transfected TS/A cells. Groups of ten BALB/c mice were challenged in the left flank with $1x10^5$ cells from TS/A untreated parental cells (TS/A); TS/A cells preincubated with Mitomycin-C (Mit-C); TS/A cells transfected with the p220.2 plasmid containing the hygromycin B resistance gene only (A0.0); A1.200 IL-1ß producing clone. Dotted columns: % tumor takes; black columns: amount of IL-1ß produced

C. TS/A cell transfection with murine IL-2 gene

Cells were treated as described for IL-1ß transfection by putting $7.5x10^6$ TS/A cells/ml in the presence of 20 μg of linearized BCMGneo or BCMGneo-IL-2 plasmids (24), kindly provided by Dr. Karasuyama. They were subjected to a single electrical pulse (0.4 kV, 960 μFD) and selected in the presence of 300 μg/ml geneticin (G418, Sigma). Six clones were isolated from the bulk culture of the transfected cells and the amount of IL-2 produced by each clone ($5x10^4$ cells/48 hours/1 ml of RPMI-FBS) was evaluated by CTLL proliferation assay as previously described (13). The in vitro growth rate of the IL-2 transfected clones, cells transfected with BCMGneo plasmid containing the neomycin resistance gene only (B0.0) and TS/A was similar. When BALB/c mice were challenged as described in section B, TS/A and B0.0 cells grew and killed all the mice, whereas Mit-

C did not. The B1 clone released about 15 U of IL-2 (B1.15) and was still able to grow and kill 60% of the mice, whereas all the other clones producing higher amounts of IL-2 were not (Figure 2). With IL-2, too, the overall picture obtained with repeated rIL-2 injections and IL-2 transfected TS/A cells is similar.

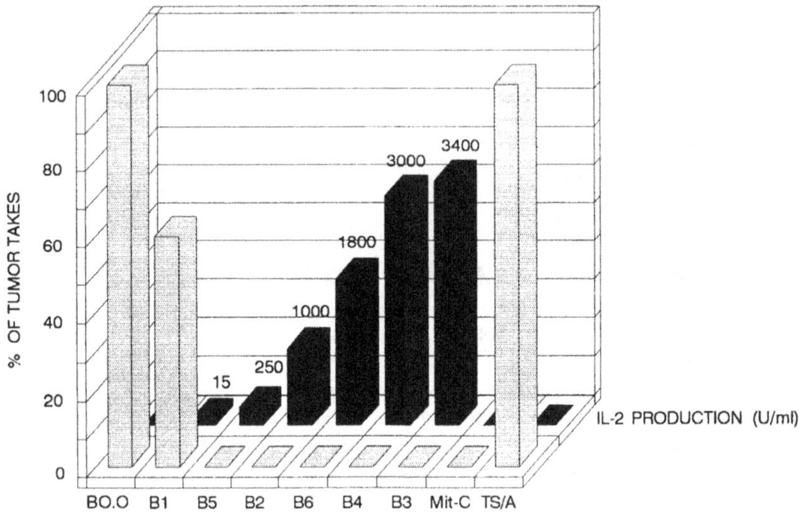

Fig. 2 In vivo growth of IL-2 transfected TS/A cells. Groups of ten BALB/c mice were challenged in the left flank with $1x10^5$ cells from: TS/A untreated parental cells (TS/A); TS/A cells preincubated with Mitomycin-C (Mit-C); TS/A cells transfected with BGMBneo plasmid containing the neomycin resistance gene only (BO.0); B1-B6 TS/A cell clones transfected with IL-2 gene. Dotted columns: % takes; black columns: amount of IL-2 produced

D. Immunogenicity of TS/A transfected cells

As cancer is a metastasizing disease, local reactions and their immediate consequences may have little or no effect on its progression. Release of ILs at the tumor growth site with subsequent induction of tumor cell oncogenicity acquires significance only if it elicits a systemic, long-lasting immune memory. To see whether the rejection of ILs-transfected TS/A cells leads to a specific immune memory, untreated mice, mice injected with Mit-C TS/A cells, and mice that had rejected IL-2 transfected TS/A cells 60 days earlier in the left flank were challenged contralaterally with the minimal 100% lethal dose of TS/A cells.

As previously observed (15,17), no protection was conferred by pre-treatment with Mit-C TS/A cells (Table 1). By contrast, a substantial number of mice surviving the first challenge of IL-1ß- or IL-2-transfected TS/A cells were protected. With the IL-2-transfected clones, the efficacy of protection is nil in the mice surviving a challenge with the B1.15 clone, a poor producer of IL-2, whereas it is good in those surviving a challenge with the B6.1800 clone, a good IL-2 producer. In this case, the disappearance of tumorigenicity and the acquisition of immunogenicity appear to be connected events.

Table 1
Oncogenicity and immunogenicity of TS/A cells transfected with IL-1ß and IL-2 genes when injected alone or in combination

Oncogenicity First challenge[a] (left flank)		Immunogenicity Second challenge[b] (right flank)	
Challenging cells	Takes/Total challenged mice	Challenging cells	Takes/Total challenged mice
TS/A	20/20 (0%)[c]	--	--
Mit-C-TS/A	0/20 (100%)	TS/A	20/20 (0%)[c]
A0.0	8/10 (20%)	TS/A	2/2 (80%)
A1.200	0/10 (100%)	TS/A	2/10 (0%)
BO.0	10/10 (0%)	TS/A	--
B1.15	6/10 (40%)	TS/A	4/4 (0%)
B6.1800	0/10 (100%)	TS/A	3/10 (70%)
A1.200+B1.15[d]	2/10 (80%)	TS/A	0/10 (100%)
A1.200+B6.1800[d]	0/10 (100%)	TS/A	0/10 (100%)

[a]Performed with $1x10^5$ cells
[b]Performed with $4x10^4$ cells 60 days after first challenge
[c]% survival between brackets
[d]Performed with $1x10^5$ cells of each population

The possibility that IITI is enhanced by the association of IL-1β and IL-2 producing clones was also investigated. A group of mice was challenged with $1x10^5$ cells of B1.15 clone admixed just before injection. Although these mice received a double dose of transfected tumor cells, their survival was equal to or higher than that of mice challenged with either the clone alone. Most important, when the efficiency of the memory established in surviving mice was tested, it was found that all surviving mice were able to reject a contralateral challenge of parental TS/A cells 60 days later. All mice survived a combined challenge of $1x10^5$ cells of A1.200 and $1x10^5$ cells of B6.1800 clones and became resistant to a subsequent TS/A challenge (Table 1).

IV THE POTENTIAL OF IITI

In recent years, the strategies of IL utilization in tumor immunotherapy have changed rapidly (25). Initially, the unlimited availability of rIL-2 offered a new way of boosting NK cells and allowed the generation of LAK cells. The adoptive transfer of LAK cells and the systemic administration of high doses of IL-2 are now a medical strategy for a few tumors (2). This non-specific approach assumed that human tumors do not express TAA (26). The possibility of using rIL-2 to expand tumor-infiltrating lymphocytes (TIL) in vitro was next tested in both an experimental system and clinical studies. In addition to their higher ability to reach the tumor and cause its regression, TIL can also establish a long-lasting immune memory. However, a semantic ambiguity is intrinsic in rIL-2 activated TIL, which display a variable degree of both LAK and TAA-specific lytic activity (25,27). The progressively growing evidence that human tumors often express TAA and thus are potentially immunogenic justifies this specific immunotherapeutic approach (5,6,28).

The evidence that IITI follows LATI of apparently non-immunogenic tumors further supports the notion most tumors are potentially immunogenic, and prospects a completely new strategical use of ILs as a component of a new tumor vaccine (29).

Local application of exogenous rILs to increase tumor immunogenicity is a powerful tool that could become a medical procedure, since the peritumoral administration of low doses of rILs is simple and inexpensive. By avoiding most of the side-effects associated with systemic injection of large IL doses, it can be easily undertaken with tumor patients. Clinical trials have shown that the infiltration of low doses of IL-2 around lymph nodes draining primary and recurrent squamous cell carcinomas of the head and neck (30,31), and IFN-γ around lymph nodes draining bladder carcinomas (32), induces a substantial number of complete or partial responses.

In this perspective, it should be carefully explored if IITI stemming from the presence of a single IL or from the combination between a few ILs is powerful enough to activate an efficient immune response, even in conditions of minimal residual disease or initial metastatic spread. The efficacy and feasibility of IITI obtained through peritumoral IL administration or by using IL-gene engineered tumor cells is currently being compared.

Acknowledgements

We wish to thank Dr. John Iliffe for careful review of the manuscript. This paper was supported by grants from the Italian Association for Cancer Research (AIRC), the ISS Italy-USA project for tumor therapy and PF ACRO.

References

1. Smith KA. Cytokines in the nineties. Eur Cytokine Net 1990; 1:7-13.
2. Rosenberg SA, Lotze MT, Muul LM, Chang AE, Avis FP, Leitman S, Linehan WM, Robertson CN, Lee RE, Rubin JJT, Seipp CA, Simpson CG, White DE. A progress report on the treatment of 157 patients with advanced cancer using lymphokine-activated killer cells and interleukin-2 or high-dose interleukin-2 alone. New Eng J Med 1987; 316:889-95.
3. Janeway CA, Bottomly K, Horowitz J, Kaye J, Jones B, Tite J. Modes of cell-cell communication in the immune system. J Immunol 1985; 135:739-42.
4. Forni G, Fujiwara H, Martino F, Hamaoka T, Jemma C, Caretto P, Giovarelli M. Helper strategy in tumor immunology: expansion of helper lymphocytes and utilization of helper lymphokines for experimental and clinical immunotherapy. Cancer Met Rev 1988; 7:289-95.
5. Forni G, Santoni A. Immunogenicity of non-immunogenic tumors. J Biol Resp Modif 1984; 3:128-31.
6. Van Der Eynde B, Hainaut P, Herin M, Knuth A, Lemoine C, Weinants P, Van Der Bruggen P, Fauchet R, Boon T. Presence on a human melanoma of multiple antigens recognized by autologous CTL. Int J Cancer 1989; 44:634-40.
7. Forni G, Varesio L, Giovarelli M, Cavallo G. Dynamic state of spontaneous immune reactivity towards a mammary adenocarcinoma. In: Spreafico F, Arnon R, eds. Tumor antigens and their specific immune response. London: Academic Press. 1979; 167-92.
8. Forni G, Giovarelli M, Santoni A, Modesti A, Forni M. Tumor inhibition by interleukin-2 at the tumor host interface. Biochim Biophys Acta 1986; 865; 307-27.
9. Carbone G, Colombo M, Sensi ML, Cernuschi A, Parmiani G. In vitro detection of cell mediated immunity to individual tumor specific antigens of chemically induced BALB/c fibrosarcomas. Int J Cancer 1983; 31:483-90.
10. Marks TA, Woodman RJ, Germain RI, Billupus LH, Maddison RM. Characterization and responsiveness of the Madison 109 lung carcinoma to various antitumor agents. Cancer Treat Rev 1977; 61:1459-64.
11. Di Marco A, Dasdia T, Giuliana F, Necco A, Casazza AM, Mora PT, Luborsky SW, Waters L. Biological properties of cell lines derived from Moloney virus-induced sarcomas. Tumori 1976; 62:415-20.

12. Nanni P, De Giovanni C, Lollini PL, Nicoletti G, Prodi G. TS/A: a new metastasizing cell line originated from a BALB/c spontaneous mammary adenocarcinoma. Clin Expl Met 1983; 1:373-80.
13. Forni G, Giovarelli M, Santoni A. Lymphokine-activated tumor inhibition in vivo. I. The local administration of interleukin 2 triggers nonreactive lymphocytes from tumor bearing mice to inhibit tumor growth. J Immunol 1985; 134:1305-11.
14. Giovarelli M, Cofano F, Vecchi A, Forni M, Landolfo S, Forni G. Interferon-activated tumor inhibition in vivo. Int J Cancer 1986; 37:141-7.
15. Bosco MC, Giovarelli M, Forni M, Modesti A, Scarpa S, Masuelli L, Forni G. Low doses of IL-4 injected perilymphatically in tumor-bearing mice inhibit the growth of poorly and apparently nonimmunogenic tumors and induce a tumor-specific immune memory. J Immunol 1990; 145:3136-43.
16. Forni G, Musso T, Jemma C, Boraschi D, Tagliabue A, Giovarelli M. Lymphokine activated tumor inhibition (LATI) in mice: ability of a nonapeptide of the human interleukin-1 to recruit antitumor reactivity in recipient mice. J Immunol 1989; 142:712-20.
17. Forni G, Giovarelli M, Santoni A, Modesti A, Forni M. Interleukin-2 activated tumor inhibition in vivo depends on the systemic involvement of host immunoreactivity. J Immunol 1987; 138:4033-41.
18. Forni G, Giovarelli M, Bosco MC, Caretto P, Modesti A, Boraschi D. Lymphokine activated tumor inhibition: combinatory activity of a synthetic nonapeptide from interleukin-1, interleukin-2, interleukin-4, and interferon-γ injected around tumor-draining lymph nodes. Int J Cancer 1989; S4:62-5.
19. Tepper RI, Pattengale PK, Leder P. Murine interleukin-4 displays potent anti-tumor activity in vivo. Cell 1989; 57:503-12.
20. Fearon ER, Pardoll DM, Itaya T, Golumbek P, Levitsky HI, Simons JW, Karasuyama H, Vogelstein B, Frost P. Interleukin-2 production by tumor cells bypasses T helper function in the generation of an antitumor response. Cell 1990; 60:397-403.
21. Gansbacher B, Zier K, Daniels B, Cronin K, Bannerji R, Gilboa E. Interleukin 2 gene transfer into tumor cells abrogates tumorigenicity and induces protective immunity. J Exp Med 1990; 172:1217-24.
22. Watanabe Y, Kuribayashi K, Miyatake S, Nishihara K, Nakayama E, Taniyama T, Sakata T. Exogenous expression of mouse interferon-gamma cDNA in mouse neuroblastoma C1300 cells results in reduced tumorigenicity by augmented anti-tumor immunity. Proc Natl Acad Sci (USA) 1989; 86:9456-61.
23. Colombo MP, Ferrari G, Stoppacciaro A, Parenza M, Fodolfo M, Mavilio F, Parmiani G. Granulocyte Colony-stimulating factor gene transfer suppresses tumorigenicity of a murine adenocarcinoma in vivo. J Exp Med 1991; 173:889-97.
24. Karasuyama H, Tohyama N, Tada T. Autocrine growth and tumorigenicity of interleukin 2-dependent helper T cells transfected with the IL-2 gene. J Exp Med 1989; 169:13-25.

25. Forni G, Bosco MC, Vai S, Giovarelli M. Interleukin 2: In vivo induction of effector cells. In: Mertelsmann R. ed. Lymphohaematopoietic growth factors in cancer therapy. Berlin: Springer-Verlag 1990, 37-46.
26. Hewitt HB. Animal tumor models for tumor immunology. J Biol Resp Modif 1982; 1:107-19.
27. Rosenberg SA, Packard BS, Aebersold PM, Solomon D, Topalian SL, Toy ST, Simon P, Lotze MT, Yang JC, Seipp CA, Simpson C, Carter C, Bock S, Schwartzentruber D, Wei JP, White JP. Use of tumor-infiltrating lymphocytes and interleukin-2 in the immunotherapy of patients with metastatic melanoma. New Engl J Med 1988; 319:1676-80.
28. Anichini A, Fossati G, Parmiani G. Heterogeneity of clones from human metastatic melanoma detected by autologous cytotoxic T-lymphocyte clones. J Exp Med 1986; 163:215-20.
29. McCune CS, Marquis DM. Interleukin 1 as an adjuvant for active specific immunotherapy in a murine tumor model. Cancer Res 1990; 50:1212-5.
30. Cortesina G, De Stefani A, Giovarelli M, Barioglio MG, Cavallo GP, Jemma C, Forni G. Treatment of recurrent squamous cell carcinoma of head and neck with low doses of interleukin-2 (IL-2) injected perilymphatically. Cancer 1988; 62:2482-90.
31. Musiani P, De Campora E, Valitutti S, Castellino F, Calearo C, Cortesina G, Giovarelli M, Jemma C, De Stefani A, Forni G. Effect of low doses of interleukin-2 injected perilymphatically and peritumorally in patients with advanced primary head and neck squamous cell carcinoma. J Biol Resp Modif 1989; 8:571-8.
32. Forni G, Giovarelli M, Jemma C, Bosco MC, Caretto P, Modesti A, Santoni A, Forni M, Cortesina G, De Stefani A, Cavallo GP, Galeazzi E, Musiani P, De Campora E, Valitutti S, Castellino F, Calearo CV, Fontana G, Sesia G. Perilymphatic injections of cytokines: A new tool in active cancer immunotherapy. Experimental rationale and clinical findings. Ann Ist Super Sanita 1990; 26:397-410.

17

Biological Significance of Autologous Tumor Killing

Atsushi Uchida
Yoshitaka Kariya
Naoya Inoue
Norihiko Okamoto
Katsuji Sugie
Radiation Biology Center
Kyoto University
Kyoto, Japan

I INTRODUCTION

Cell-mediated cytotoxicity is considered to be one expression of host immune defense mechanisms against tumors. Most in vitro studies on cytotoxicity against tumor were performed by the use of tumor cell lines as targets. Tumor cells, however, alter their susceptibility to cell-mediated lysis when cultured in vitro (1). For a better evaluation of cytotoxicity of lymphocytes against tumor cells in cancer patients, autologous combinations of fresh effector and target cells have been used. In these studies, peripheral blood lymphocytes of cancer patients expressed lysis ranging from 10-50%, depending on tumor types and metastases, of tumor cells freshly isolated from the same patients (2-6). The results we have obtained from the studies of cell populations and single cells indicate that $CD3^-CD16^+$ large granular lymphocytes (LGL) from the blood and tumor tissues of cancer patients lyse autologous, freshly isolated tumor cells (5,6) and release a novel cytotoxic factor with lytic effects on autologous and allogeneic fresh human tumor cells (7,8). In addition, results of a previous study showed that autologous tumor cell killing (ATK) was also mediated by $CD3^+$ T lymphocytes; this killing was performed by T lymphocytes in patients with localized neoplasms and primarily by LGL in patients with metastatic tumors.

The autologous mixed lymphocyte-tumor culture reaction (AMLTR) has been used to investigate the recognition by lymphocytes of autologous tumor cells in mixed culture

(9,10). If cancer patients are sensitized against tumor-associated antigens, the tumor cells are expected to stimulate the proliferation of the lymphocytes in vitro. Indeed, proliferation of T lymphocytes is induced in response to autologous tumor cells, while the proliferation is minimal by stimulation with allogeneic tumor cells. We have recently reported that the T-cell subset that responds to autologous tumor cells differs from the subset that is reactive to autologous non-malignant non-T cells (11). The data suggest that the tumor-specific event may be observed in the AMLTR.

Cancer patients are generally treated with some combination of surgery, chemotherapeutic agents, and radiation. While some patients respond in varying degrees to these treatment modalities, others remain unresponsive. Immunotherapy has been proposed as a fourth modality of cancer treatment. Clinical studies with a variety of biological response modifiers (BRMs), however, demonstrated response rates of 5-25%, depending on BRM used and the type of cancer treated. The response rates for most protocols have not been markedly elevated by alterations in doses and schedules of treatment. Thus, it remains unclear why a subset of patients responds to a given treatment modality, while the majority remains unresponsive. It is important to search for some immunological parameter or tumor characteristic that could be used prospectively to predict patient response to a given treatment modality or to provide definitive answers regarding the mechanisms responsible for an antitumor response.

Here we describe the biological and clinical significance of blood ATK activity in human cancer patients (12). We suggest that in vivo induction of ATK activity prior to surgery by administration of BRM may improve the clinical outcome in patients who naturally have no such potential.

II BIOLOGICAL SIGNIFICANCE OF AUTOLOGOUS TUMOR KILLING (ATK)

We have studied lymphocyte reactivity against autologous freshly isolated tumor cells by the use of cytotoxicity and proliferation assays in more than 1,500 patients with various types of neoplasms, including adenocarcinomas or squamous cell carcinomas of the lung, breast, stomach, liver, colon, ovary, and uterus. Blood samples were obtained from each patient prior to anesthesia for surgery, and non-adherent lymphocytes consisting mainly of T lymphocytes (60-95%) and large granular lymphocytes (5-40%) were used as effectors (5-9). Tumor cells were isolated from tumor specimens obtained from cancer patients at the time of surgery by an enzymatic treatment, followed by centrifugation on discontinuous three-step Percoll gradients and two-step Ficoll-Hypaque gradients and by adherence to plastic surfaces (5-9). the tumor-enriched fractions contained more than 90% tumor cells which were more than 90% viable.

At the time of surgery, blood lymphocytes from patients with a variety of primary localized solid neoplasms demonstrated varying levels of cytotoxicity against autologous, freshly isolated tumor cells in a 6-h ^{51}Cr-release assay. Significant levels of cytotoxicity above baseline ($\geq$10%) were observed in 5-60% of patients, depending on tumor types

(Table 1). Tests of ATK activity were positive in approximately 60% of patients with stage I and 40% with stage II tumors. Chi-square analysis showed no statistical differences among the groups.

The lack of ATK activity is unlikely to be due to the inability of certain cancer patients to mount an immunologic response, since fresh tumor target preparations from these patients were lysed by lymphokine-activated killer (LAK) cells or killer cells activated by the streptococcal preparation OK432 or ß-1-3-D-glucan sizofiran. The frequency of positive results in ATK tests is comparable to our previous observations (1) and the results of other studies (3,4). Although the number of samples that were positive for ATK activity decreased when patients developed metastases, the prognosis of metastasis was not indicated by the clinical and pathological TNM stage while the primary tumor was localized. In patients with localized neoplasms, $CD3^-$ LGL and/or T-lymphocytes expressed ATK activity, while ATK was mediated primarily by LGL, when patients developed metastases (1). LGL of patients with metastatic cancer suppressed the induction of ATK potential in the AMLTR, which may explain the low frequency of T-cells with ATK activity in these patients (10).

Table 1
Frequency of ATK activity in a variety of cancer patients

Tumor	% Positive ATK tests
Squamous cell carcinoma of lung	56
Adenocarcinoma of lung	53
Gastric carcinoma	60
Carcinoma of colon and rectum	52
Hepatocellular carcinoma	5
Breast carcinoma	40
Ovarian cancer	42
Carcinoma of uterine cervix	56
Melanoma	38

Blood lymphocytes from each patient were tested for ATK activity in a 6-h 51 Cr-release assay at the time of surgery, and % cytotoxicity of 10% or more was statistically significant and considered to be positive.

The patients who had received curative surgery were retrospectively evaluated for the postoperative clinical course. Pathological examinations showed no tumor cells in the margins of tumor resections in all patients. No adjuvant anticancer therapy was performed after surgery. When local or distant recurrences developed, patients received chemotherapy, radiation therapy, or biological therapy. Clinical parameters assessed prospectively included age, sex. weight loss, other organ system diseases, and use of medications. Approximately 80% patients had performance status of 0, and 20% had performance status of 1, according to the ECOG performance status scale. The patients who are alive have had follow-up for 60 months. More than 80% of patients who were positive for ATK at the time of surgery remained tumor-free and were alive more than 5 years after the operation (Figure 1A,B). The other patients with ATK, however, developed metastases by 18 months and died by 37 months. In contrast, all patients without ATK activity relapsed within 18 months and died within 42 months. When the disease-free interval and total survival time were estimated by Kaplan-Meier analysis, the differences observed in curves for postoperative survival (disease-free interval and total survival) for patients with or without ATK activity were statistically highly significant according to the Cox-Mantel test ($P<0.00003$) and the generalized Wilcoxon test ($P<0.00003$). The correlation coefficient for ATK and postoperative clinical course was also high. The data strongly indicate that the potential of blood lymphocytes to kill autologous fresh tumor cells, tested at the time of surgery, may predict good prognosis of patients with primary localized tumors. The results also suggest that the measurement of ATK function at the time of surgery in cancer patients will provide valuable information for the probability of disease recurrence. Although all patients received curative surgery, only those with ATK ability are free from tumor and are alive after more than 5 years. Updated clinical data obtained indicate that none of the patients who have remained tumor-free for the observation period of 60 months develops recurrence and die after that period. No differences were observed in performance status, TNM classification, age or sex between groups that were positive for ATK activity and those that were negative, suggesting that the measurement of ATK activity may represent an independent prognostic parameter. All patients without ATK activity developed recurrences and died within 5 years, showing that a negative result in ATK tests indicates a poor prognosis. Our results suggest that ATK lymphocytes may be the main effectors in the immunological defense against growth and metastasis of tumor. The test, however, has no absolute prognostic value, since some patients with short disease-free interval and short total survival showed positive results. Since AMLTR and NK cell activity were also positive in these patients, the immunological control may not be operative for some types of cancer.

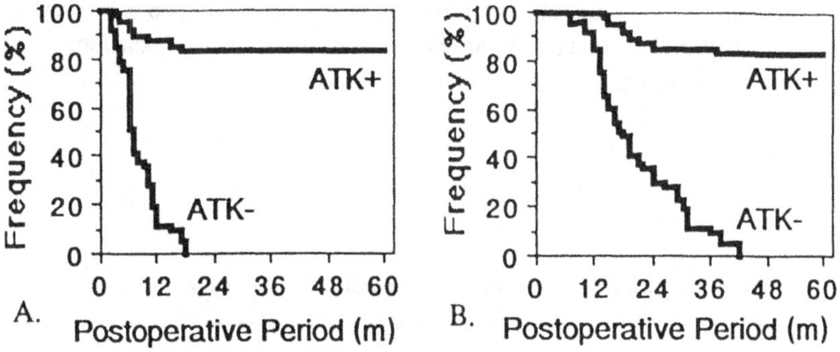

Fig. 1A,B Curves for postoperative disease-free survival(A) and total survival(B) of 48 ATK-positive patients and 53 ATK-negative patients. Blood lymphocytes from each patient were tested for ATK activity in a 6-hr 51CR-release assay at the time of surgery, and % cytotoxicity of 10% or more was statistically significant and considered to be positive. Patients were retrospectively evaluated for disease-free survival and total survival after 60 months of follow-up after curative operation. Duration of tumor-free period was calculated from the date of surgery to the last noted physician contact. Tumor-free survival and total survival was estimated by Kaplan-Meier analysis

We next examined the possibility that patients with lymphocytes expressing ATK activity have better general immune status and thus survive longer than those without the activity by concomitantly testing other immunological factors. ATK activity was not correlated with T cell proliferation induced by autologous tumor cells in mixed cultures or LGL-mediated NK cell activity against the NK prototype target K562 (Table 2). In fact ATK was mediated by heterogeneous populations: LGL in 35% of the patients, T-lymphocytes in 30%, and both types of lymphocytes in the other 35% (1). In addition, there were no correlations between ATK activity and other immune functions, including mitogenic response, autologous and allogeneic mixed lymphocyte reactions, and production of interferon, interleukin (IL)-1 and IL-2. It is thus evident that the absence of ATK activity does not reflect general low immunity of the patients. In retrospective evaluation, there were no statistical differences in postoperative survival curves for patients who were positive in AMLTR tests and NK tests and those who were negative in these tests, though the former group had a slightly longer disease-free interval than the latter. Previous reports also suggested the prognostic significance of ATK tests in cancer patients (3,4). In those studies, however, no statistical analysis of survival curves was performed, the follow-up period was shorter, and other immunological functions were not concomitantly tested.

Table 2
ATK and other immunological functions and their association with clinical course

Immunological functions	Correlation with ATK	Correlations with clinical course
ATK activity	–	Yes
AMLTR	No	No
NK activity	No	No
PHA response	No	No
Con A response	No	No
Allogeneic MLR	No	No
Autologous MLR	No	No
IFN production	No	No
IL-1 production	No	No
IL-2 production	No	No

A recent report has shown correlation between survival and amplification of oncogene-coamplification units in breast cancer patients. However, in our study, ATK was independent of oncogene amplification of tumor cells and expression of the major histocompatibility complex (MHC) class I and II antigens. No correlation has been demonstrated between NK cells activity and postoperative prognosis in melanoma patients (13). The findings of other studies, however, supported the prognostic value of NK activity in patients with head and neck cancer (14).

This study includes only patients who had primary localized tumors and underwent curative surgery and whose tumor specimens were suitable for cytotoxicity and proliferation assays. Such specimens were obtained in approximately 30-50% of patients with solid tumors of different origins.

III MECHANISMS OF AUTOLOGOUS TUMOR KILLING

The mechanism by which $CD3^-$ LGL and $CD3^+$ T lymphocytes recognize and kill autologous, freshly isolated tumor cells is somewhat different from those responsible for lysis of allogeneic tumor cells and tumor cell lines. An analysis by the use of monoclonal antibodies (mAb) revealed that freshly isolated tumor cells from human cancer patients lack the expression of HLA class I and/or class II molecules on their surface in 10-50% tumor samples, depending on the tumor type. Recent evidence indicates that NK cells preferentially recognize and lyse target cells which do not express MHC molecules. However, there was no correlation between MHC class I and II expression on tumor cells and their susceptibility to lysis by autologous LGL. In addition, blocking of MHC class I antigen expression by anti-MHC class I mAb did not affect their sensitivity to lysis by autologous LGL in approximately two-thirds of our cases and suppressed it in other 10% cases, though the treatment enhanced or induced sensitivity to lysis by LGL only in one-fourth of the cases. Furthermore, the induction of MHC class I molecules by pre-treatment of tumor cells with interferon (IFN)-γ or tumor necrosis factor (TNF)-α did not make them resistant to lysis by autologous LGL in the majority of cases. These results suggest that LGL may lyse autologous tumor cells and NK target cells through different mechanisms. This is confirmed at the clone level (Table 3). $CD3^-$ LA1.1 clone which is established by repeated stimulation with autologous tumor cells expressed autologous tumor-restricted lysis, without killing of allogeneic fresh tumor cells or K562. In contrast, LK1.1 clone that is obtained by repeated stimulation with K562 killed only K562 but not autologous tumor cells. We next examined a possible role of adhesion molecules in the ATK system. Freshly isolated human tumor cells expressed varying levels of the intercellular adhesion molecule 1 (ICAM-1; CD54) on their surface when analyzed by flow cytometry with anti-CD54 mAb. The expression of ICAM-1 was more frequently observed with squamous-cell carcinoma than with adenocarcinoma. There was no correlation between the ICAM-1 expression of tumor cells and susceptibility to lysis by autologous $CD3^-$ LGL. Tumor cells of each patient were then fractionated according to ICAM-1 expression. Both CD54-positive and -negative fractions of adenocarcinomas were lysed by autologous LGL, while LGL killed predominantly CD54-positive fractions of squamous cell carcinomas. Treatment with anti-CD54 mAb of tumor cells produced no effect on the sensitivity of adenocarcinomas, whereas it partially inhibited that of squamous cell carcinomas. ATK LGL that were isolated by binding to autologous tumor cells expressed varying levels of adhesion molecules, including CD11a (α chain of leukocyte function-associated antigen 1; LFA-1), CD11b (α chain of Mac-1), CD11c (α chain of p150/95), CD18 (β chain of the β2-integrin family) and CD54 on their surface. Treatment of LGL with mAb against these adhesion molecules did not abrogate cytotoxicity against autologous tumor cells (Table 4). The mixture of these mAb was also ineffective, while it inhibited binding and killing of the NK prototype target K562 cells. Autologous tumor killing activity was also resistant to anti-CD54 mAb treatment of effector cells, which inhibited NK activity of $CD3^-$ LGL. These results indicate that $CD3^-$ LGL may kill autologous tumor cells and K562 by different interactions and that the former could be independent of LFA-1/ICAM-1 systems and the latter is dependent thereof.

Table 3
Lysis by $CD3^-$ LGL clone of autologous tumor cells and K562

Clones	% Cytotoxicity against		
	Autologous tumor	Allogeneic tumor	K562
LA1.1	36.2	1.4	3.5
LA1.2	30.8	29.6	47.9
LA2.1	20.4	11.8	36.1
LK1.1	1.2	2.2	48.1
LK2.1	18.9	18.3	37.2

LGL clones were obtained by repeated stimulation with autologous tumor cells (LA clones) or K562 (LK clones). All clones are $CD2^+CD3^-CD16^+CD56^+$ and lacked rearrangement of T-cell receptor (TCR)α/β and TCRγ/δ. Results are expressed as % cytotoxicity at an E:T ratio of 5:1 in a 6-h 51Cr-release assay.

IV INDUCTION OF ATK ACTIVITY AND ITS BIOLOGICAL SIGNIFICANCE

The strong correlation of blood ATK activity with long-term, tumor-free status is consistent with the hypothesis that the immune system may play a beneficial role in the eventual rejection of at least some human tumors. The evidence also suggests that ATK effector cells may play a critical role in the interaction between cancer patients' immune cells and their tumors. Thus, basing therapeutic strategies on the activity of other effector cell types may be misleading. Biological effects of BRM have been assessed for the ability to augment NK cell activity and induce LAK cell activity which is largely derived from NK cells. However, since there was no positive correlation between NK cell activity and tumor-free and total survival, conclusions drawn from such parameters may be irrelevant for ATK activity and clinical course.

Table 4
Unlikely involvement of CD11 and CD18 adhesion molecules in ATK by LGL

Treatment	% Cytotoxicity to	
	Autologous tumor	K562
None	25.7	42.6
Control serum	27.2	43.2
Anti-CD11a	23.7	32.2*
Anti-CD11b	23.5	39.3
Anti-CD11c	26.6	37.9
Anti-CD11a + anti-CD11b + anti-CD11c	25.9	19.4*
Anti-CD18	28.2	21.8*
Anti-CD11a + anti-CD11b + anti-CD11c + anti-CD18	23.1	14.8*

LGL were treated with medium or anti-CD11a mAb, anti-CD11b mAb, anti-CD11c mAb, anti-CD18, or mixtures of the mAb and tested for lysis of autologous tumor cells and K562 in a 6-h 51 Cr-release assay.
*Value is significantly different from that of none at $P<0.05$.

Adoptive immunotherapy with various types of cytotoxic cells has been beneficial for a subset of patients with melanoma and renal cell cancers (13). Adoptive immunotherapy with activated tumor-infiltrating lymphocytes (TIL) is based on the concept that TIL sensitized in vivo by autologous tumor cells could become differentiated in vitro to mature ATK lymphocytes by re-stimulation with autologous tumor cells and IL2 and that those effector cells might be involved in the therapeutic benefit observed with this treatment. Those TIL may be more analogous to the putatively nonprognostic lymphocytes that respond in AMLTR than to the predictive ATK lymphocytes. When compared with blood lymphocytes, TIL showed little or no ATK activity. In addition, ATK activity of TIL was not associated with long-term survival. Thus, the most important effector cells may actually be found in the blood rather than in tumor tissues.

It is of clinical and biological importance to ascertain whether in vivo induction of ATK activity prior to surgery could reduce the incidence of tumor metastasis formation after curative operation and prolong the postoperative clinical course in patients who naturally have no such potential. We have previously demonstrated that intrapleural administration of the streptococcal preparation OK432 to patients with carcinomatous pleural effusions results in an induction of ATK activity of TIL and that this is strongly associated with a reduction or complete disappearance of tumor cells in the effusions (17). On the basis of these findings, patients with localized hepatocellular carcinoma were treated with daily intradermal or intravenous administration of OK432 for successive 7 days prior to curative operation.

Approximately 50% of patients who naturally had no blood lymphocytes with ATK activity responded to the biological therapy by acquiring ATK activity by the time of surgery. The BRM-induced ATK activity was maintained for more than 4 weeks after surgery by repeated injections of the agent. By contrast, the other patients failed to acquire ATK activity in response to the BRM therapy. In vitro treatment of blood lymphocytes with OK432 induced ATK activity in patients who subsequently responded to in vivo OK432 treatment. Thus, there was good correlation between in vitro and in vivo response to BRM.

The patients who had received curative surgery and the biological therapy were evaluated for the postoperative clinical course. Pathological examinations showed no tumor cells in the margins of tumor resections in all patients. No adjuvant anticancer therapy was performed after surgery. When local or distant recurrence developed, patients received chemotherapy, radiation therapy, or biological therapy. The patients who were induced to express ATK activity at the time of operation by the biological therapy remained disease-free after being observed for more than 2 years (Figure 2). By contrast, all patients who had failed to acquire ATK activity in spite of the therapy developed local or distant recurrences within 1 year and died by 2 years. As was the case with endogenous ATK activity, there was a strong correlation between the presence of ATK activity at the time of surgery and postoperative clinical course. Our data imply that the biological therapy with ATK-inducing effects before surgery may be of clinical benefit to cancer patients who naturally have no ATK activity.

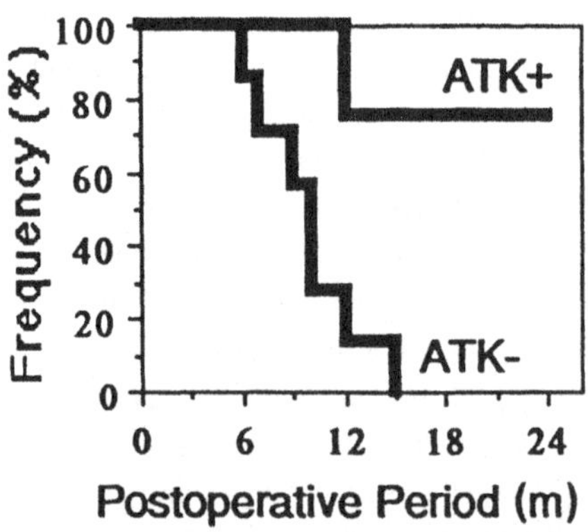

Fig. 2. Curves for postoperative total tumor-free survival of ATK-positive and -negative patients with hepatocellular carcinoma. Patients received OK432 treatment before curative operation. ATK activity was determined at the time of surgery. Patients were retrospectively evaluated for total survival after 48 months of follow-up after operation. Patients were retrospectively evaluated for total survival after radical surgery, and survival was estimated by Kaplan-Meier analysis

V CONCLUSION

The data presented in this report strongly indicate that positive ATK activity of blood lymphocytes at the time of surgery predicts a favorable clinical course in patients with primary localized solid tumors. The strong correlation of disease-free interval and total survival indicates that ATK activity is a meaningful prognostic indicator and provides evidence for immunological control of tumor growth and metastasis. According to our data, it is unlikely that cancer patients who remain tumor free after 60 months of follow-up will develop recurrence or die from the disease. On the basis of these clinical data we have been conducting a study to determine whether induction of ATK activity before surgery, by treatment with BRM, can improve the clinical outcome in patients who do not naturally have this potential. Our preliminary data clearly indicate that the presence of natural and BRM-induced ATK activity in the peripheral blood is strongly associated with long-term survival.

References

1. Uchida A, Mizutani M. Autologous tumor killing activity in human mechanisms and biological significance. In:Torisu Y, Yoshida T,ed. New horizons in tumor immunotherapy. Amsterdam:Elsevier. 1989; 201-213.
2. Klein E, Vanky F. Natural and activated cytotoxic lymphocytes which act on autologous and allogeneic tumor cells. Cancer Immunol Immunother 1981; 11:183-188.
3. Vanky F, Williams J, Kreichbergs A, et al. Correlation between lymphocyte-mediated auto-tumor reactivities and clinical course. I. Evaluation of 46 patients with sarcoma. Cancer Immunol Immunother 1983; 16:11-16.
4. Vank F, Peterffy A, Book K, et al. correlation between lymphocyte-mediated auto-tumor reactivities and clinical course. I. Evaluation of 69 patients with lung carcinoma. Cancer Immunol Immunother 1983; 16:17-22.
5. Uchida A, Micksche M. Lysis of fresh human tumor cells by autologous large granular lymphocytes from peripheral blood and pleural effusions. Int J Cancer 1983; 32:37-44.
6. Uchida A, Ynagawa E. Natural killer cell activity and autologous tumor killing activity in cancer patients: Overlapping involvement of effector cells as determined in two-target conjugate cytotoxicity assay. J Natl Cancer Inst 1984; 73:1093-1100.
7. Uchida A, Klein E. Generation of cytotoxic factor by human large granular lymphocytes during interaction with autologous tumor cells: Lysis of fresh human tumor cells. J Natl Cancer Inst 1988; 80:1398-1403.

8. Uchida A, Fujimoto T, Mizutani Y. Lysing of fresh human tumor by a cytotoxic factor derived from autologous large granular lymphocytes independently of other known cytokines. Cancer Immunol Immunother 1990; 31:60-94.
9. Uchida A, Klein E. Suppression of T-cell response in autologous mixed lymphocyte-tumor culture by large granular lymphocytes. J Natl Cancer Inst 1986; 76:389-398.
10. Vose BM, Bonnar GD. Human tumor antigens defined by cytotoxicity and proliferative responses of cultured lymphoid cells. Nature 1982; 296:359-361.
11. Uchida A, Moore M, Klein E. Autologous mixed lymphocyte-tumor reaction and autolgous mixed lymphocyte reaction. II. Generation of specific and non-specific killer T cells capable of lysing autologous tumor. Int J Cancer 1988; 41:651-656.
12. Uchida A, Kariya Y, Okamoto N, Sugie K, Fujimoto T, Yagita M. Prediction of postoperative clinical course by autologous tumor-killing activity in lung cancer patients. J Natl Cancer Inst 1990; 82:1697-1701.
13. Hersey P, Edwards A, Milton GW, et al. No evidence for an association between natural killer cell activity and prognosis in melanoma patients. Nat Immun Cell Growth Regul 1983/84; 3:87-94.
14. Schantz SP, Shillitoe EJ, Brown B, et al. Natural killer cell activity and head and neck cancer: A clinical assessment. J Natl Cancer Inst 1986; 77:869-875.
15. Rosenberg SA< Packard BS, Aebersold PM, et al. Use of tumor-infiltrating lymphocytes and interleukin-2 in the immunotherapy of patients with metastatic melanoma. A preliminary report. N Engl J Med 1988; 319:1676-1680.
16. Kradin RL, Boyle LA< Preffer FI, et al. Tumor-derived interleukin-2-dependent lymphocytes in adoptive immunotherapy of lung cancer. Cancer Immunol Immunother 1987; 24:76-85.
17. Uchida A, Micksche M. Lysis of fresh human tumor cells by autologous peripheral blood lymphocytes and pleural effusion lymphocytes activated by OK 432. J Natl Cancer Inst 1983; 71:673-80.
18. Uchida A, Moore M, Hoshino T. Intrapleural administration of OK432 in cancer patients: Augmentation of autologous tumor killing activity of tumor-associated large granular lymphocytes. Cancer Immunol Immunother 1984; 18:5-12.
19. Uchida A, Kariya Y, Okamoto N. et al. Biological significance of autologous tumor killing in human cancer patients and its modulation by biological therapy. In: Pastorino H, Hong WK, ed. Chemoimmuno Prevention of Cancer. New York: Georg Thieme Verlag. 1991:9-14.

18

Biological Response Modifiers and Chemotherapeutic Agents that Alter Interleukin 2 Activities

William L. West
Allen R. Rhoads
Clement O. Akogyeram
Howard University College of Medicine
Washington, D.C.

I INTRODUCTION

Immunopharmacology has as its goal the identification of agents that act on specific components of the immune system to either selectively enhance or suppress their activities. In a cancer-bearing host, agents of interest that modulate the immune system are of two general types: (a) immunosuppressants which are used as cytotoxic agents in cancer therapy and (b) biological response modifiers (BRMs) which enhance the immune response. More generally, BRMs are defined as those agents which modify a biological response to tumor cells with resultant therapeutic benefits. BRMs, whether biologic or chemical agents, are capable of restoring an immunologic balance in tumor-bearing hosts by enhancing antitumor functions of the immune system. In this respect, BRMs differ from classical modalities (surgery, radiation, and chemotherapy) used in therapy of cancer. Although most BRMs are investigational drugs, they present approaches that expand the possibility for successful treatment of conditions associated with altered immunological functions including cancer.

Three classes of agents which may be considered BRMs, as they restore balance among components of the immune system, will be discussed. Historically, the first source of BRMs were microbes and microbial products which were identified early on as natural general potentiators of immunity. Opportunistic infections with certain microorganisms in cancer-bearing hosts were associated with either regression or remission of cancer in some cases. Infections with organisms such as Bacillus Calmette and Guerin (BCG), Corynebacterium paquam, Salmonella enteritides, or the protozoan like Besnoita jellisone or antigenic components isolated from them, were reported to enhance resistance of experimental animals against neoplastic disease. Clinically, the use of such crude preparations in patients with cancer has been disappointing. More recently, the availability of well-defined microbial preparations with immunostimulatory properties have renewed our interest in preparations such as OK432 (Picibanil) made from Streptoccoccus pyogenes, which stimulates cellular immunity in cancer patients (1,2) or ImmuVert (ribosomes and membrane vesicles) from Serratia marcesces, which stimulates natural killer (NK) cell activity of peripheral blood mononuclear cells and has been reported to be effective against mouse fibrosarcoma and human brain tumors (3).

A second class of BRMs is comprised of a number of structurally unrelated synthetic or natural compounds that appear to act as adjuvants: they enhance antitumor as well as other responses and modulate various components of the immune system. Synthetic levamisole has been used as an adjuvant to chemotherapy in cancer patients as well as in chronic infections, i.e., chronic polyarthritis and rheumatoid arthritis. Levamisole is an adjuvant with efficacy in the immunotherapy of melanoma and colon cancer (4). Ketoconazole is also of value in the immunosuppressed individual. Moreover, synthetic phorbol myristate acetate (PMA) and natural bryostatins (macrocyclic lactones isolated from Bugula neritina) are important in lymphocyte activation (5). Unlike tumor promoting phorbol esters, bryostatins are also antineoplastic. It is likely that among these agents are adjuvants that may increase efficacy and/or reduce toxicity of other therapeutic agents. Moreover, phorbol esters have the disadvantage of causing inhibition of effector cell cytotoxicity (6).

The third class of BRMs includes cytokines, which have been extensively studied recently for the ability to enhance immune responses (see Table 1). Cytokines produced using recombinant DNA technology are now available in sufficient quantities and acceptable purity to allow their clinical evaluation. As cytokines are naturally produced by a variety of cells in normal hosts, they may be considered to be endogenous BRMs. The spectrum of cytokine activities is broad, since they are multifunctional, so that multiple effects can be attributed to the same cytokine. Cytokine networks are complex, and it is well known that more than one cytokine may act on the same target cell.

This review focuses on selected pharmacologic agents, including both BRMs and chemotherapeutic drugs, which modulate the cytokine, interleukin-2 (IL2). It is now well established that cell-to-cell signals culminating in effective antitumor immunity are mediated by helper T-lymphocytes and IL2. Since IL2 plays a pivotal role in antitumor responses, agents which modulate its release, binding to its receptors and its catabolism are important. A better definition of these interactions raises the possibility of improved clinical efficacy through the combined use of IL2 and certain other pharmacologic agents.

Table 1
A List of Selected Cytokines and Their Effects

Cytokines	Source	Function
I. Factors that affect macrophages		
Interferon- (INFα)	Leucocytes	Antiviral and antiproliferative actions
Interferon- (INFß)	Fibroblast	Antiviral and antiproliferative actions
Interferon- (INFγ)	T-lymphocytes	Immunomodulator and upregulator B and T cells; macrophage activating factor (MAF)
Macrophage CSF		Proliferation of monocytes
Macrophage-inhibitory factor (MIF)	T-lymphocytes	Inhibits macrophage movement at inflammatory sites
II. Factors that affect lymphocytes		
Interleukin-1 (IL-1)	Macrophages/ monocytes	Proliferation B and T cells, hematopoleses, inflammation; IL-2 production; endogenous pyrogen; tumoricidal
Interleukin-2 (IL-2)	T-lymphocytes	Proliferation T- and B-lymphocytes, activation of natural killer (NK) cells
Interleukin-3 (IL-3)	Lymphocytes	Proliferation of bone marrow progenitors
Interleukin-4 (IL-4)	T-lymphocytes	Proliferation of antigen-primed B and T cells; mast cell growth factor; (stimulates IgE)
Interleukin-5 (IL-5)	T-lymphocytes	Proliferation and differentiation of eosinophils; B-cell growth and differentiation (stimulates IgA)
Interleukin-6 (IL-6)	Macrophages T-cell, B-cells	Proliferation of B cells and bone marrow progenitors; differentiation B-cell and macrophages; produces inflammation
Interleukin-7 (IL-7)	Progenitor B cells (stem cells)	Proliferation and differentiation of B-cells progenitors
III. Other Factors		
Tumor Necrosis Factor (TNF)	Macrophages, T-cells, NK-cells	Antiparasitic action, inflammation, tumorcidal, proliferation of B-cells, procoagulant, inhibits
lipoprotein lipase,		ICAM expression mobilizes calcium from bone
Granulocytes-macrophage colony stimulating factor		Stimulates granulocytes, neutrophil, monocytes and eosinophils, differentiation of granulocytes and macrophages
Granulocyte CSF (G-CSF)		Proliferation and differentiation of granulocytes
Neutrophil chemotactic factor	Basophils and mast cells	Chemotactic for neutrophils
Platelet-activating factor (PAF)	Basophils and mast cells	Platelet aggregation; activation factor chemotactic for neutrophils
Eosinophil chemotactic factor (ECF)	Basophils and mast cells	Chemotactic for eosinophils

II AGENTS WHICH MODIFY EFFECTS OF IL2

In the course of normal immune response, antigen-presenting cells (APCs) not only present antigen to lymphocytes, but they also produce and secrete a variety of cytokines, including interleukin-1 (IL1). This cytokine acts as a second signal necessary for the antigen-activated lymphocytes to produce IL2, proliferate and differentiate into antigen-specific effector cells. It is highly likely that administration of antineoplastic therapy at the time antitumor immune response is operative results in destruction of a large number of selected proliferating T cell clones within one generation. Thus, although a combination of antineoplastic and immunostimulatory agents in therapy of cancer may be attractive due to the possibilities for achieving both a decrease in proliferating tumor cell number and augmentation of antitumor immune function, it is obvious that these agents have to be used with a great deal of caution. Furthermore, not only the particular combination but also the logistics of dosages and schedules for each agent have to be based on preliminary data obtained in animal models of tumor metastases. In cancer, the usual approach has been to administer high-dose antineoplastic chemotherapy in cycles separated by a rest period in order to allow the immune system to recover. More recently, cytokine therapy has been used during the period of recovery to rescue the bone marrow cells, an example of how the two types of agents can be successfully combined for greater effectiveness. Experiments with various combinations of pharmacologic agents and cytokines in animal models of tumor growth indicate that other desirable antitumor effects may be achieved using creative combinations of these agents. Below, we describe several of the pharmacologic agents which show therapeutic promise when used in combination with IL2.

III INTERLEUKIN-2

Interleukin-2 (IL2) is a glycoprotein with a molecular weight of 15,000 daltons. It was originally described as an in vitro growth factor for T-lymphocytes (7) and later shown to be produced and released following mitogen or antigen stimulation of a subclass of helper T-cells. IL2 not only stimulates the proliferation of T-cells but facilitates growth and activities of B-lymphocytes, enhances the NK-cell activity (8), increases production of other cytokines, and induces generation of lymphocyte activated killer (LAK) cells from resting PBMNC (9). In terms of the rationale for the use of IL2 in immunodepressed disease states such as cancer, IL2 has been shown to restore immune deficiency seen in athymic nude mice (10), cytotoxic T-cell (CTL) responses in vivo in cyclophosphamide-treated mice or allograph responsiveness in rats. The administration of IL2 facilitated the formation of CTL in alloimmunized mice and resulted in the regression of established tumors metastases in mice. In vitro, lymphoid cells induced with IL2 become lytic for fresh autologous and allogenic tumor cells (9). IL2 also enhanced NK cell activity both in vitro and in vivo. Either IL2 alone or IL2 plus in vitro generated LAK cells administered in vivo result in objective signs of tumor regression in some patients with refractory metastatic disease (11,12). Although primarily considered a cytokine that enhances lymphocyte functions, IL2 may also increase cytotoxicity and microbial activity of human monocytes (13). Clonal expansion of IL2-dependent lymphocyte populations requires the synthesis of IL2 as well as the expression of IL2 receptors. IL2 mediates its activity via the surface IL2 receptors (IL2Rα, β and γ) on target cells (14). Once IL2 is bound to its receptor, the receptor complex is internalized with a $t_{1/2}$ of 25-30 minutes. Further, IL2 signal advances the cell cycle from late G_1 into S phase and increases expression of transferrin receptors required for initiation of DNA synthesis in target cells. The above brief summary indicates that IL2 plays a pivotal role in cell mediated immune responses, and because of its biologic activities has an important role in immunotherapy of cancer and/or recovery of cancer patients receiving traditional forms of therapy.

Systemic or local regional administration of IL2 alone to patients with cancer has been shown to have antitumor effects in a subset of patients with metastatic melanoma or renal cell carcinoma (11,12). IL2 therapy alone has been administered by a series of bolus injections or by intravenous infusion over several hours. Doses range from 10^3 to 10^6 U/m^2 of body surface area or 10^4 to 10^5 U/kg of body weight. The plasma half-life of IL2 is 5-7 minutes and this is not altered significantly by other routes of administration (15). This rapid clearance by catabolism in the kidney, and plasma neutralization by inhibitors contribute in part to the necessity to activate the cells in vitro (16). Administration of IL2 to patients with cancer has been associated with considerable toxicity (15). As IL2 has also been administered in conjunction with autologous activated lymphocytes (LAK cells), it has been uncertain how much of this toxicity could be attributed to large numbers of adoptively-transferred LAK cells. In the initial clinical trials, the highest toxicity occurred during LAK therapy with multiple cycles of IL2 together with LAK cells (15,14). Subsequent experience has suggested that toxicity is related to high doses of IL2 administered as a bolus rather than LAK cells.

Administration of IL2 to patients results in an initial decline in the circulating WBC, but the numbers return to normal within 24 h. Eosinophilia is frequently found in the blood and may be related to the IL2-induced release of IL5. The adverse reactions associated with administration of natural or recombinant IL2 include fever, chills, fatigue, malaise, nausea and vomiting. One or more of these symptoms occur in approximately 25% of patients undergoing IL2 therapy. Lee and colleagues (11) demonstrated that IL2 causes a hyperdynamic cardiac state and fall in vascular resistance. Higher doses may cause leakage of capillaries that leads to fluid retention and sudden, excessive weight gains. Many of adverse reactions of IL2 such as fever, chills, nausea and vomiting can be controlled with drugs, but fluid retention is a very serious problem causing edema, respiratory distress, hypotension, and acute renal failure. Because of these adverse reactions high-dose IL2 is no longer administered to patients with cancer. It has been replaced by moderately-dosed, much less toxic regimens, which can be safely administered even in the outpatient clinic.

The results of many clinical trials with LAK cells and IL2 have been now published (11,12,18). These phase I trials indicated that IL2 and LAK cells may cause tumor reduction (i.e., >50% reduction of tumor volume) or even regression in about 35% of patients refractory to conventional therapy. Patients generally received multiple cell infusions of $2x10^{11}$ autologous LAK cells and IL2. In general, clinical protocols involving generation and infusion of LAK cells to patients with cancer are very difficult and costly, because of a requirement for several leukophereses and for multiple infusions of activated LAK cells. Overall, partial regression of metastatic tumors was seen in a number of patients, and about one third of over 200 patients with metastatic cancer treated at the Surgery Branch, NIH experienced complete regression or partial responses, some of which were durable (>2 years). Tumors most sensitive to LAK cell and IL2 therapy are renal cell carcinoma, malignant melanoma, colorectal carcinoma and non-Hodgkin's lymphoma (11).

As indicated above, substantial antitumor effects were seen with high doses of systemic IL2 in early clinical trials, but these desirable effects were accompanied by severe toxicities. For this reason, other routes or methods of administration are being explored in order to be able to define an optimal protocol and to dissociate desirable from undesirable effects of therapy with IL2. For example, pharmacokinetic studies suggest that in place of i.v. infusions, subcutaneous and intramuscular administration might reduce toxicity and improve efficacy (12,19,20,21). On the other hand, subcutaneous administration of IL2 has been linked to development of anti-IL2 antibodies (21,22). Variations in the dosage and routes of administration of IL2 and LAK cells are currently being evaluated and hopefully new protocols for their administration will increase the success rate of Adoptive Immuno-Therapy (AIT) in patients with cancer.

IV ALKYLATING AGENTS

A. Cyclophosphamide

Cyclophosphamide, following its metabolic activation, has been studied extensively for effects on humoral and cell-mediated immunity. It was shown initially to be selectively toxic to B cells, and more recently, to eliminate certain suppressor T-cell populations. In addition, and depending on the dose, effects of cyclophosphamide on B cells may be reversible, whereas those on suppressor T-cell function are not. Several explanations have been suggested for this irreversible effect of the drug e.g., certain suppressor T-cell populations may not contain enzyme-mediated DNA repair mechanisms, and/or there are no reserve suppressor T cells. The exact mechanism for these desirable and selective effects on immunity are not known. Also, suppressor T-cell functions are inhibited by doses of cyclophosphamide that have no measurable effects on B cells; hence inhibition of clonal proliferation may not be the only mechanism for its effects, assuming proliferative responses of both cell types should be equally sensitive. Thus, from this brief discussion cyclophosphamide appears to be one of the most suitable drugs for exploration of selective effects on the immune system. Cyclophosphamide affects all phases of the immune response, as shown in Table 2 and depending on the dose used may be able to restore balance among various components of the immune system.

Table 2
Effects of Drugs on Phases of Immune Response

Phases of Immune Response (cells involved)	Effect of Drug[a]	
	Suppressant	Stimulant
I. Antigen recognition and/or processing (macrophages, activated B-cells,	Cyclophosphamide Cytimun Corticosteroids	BCG C. parvum Levamisole [b]Doxorubicin
II. Amplification (lymphocytes in blastogenesis, macrophages, B-cells, T-cells)	[b]Cyclophosphamide Cytimun Antimetabolites Antimetabolites 5-FU 6-MP Cytarabine L-asparaginase Corticosteroids	Levamisole Concanavalin A [b]Doxorubicin Bryostatins Phorbol esters Ketoconazole Swainsonine
III. Antibody Production Lymphokine Production	Cyclophosphamide Cytmun Corticosteroids	Lipopoly-saccharides Levamisole
IV. Immune effector responses (plasma cells, small lymphocytes, B-cells, T-cells)	Cyclophosphamide Cytimun Corticosteroids Methotrexate Cytarabine	Levamisole C. parvum

[a]5-FU, 5-fluorouracil; 6-MP, 6-mercaptopurine; C. parvum, Corynebacterium parvum; BCG, Bacillus Calmette-Guerin

[b]Selectively inhibits formation of T-suppressor cells; increases IL-2 production

In addition to cyclophosphamide, many active metabolites such as 4-hydroxycyclophosphamide, aldophosphamide, phosphoramide mustard, acrolein, and possibly nornitrogen mustard are believed to have both antineoplastic and immunosuppressive effects. Alkylating agents in general are known to be cytotoxic to lymphocytes by alkylating nucleic acids and forming cross-links between macromolecules (polypeptide or polynucleotide chains) with relatively stable bonds. In addition, they are known to inhibit protein and nucleic acid (DNA) biosynthesis, and to selectively alkylate purine bases, which not only results in depurination but also introduces errors during transcription and replication.

Cyclophosphamide, or its active metabolites, though promising for selective effects throughout the immune system, must await further evaluation. In this regard, the report of Mitchell, et al. (23), in which pretreatment with a low dose of cyclophosphamide prior to low doses of IL2 not only increased efficacy but reduced toxicity in patients with disseminated melanoma. Prior to AIT with tumor-infiltrating lymphocytes, patients at the Surgery Branch, NCI are now routinely treated with cyclophosphamide. Further, as a general immunosuppressant, cyclophosphamide therapy has yielded clinically desirable results in several diseases with known immunologic associations. In patients who are refractory to steroid therapy or whose disease process exacerbates during withdrawal from steroid therapy, cyclophosphamide is known to be beneficial. In addition, clinical improvements as a result of cyclophosphamide therapy were seen in: (a) Wegener's granulomatosis, (b) selected patients with rheumatoid arthritis, and (c) nephrotic syndrome.

Cyclophosphamide is generally given orally in 2-mg/kg doses in the diseases mentioned above. A rigid monitoring program should be established before the treatment begins, as a moderate leukopenia (WBC count of 2500-4000/cm^3) is induced and maintained throughout the course of therapy. Also, a substantial fall in granulocytes to less than 1000/cm^3 often predisposes the patient to infections and a fall in the platelet count below 10^5/cm^3 is a signal to discontinue therapy. When cyclophosphamide is used as a prelude to AIT, a single high dose (25 mg/kg body weight) is generally given one day prior to the transfer of cells and IL2 (24).

As with other cytotoxic drugs, cyclophosphamide has an enormous potential for adverse effects on the gonads and blood-forming organs (reticuloendothelial system). Testicular atrophy and azoospermia have been reported in male patients. Severe hair loss (alopecia), occasional hemorrhagic cystitis (increase in fluid intake may reduce incidence), gastrointestinal intolerance, and mucositis have been reported in male and female patients. Also, cyclophosphamide is in part activated to alkylating molecular species in the liver smooth endoplasmic reticulum, and toxicity may be altered by other drugs or conditions that activate or inhibit these enzymes.

V ANTIBIOTICS

A. Anthracyclines

Doxorubicin, daunorubicin and aclacinomycin are examples of naturally occurring anthracycline glycoside antibiotics. These molecules consist of an aromatic tetracyclic aglycone and an amino sugar. The sugars and amino sugars found in the three glycosides include daunosamine, rhodosamine, 2-deoxy-L-fucose, and rhodonose. Doxorubicin and daunorubicin contain amino sugar daunosamine, whereas aclacinomycin contains rhodosamine and a trisaccharide. The aglycone region of each molecule is hydrophobic whereas the amino sugar is hydrophilic. These compounds are amphipathic, containing both basic and hydrophobic regions. Doxorubicin (pK_a 7.2-7.4) is partially ionized at pH 7.4 by protonation of the amino group. The pKa of aclacinomycin is lower resulting in even less of this compound being ionized at physiological pH.

Daunorubicin and doxorubicin possess vesicant-like properties, and are always administered intravenously. The usual dose of doxorubicin is approximately 45-60 mg/m^2 of surface area, given every three weeks or 15-20 mg/m^2 given weekly. The recommended dose of daunorubicin is 30 to 60 mg/m^2 daily for 3 days or once weekly. The total cumulative dose range is between 500 and 550 mg/m^2.

The pattern of distribution for doxorubicin and daunorubicin is similar. After an intravenous administration, they are cleared rapidly from the plasma, and clearance kinetics appear to be triphasic, displaying an initial half-life of 12 min, an intermediate value of 3.3 to 8 h, and a final value of approximately 30 h. The drugs are rapidly taken up into tissues, especially the heart, lungs, liver and kidneys. Neither doxorubicin nor daunorubicin has been shown to cross the blood-brain barrier in significant amounts.

Doxorubicin and daunorubicin are actively metabolized by cytoplasmic aldoketoreductases to doxorubicinol and daunorubicinol, respectively, and by reductive deglycosylation to a 7-deoxyaglycone, but significant amounts are excreted unchanged and as multiple metabolites. The liver appears to be the major site for metabolism. In patients with liver disease, plasma levels of doxorubicin, daunorubicin and their metabolites are markedly elevated and are associated with exaggerated clinical toxicity (25).

Urinary excretion is low, rarely accounting for more than 10% of the administered dose. In contrast, biliary excretion is high, accounting for 40-50% of the administered dose after seven days.

Aclacinomycin (ACM) and its metabolites were studied in twelve patients treated with 60-120 mg/m^2 during a phase 2 clinical trial (25). Total plasma drug fluorescence initially declined very rapidly, but from 2 to 24 h after injection, fluorescence rose progressively to intensities greater than those measured one minute after aclacinomycin injection. Total drug fluorescence in the plasma slowly declined from 24-72 h after ACM administration. These events reflected the rapid disappearance of ACM and the subsequent appearance of two highly-fluorescent metabolites. One of these metabolites co-chromatographed with bis-anhydro-aklavinone and had a fluorescence spectrum identical to that of bis-anhydro-aklavinone (26).

Comparison of the cellular accumulation and disposition of the anthracycline antibiotics indicated that although daunorubicin and ACM were both avidly accumulated by cells, intracellular concentrations of ACM were 2-3 times those of daunomycin. The intracellular accumulation of both drugs is temperature dependent. Both daunomycin and ACM were released from the cells placed in a drug-free medium, a process that was also temperature dependent. Unlike daunorubicin, which localized in cell nuclei, ACM localized in the cytoplasm with no detectable nuclear fluorescence. Although both drugs produced dose-dependent inhibitions of (^{3}H)-thymidine and (^{3}H)-uridine incorporation by L1210 and P388 cells, ACM inhibited both processes at lower concentrations than did daunorubicin. While daunorubicin inhibited (^{3}H)-thymidine incorporation more effectively than (^{3}H)-uridine incorporation, the reverse was observed for ACM. these findings have been used to classify the anthracyclines, doxorubicin and daunorubicin, in one class and ACM in another, separate class of antibiotics.

Anthracyclines inhibit cell division and growth in tumors and other cells with high mitotic rates, in part, by intercalation between the base pairs and subsequent inhibition of DNA, RNA and protein synthesis (see Table 2).

However, studies of the host response to tumors have suggested that doxorubicin may also selectively modulate cell mediated immune responses (see Table 2). More specifically, doxorubicin was shown to be more effective in murine models of strongly immunogenic than non-immunogenic tumors. Doxorubicin treatment resulted in an increase in cell-mediated cytotoxic responses (27-29). These findings were shown to be due to increased T-cell and macrophage activities. Simultaneously, there was an inhibition of humoral antibody formation and NK-cell response. Tissue macrophages are relatively insensitive to doxorubicin in comparison to monocytes. Further, doxorubicin appears to increase prostaglandin E_2 (PGE_2) production as well as that of IL2. Whether this increased IL2 production represents a direct effect on lymphocytes or is indirectly mediated via IL1 production by monocytes is unclear. ACM also shows strong cytotoxicity for lymphocytes stimulated by lectins such as Con-A, PHA and LPS. However, ACM also selectively inhibits generation of suppressor cells involved in the regulation of both antibody production and delayed hypersensitivity reactions.

Doxorubicin was used in conjunction with adoptive transfer of IL2-activated cytotoxic lymphocytes plus IL2 in treatment of established murine renal cell carcinoma (30). In the adjuvant setting, this combination of chemoimmunotherapy resulted in the cure of 67% of mice bearing established renal cancer (30). These studies demonstrated that the combination of doxorubicin and IL2-stimulated cytotoxic lymphocytes acted synergistically for treatment of established renal cancer (30). The observed therapeutic effect may be due to sublethal damage of tumor cells by the drug, making them more susceptible to the effect of transferred cytotoxic cells.

This modulation of immune cell functions by anthracyclines may be related to activation of phosphoinositol pathways, calcium-dependent enzymes, and the protein kinase C pathway (PKC). PKC is a pivotal enzyme in the transduction of signals which regulate cellular proliferation and differentiation, especially those involving receptor-mediated hydrolysis of inositol phospholipids. PKC or its isozymes may be involved in the stimulation and activation of T-lymphocytes. The specific events are the hydrolysis of phosphatidylinositol 4,5-biphosphate (PIP_2) to yield second messengers, diacylglycerol (DAG) and inositol 1,4,5-triphosphate (IP_3). DAG activates PKC and IP_3 mobilizes calcium ions from the endoplasmic reticulum. Anthracyclines cause an elevation of IP_3 and a sustained rise in DAG. Moreover, PKC was shown to influence phosphorylations involving the multidrug resistance (MDR)-associated proteins (20-kD and 170kD). Additional, messenger systems are also modulated by anthracyclines, i.e., adenylate cyclase and guanylate cyclase are both inhibited as are phosphodiesterases and calmodulin-mediated events.

Lastly, alterations in membrane fluidity may be an explanation for the role of anthracyclines in signal transduction. Doxorubicin was shown to affect membrane fluidity, and anthracyclines which do not evoke the fluidity response are inactive as cytotoxic agents (31).

Doxorubicin is used in the treatment of malignant lymphomas, acute lymphoblastic and granulocytic leukemias, Wilms tumor, osteogenic sarcoma, Ewings sarcoma, soft tissue sarcoma and neuroblastoma. In acute non-lymphoblastic leukemia in adults daunorubicin is the single most active drug and, when given with cytarabine, it is the treatment of choice. Doxorubicin is also considered to be the drug of choice for progressive metastatic thyroid carcinoma, because of its known activity against this tumor and lack of effect by other agents.

Daunorubicin and doxorubicin produce myleosuppression with leukopenia and thrombocytopenia (60 to 80 percent of patients), stomatitis and gastrointestinal disturbances (80 percent), nausea, vomiting or both (20 to 55 percent) and alopecia (85 to 100 percent). These effects do not limit the use of these drugs in the treatment of cancer. The major dose-limiting toxicity of doxorubicin and daunorubicin is the development of cardiomyopathy. The effects of doxorubicin and daunorubicin on the heart may be classified as acute, i.e., transient and usually benign electrocardiographic changes and chronic, i.e., cumulative dose dependent and potentially irreversible degenerative changes leading to congestive heart failure.

VI BACILLUS CALMETTE-GUERIN (BCG)

Bacillus Calmette-Guerin (BCG) vaccine has been used in the treatment of melanoma and acute lymphocytic leukemia in human and in a variety of experimental tumors. BCG is widely accepted as an immunological enhancer or a nonspecific adjuvant, and it has been shown to contain antigens in common with human and animal tumor cells. Hence, it can stimulate specific and nonspecific antitumor immune responses in humans.

BCG is known to directly upregulate macrophage functions, resulting in significant increases in macrophage cytotoxicity against neoplastic cells, phagocytic activity, lysosomal enzymes, activated phospholipase A_2, production of IL1 that enhances T-cell helper function and T-cell cytotoxicity, and chemotaxis. The main effect of BCG is on the phase 1 of the immune response see Table 2. The route and relative time of administration are crucial, and BCG contact with the tumor is important. For example, injection of BCG intralesionally at the sites drained by lymph nodes increases the potential for a favorable response.

BCG alone or in combination with antigen preparations has shown exciting results in several diseases. The FDA approved two preparations made from substrains of Bacillus Calmette and Guerin, a live, weakened strain of Hycobacterium bovis. They are: BCG live (Intravesical), trade name Tera Cys and BCG Vaccine (for intrvesical or percutaneous use), trade name Tice BCG.

It is necessary to remember that patients whose immune systems is suppressed or who are immunologically deficient should not receive BCG products, because of the high risk of systemic mycobacterial infection. Similarly, therapeutic combinations of BCG with agents which cause bone marrow suppression should be avoided. On the other hand, the potential of BCG for stimulation and activation of immune cells raises new possibilities for its use in treatment of patients with cancer.

Adverse effects of BCG products include hematuria, urinary frequency, dysuria, bacterial urinary tract infection as well as genitourinary symptoms. Side effects include malaise, fever, chills, anemia, nausea, vomiting and anorexia. When used locoregionally, as in therapy of a primary and relapsed carcinoma-in situ (CIS) of the urinary bladder, with or without papillary tumor, BCG has only mild toxicity and has been effective.

VII FK-565

FK-565, developed in the research laboratories of the Fujisawa Pharmaceutical Company, is a low molecular weight (502 daltons) synthetic acylpeptide with potent biological response modifying activity as an anti-infectious and antitumor agent (32). FK-565 is a synthetic derivative of the immunostimulatory FK-156 dipeptide isolated from the fermentation broth of Streptomyces (30). FK-156 has structural similarity to the bacterial cell wall peptidoglycan; FK-156 has immunological activity even though it lacks the sugar moiety of peptidoglycan and is a compound that illustrates that the glutamyldiaminopimelic acid moiety is a minimal and essential structure for activity. FK-565 represents an interesting and effective acyltripeptide derivative of FK-156 with enhanced antitumor host defense activity. It has been reported that peritoneal macrophages harvested from C57BL/6 mice and beige mice are cytotoxic against syngeneic B 16 melanoma cells following in vitro activation with FK-565; it has therefore been suggested that FK-565 has potential as an effective immunomodulator in immunotherapy (30). Moreover, FK-565 has also been found to augment NK-cell activity in vivo (33). FK-565 has therefore been shown to stimulate both NK cells and macrophages in mice and to have therapeutic effects on experimental tumor metastases in mice.

In vitro effects of FK-565 on NK activity, lymphokine-activated killer (LAK)-cell generation and cytokine production of normal human peripheral blood mononuclear cells (MNC) were studied (34). FK-565 used at concentrations ranging from 0.1 to 100 μg/ml enhanced NK-cell activity only if adherent MNC were removed. The optimal NK-cell enhancing dose was 2 μg/ml FK-565. At the same range of concentrations, FK-565 activated adherent MNC to induce suppression of NK-cell activity in autologous non-adherent MNC preparations. FK-565 also potentiated both the generation of LAK-cell activity in the presence of IL2 and FK-565 on LAK-cell activity was observed for all drug concentrations used. The effects of FK-565 on cytotoxic cells could not be attributed to IL2, interferon gamma or tumor necrosis factor-alpha, because FK-565 alone had no detectable influence on in vitro production of these cytokines by MNC. The ability of FK-565 to modulate effector cells of natural antitumor immunity indicates that it may have promise as a biological response modifier in humans (34).

In addition, FK-565 showed in vitro effects on human monocytes and granulocytes (35). Monocyte cytotoxicity against A375 melanoma targets was significantly increased following pretreatment with FK-565 at concentrations of 1 μu/ml or more. The tripeptide also upregulated antitumor cytostasis by monocytes and showed a strong stimulatory effect on superoxide generation by resting monocytes over a wide range of FK-565 concentrations after 18 h preincubation. The monocyte preparations contained an average of 76% LeuM3^{+}HLA-DR^{-} and 10% LeuM3-HLA^{-}DR^{+} cells, and this phenotype distribution was not altered after incubation with FK-565. At concentrations above 1 μg/ml and after 2h preincubation, FK-565 also increased superoxide generation by resting but not stimulated granulocytes. Pre-exposure of cultured bovine endothelial cells to the peptide resulted in a significant inhibition of fMLP-stimulated granulocyte adherence to these cells. These data indicate that in vitro incubation of human monocytes and granulocytes with FK-565 (0.1-100 μg/ml) had resulted in simultaneous upregulation of several antitumor functions mediated by these cells (35).

VIII LEVAMISOLE

Levamisole, a phenylimidazole derivative, is a potent antihelmintic agent that is known to stimulate the immune system in immunologically incompetent animals and humans. Cellular functions said to be enhanced by levamisole include phagocytosis, chemotaxis, random migration and adherence, as well as mitogen and antigen-induced proliferative responses. Levamisole affects all phases of the immune response as shown in Table 2.

Levamisole has no intrinsic antibacterial or antiviral properties, and hence these effects are mediated via immunoenhancement.

Levamisole (see chemical structure in Figure 1) is known to act on cyclic nucleotide metabolism, increasing the breakdown of cyclic adenosine monophosphate (cAMP) and decreasing the breakdown of cyclic guanosine monophosphate (cGMP). Levamisole also exerts cholinergic effects, i.e., increasing cGMP levels and formation. Increased levels of cGMP are correlated with lymphocyte proliferation and augmentation of chemotactic responses after administering this drug to laboratory animals. Overall, differentiation of precursor T lymphocytes into mature effector cells, is facilitated by levamisole, although the exact mechanism of this process is unknown.

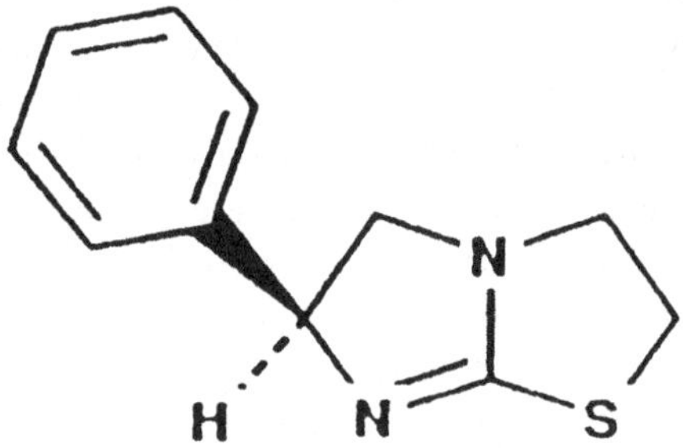

Levamisole

Fig. 1 Chemical structure of levamisole (L-2,3,5,6-tetrahydro-6-phenyl-imidazol[2,1-b]thiazole

Levamisole is rapidly absorbed from the gastrointestinal tract in humans. An average dose for immunoenhancement in adults may be 150 mg/kg (orally), 50 mg three times a day. Peak plasma levels of 0.7 mg/1 are obtained in 1-2 hr, and the plasma half-life is equal to 4 hr. Levamisole is metabolized in the liver. Parenteral administration (e.g., i.m.) of the drug will give higher (2x) peak plasma levels. Urinary excretion of levamisole (parent drug) is slow and is influenced by pH (low pH increasing elimination). Overall elimination of a single dose is complete within 48 hr. Adverse effects from levamisole occur in about 20% of the patients. Rash is the most common but leukopenia, agranulocytosis, and thrombocytopenia have been reported. Also an influenza like illness, nausea and vomiting and oral ulcerations may occur. Reversible immune complex glomerulonephritis may also be a problem in certain patients. Discontinuation of therapy will reverse most of adverse effects (36).

Levamisole hydrochloride in combination with fluorouracil has been approved for treatment of Dukes C colon cancer (37). Levamisole is an adjuvant with efficacy in immunotherapy of melanoma (4). Adverse effects from this combination include nausea and vomiting, diarrhea, stomatitis and anorexia in addition to those listed for Levamisole alone. Moreover, CNS symptoms such as dizziness, ataxia, depression and confusion may occur.

Immunodeficiency diseases such as Job's syndrome or hyperimmunoglobulin E syndrome, lazy leucocyte syndrome, Wiskott-Aldrich syndrome, ataxia telangiectasis, chronic granulomatous disease, and cyclic neutropenia all have been treated with some success. Infectious diseases such as recurrent herpes infections, recurrent furunculosis, acne conglobata, and chronic pyogenic skin infections have been improved by treatment with levamisole as indicated by an improvement of the lesions. Diseases associated with immunological abnormalities such as rheumatoid arthritis, Reiter's syndrome, systemic lupus erythematosus, aphthous stomatitis, and Crohn's disease have in some cases been improved following treatment with levamisole.

IX BRYOSTATINS

Bryostatins are macrocyclic lactones extracted and chemically purified from marine bryozoans, sea mosses (38). Synthetic phorbol myristate and natural bryostatins activated protein kinase C (PKC), an intracellular enzyme important in lymphocyte activation (39,40). Unlike tumor-promoting phorbol esters, bryostatins are antineoplastic. Bryostatins induce transcriptional activation of the IL2R gene and down-regulate certain specific T-cell surface markers. Bryostatins are also able to block some biochemical responses caused by phorbol esters.

The nontumor-promoting activators of PKC such as bryostatins, appear to be excellent candidates for further evaluation as immunomodulating agents. They seem to modulate lymphocyte activation, via PKC and since PKC is not one enzyme but a group of isoenzymes, this raises the possibility for selective pharmacologic intervention in the lymphocyte activation pathway.

Bryostatins when combined with a calcium ionophore can effectively induce IL2 production and proliferation of lymphocytes. Even more interesting is that observation that low concentrations (5-70 µg/kg _in vivo_ or 0.89 µg/ml _in vitro_) of bryostatins, potentiate effects of IL4 in stimulating resting T-cells to proliferate and differentiate into cytotoxic lymphocytes. Thus, both _in vivo_ and _in vitro_ bryostatins appear to synergize with IL4 in the generation of nonspecific cytotoxic T-cells. They also enhance the ability of recombinant IL2 to induce the development of _in vivo_ primed cytotoxic lymphocytes (CTL) during their _in vitro_ incubation. On the other hand, bryostatin seems to inhibit CTL activity against antigen-specific target cells. Bryostatins in combination with calcium ionophores also facilitate exocytosis of cytolytic granules from CTL.

Bryostatins are also useful as stimulators of normal human hematopoietic cells in that they directly stimulate bone marrow progenitor cells (in vitro) and functionally activate neutrophils. The effects reviewed above strongly suggest the use of bryostatins either alone or in combination with IL2 because an exogenous dose of IL2 may be greatly reduced and endogenous IL2 may be more effective.

X SWAINSONINE

Swainsonine, an indolizidine alkaloids, has generated interest in its potential use as an anticancer agent with reports that it: (i) inhibits tumor growth and metastasis (41), (ii) augments natural killer (NK) and macrophage-mediated tumor cell killing (42) and (iii) stimulates bone marrow cell proliferation (43). The antineoplastic activity of swainsonine can be explained, at least in part, by augmentation of immune effector mechanisms (44,45).

Swainsonine may confer protection against the cytotoxic effects of both cell cycle-specific and nonspecific cytotoxic anticancer agents. In murine models, results indicate that the intraperitoneal administration of swainsonine decreased the lethality of methotrexate, 5-fluorouracil, cyclophosphamide, and doxorubicin in nontumor bearing mice. The increased survival was found to correlate with stimulation of BM cellularity, in vivo and in vitro colony formation, and engraftment efficiency.

Swainsonine may have the potential to allow increases in the dose and/or frequency of administration of anticancer chemotherapy without increasing the complications associated with bone marrow toxicity. It may play an important role in the treatment of immunosuppressive disorders, including malignancy. It also may serve as an important probe to investigate the mechanism of normal hematopoiesis. Swainsonine has no obvious adverse side effects at the levels used in the above murine experiments; however, toxicological studies are yet to be done in humans.

XI SUMMARY AND CONCLUSIONS

IL2 immunotherapy alone or with LAK cells represents a novel approach to the treatment of metastatic cancers. Similarly, other BRMs and classical chemotherapeutic drugs through their molecular effects on the components of the immune system reveal new and exciting prospects for the better use of these agents. Both approaches converge in that they restore balance among numerous components of the immune surveillance system. This review raises the possibility of improved protocols through the judicious use of these agents and stresses the need for further investigation of combined use of chemotherapy and immunotherapy.

References

1. Uchida A, Hoshino T. Clinical studies on cell-mediated immunity in patients with malignant disease. Cancer Immun Immunother 1980; 9:153-158.
2. Nio Y, Zighelboim J, Berek J, Bonavida B. Cytotoxic and cytostatic effects of the streptococcal preparation OK-432 and its subcellular fractions on human ovarian tumor cells. Cancer 1989; 64:434-441.
3. Jaekle KA, Mittleman A, Hill FH. Phase II trial of Serratia marcescens eltract in recurrent malignant astrocytoma. J Clin Oncol 1990; 8:1408-1418.
4. Wauwe JV, Janssen PAJ. Review article on the biochemical mode of action of levamisole: an update. J Immunopharmac 1991; 13:3-9.
5. Esa AH, Boto WO, Adler WH, May WS, Hess AD. Activation of T-cells by bryostatins: Induction of IL-2 receptor gene transcription and down modulation of surface receptors. Int J Immunopharmac 1990; 12:481-490.
6. Goldfarb R, Herberman RB. Natural killer cell reactivity: Regulatory interactions among phorbol ester, interferon, cholera toxin and retinoic acid. J Immunol 1981; 126:2129-2135.
7. Morgan DA, Ruscetti FM, Gallo RC. Selective in vitro growth of T lymphocytes from human bone marrows. Science 1976; 193:1007-1008.
8. Rosenberg SA, Speiss PA, Schwartz S. In vivo administration of interleukin-2 enhances specific alloimmune responses. Transplantation 1983; 35:631-634.
9. Lotze MT, Grimm EA, Mazumder A, Strausser JL, Rosenberg SA. In vitro growth of cytotoxic human lymphocytes: IV lysis of fresh and autologous tumor of lymphocytes cultured in T-cell growth factor. Cancer Res 1981; 41:4420-4425.
10. Wagner H, Hardt C, Heeg K, Rollinghoff M, Pfizenmaier KT. Cell derived helper factor allows in vivo induction of cytotoxic T cells in nu/nu mice. Nature 1980; 284:278-280.
11. Rosenberg SA, Lotze MT, Muul LM, Chang AE, Avis FP, Leitman S, Linehan WM, Robertson CN, Lee RW, Rubin JT, Seipp CA, Simpson CG, White DE. A progress report on the treatment of 157 patients with advanced cancer using lymphokine-activated killer cells and interleukin 2 or high dose interleukin 2 alone. N Engl J Med 1987; 316:889-897.
12. West WH, Tauer KW, Yannelli JR, Marshall GD, Orr DW, Thurman GB, Oldham RK. Constant-infusion recombinant interleukin-2 in adoptive immunotherapy of advanced cancer. N Engl J Med 1987; 316:898-905.
13. Wahl SM, McCartney-Francis N, Hunt DA, Smith PD, Wahl LA, Katona IM. Monocyte interleukin-2 receptor gene expression and interleukin-2 augmentation of microbicidal activity. J Immun 1987; 139:1342-1347.
14. Waldman TA. The structure, function and expression of interleukin 2 receptors on normal and malignant lymphocytes. Science 1986; 232:727.

15. Lotze MT, Matory YL, Ettinghausen SE, Rayner AA, Sharrow SO, Seipp CAY, Custer MC, Rosenberg SA. In vivo administration of purified human interleukin 2 (IL-2) II Half-life, immunologic effects, and expansion of peripheral lymphoid cells in vivo with recombinant IL-2. J Immun 1985; 135:3865-2875.
16. Bubenik J. Mini-review: Local immunotherapy of cancer with interleukin 2. Immunology Letters 1989; 21:267-274.
17. Lee RE, Lotze MT, Skibber JM, Tucker E, Bonow RO, Ognibene FP, Carrasquillo JA, Shelhamer JH, Parrillo JE, Rosenberg SA. Cariorespiratory effects of immunotherapy with interleukin-2. J Clin Oncol 1989; 7:7-20.
18. Dutcher JP, Gaynor ER, Boldt DH, Doroshow JH, Car MH, Sznol M, Mier J, Sparano J, Fisher RI, Weiss G, Margolin K, Aronson FR, Hawkins M, Atkins M. A Phase II study of high-dose continuous infusion interleukin-2 with lymphokine-activated killer cells in patients with metastatic melanoma. J Clin Oncol 1991; 9:641-648.
19. Thompson JA, Douglas JL, Cox WW, et al.: Recombinant interleukin-2 toxicity, pharmacokinetics, and immunomodulatory effects in a phase I trial. Cancer Res 1987; 47:4202-4207.
20. Clark JW, Smith II JW, Steis RG, Urba WJ, Crum E, Miller R, McKnight J, Beman J, Stevenson HC, Creekmore S, Stewart M, Conlon K, Sznol M, Kremers P, Cohen P, Longo DL. Interleukin 2 and lymphokine-activated killer cell therapy: Analysis of a bolus interleukin 2 and a continuous infusion interleukin 2 regimen. Cancer Res 1990; 50:7343-7350.
21. Whitehead RP, Ward D, Hemingway L, Hemstreet III GP, Bradley E, Konrad M. Subcutaneous recombinant Interleukin-2 in a dose escalating regimen in patients with metastatic renal cell adenocarcinoma. Cancer Res 1990; 50:6708-6715.
22. Krigel RL, Padavic-Shaller KA, Rudolph AR, et al. A phase I study of recombinant interleukin-2 plus recombinant beta-interferon. Cancer Res 1988; 48:3875-3881.
23. Mitchell MS, Kempt RA, Harel W, et al. Effectiveness and tolerability of low-dose cyclophosphamide and low-dose intravenous interleukin-2 in disseminated melanoma. J Clin Oncol 1988; 6:409-424.
24. Aebersold P, Hyatt C, Johnson S, Hines K, Korcak L, Sander M, Lotze M, Tapalian S, Yang J, Rosenberg SA. Lysis of autologous melanoma cells by tumor-infiltrating lymphocytes: Association with clinical response J Natl Cancer Inst 1991; 83:932-937.
25. Nwankwoala RNP, West WL. Anthracyclines: In cancer chemotherapy. Nigerial Med Practitioner 1987; 13:1-2.
26. Egorin MJ, Andrews PA, Nakazawa H, Bachur NR. Purification and characterization of aclacinomycin A and its metabolites from human urine. Drug Metabolism and Disposition 1983; 2(2):167-171.
27. Vecchi A, Mantovani A, Tagliabue A, Spreafico F. A characterization of the immunosuppressive activity of Adriamycin and Daunomycin on humoral antibody production and tumor allograft rejection. Cancer Res 1976; 36:1222-1227.

28. Mantovani A, Vecchi A, Tagliabue A, Spreafico F. The effect of adriamycin and daunomycin on antitumoral effector mechanisms in an allogenic system. Eur J Cancer 1976; 12:371-380.
29. Spreafico F. The immunological activity of anthracyclines. In anthracyclines antibiotics in cancer therapy. (eds Muggia F, Young CN, Carter SK) 1982; 50-68. Nijhoff N Publ Co, The Hague.
30. Salup RR, Wltrout RH. Adjuvant immunotherapy of established murine renal cancer by interleukin 2-stimulated cytotoxic lymphocytes. Cancer Res 1986; 46:3358-3363.
31. Tritton TR. Disruption of cellular growth control and signal transduction mechanisms a target for cancer chemotherapy. In developments in cancer chemotherapy by Robert I. Glazer II:46-54, copyright 1989, CRC Press Inc., Boca Raton, Florida.
32. Inamura N, Nakahara K, Kino T, Gotoh T, Kawamura I, Aoki H, Imanaka H, Sone S. Activation of tumoricidal properties in macrophages and inhibition of experimentally-induced murine metastases by a new synthetic acyltripeptide, FK-565. J Biol Resp Modifiers 1985; 4:408-417.
33. Talmade JE and chirigos MA. Comparison of immunomodulatory and immunotherapeutic properties of biological repsonse modifiers. Springer Semin Immunopathol 1985; 17-23.
34. Wang YL, Whiteside TL, Friberg D, Herberman RB. In vitro effects of an acyltripeptide, FK565, on NK-cell activity, LAK-cell generation and cytokine production by human mononuclear cells. Immunopath 1989; 17:175-185.
35. Wang YL, Kaplan S, Whiteside T, Herberman RB. In vitro effects of an acyltripeptide, FK565, on antitumor effector activities and on metabolic activities of human monocytes and granulocytes. Immunopath 1989; 18:213-222.
36. West, WL, Pradhan, SN: Pharmacology in Medicine (Pradhan SN, Maickel RP, and Dutta, eds.) SP Press Inter 1986; 13.5.3:524-530.
37. Moertel CG, Fleming TR, MacDonald JS, et al. Levamisole and fluorouracil for adjuvant therapy of resected colon carcinoma. New Engl J Med 1990; 322:362-358.
38. Pettit GR, Herald CL, Doubek DI, Herald DL. Isolation and structure of Bryostatin 1, J Am Chem Soc 1982; 104:6846-6848.
39. Trenn G, Pettit GR, Takyama H, Hu-Li J, Sitkovsky MV. Immunomodulating properties of a novel series of protein kinase C activators: they bryostatins. J Immun 1988; 140:433-439.
40. Dale IL and Gescher A. Effects of activators of protein kinas C, including bryostatins 1 and 2, on the growth of A549 Human lung carcinoma cells. Int J Cancer 1989; 43:158-163.
41. Humphries MJ, Matsumoto K, White SL, Molyneux RJ, Olden K. An assessment of the effects of swainsonine on survival of mice injected with B16-F10 melanoma cells. Clin Exp Metastasis 1990; 8:89-102.

42. Humphries MJ, Matsumoto K, White SL, Molyneux RJ, Olden K. augmentation of natural killer cell activity by swainsonine, a new antimetastatic immunomodulator. Cancer Res 1988; 48:1410-1415.
43. White SL, Nagai T, Akiyama SK, Reeves EJ, Grzegorzewski K, Olden K. Swainsonine stimulation of the proliferation and colony forming activity of murine bone marrow. Cancer Commun 1991; 3:83-91.
44. Bowlin TL and Sunkara PS. Swainsonine, an inhibitor of glycoprotein processing, enhances mitogen induced interleukin-2 production and receptor expression in human lymphocytes. Biochem Biophys Res Commun 1988; 151:859-846.
45. Bowlin TL, McKown BJ, Kang MS, Sunkara PS. Potentiation of human lymphokine-activated killer cell activity by swainsonine, an inhibitor of glycoprotein processing. Cancer Res 1989; 49:4109-4113.

PART VI

Introduction

Immunotherapy of Tumors

Giorgio Parmiani
Instituto Nazionale Tumori
Milan, Italy

In the last few years adoptive immunotherapy studies have focused on the use of interleukin-2 (IL-2) given alone or with two different types of effectors, namely the non-specific, MHC-unrestricted, IL-2 activated lymphocytes (LAK) which included both T and NK cells and, more recently, the tumor-specific, MHC-restricted T lymphocytes. However, the lack of detailed information on the in vivo distribution of LAK cells in man and the underestimation of the crucial role which is played by tumor microenvironment, have led to a plethora of often useless clinical studies based on the simple (but essentially wrong) rationale that if LAK cells kill tumor cells in vitro, they should do the same in vivo, provided that IL-2 is available to such lymphocytes.

Nevertheless, this generation of clinical studies has provided us with important information that human tumors can regress after IL-2/LAK therapy and that in a small number of cases, patients can be cured by such treatment. Unfortunately, the mechanism(s) of such curative effects is still unknown despite many studies and even the therapeutic role of LAK cells or their contribution to tumor regression in vivo remain controversial. To me, the most important limitation of LAK cell therapy is in the insufficient influx of LAK cells into the tumor, which prevents a more effective clinical response. It is intriguing to note that T lymphocytes and not NK cells have been observed in the infiltrates of regressing lesions (Cohen et al., 1987), a finding which suggests a role for specific T cells in IL-2/LAK-based therapy (Parmiani, 1990).

In addition, the earlier studies of adoptive immunotherapy with non-specific LAK cells laid the foundation for subsequent, more recent clinical trials aimed at increasing the response rate by improving the tumor cytotoxic activity of lymphocytes, their homing and/or tumor specificity. In fact, two important developments ensued. First, the use of more selected subpopulations of LAK cells allowed for a better tumor-targeting activity (Basse et al., 1991). Second, the use of operationally tumor-specific, MHC-restricted T cells characterized by the ability to target distant neoplastic lesions seemed to require less IL-2, whose toxicity has been a major concern for many oncologists. A direct comparison of therapeutic efficacy between LAK cells and T cells from the same mice showed a superior antitumor activity of the latter (Rodolfo et al., 1991). Along these lines, clinical studies have been initiated aimed at evaluating the therapeutic activity of selected subsets of NK cells (A-NK) in various types of metastatic cancer (Kirkwood JM, Herberman R, unpublished data) and of T cells obtained from the tumor infiltrates (Rosenberg et al., 1988).

Newer clinical trials suggest that even the use of IL-2/LAK therapy, which may certainly be appropriate for subgroups of patients with advanced kidney tumors (Negrier et al., 1990), may find a more rational application in a clinical setting which allows a high concentration of LAK cells into the tumor to occur. For example, in-transit metastases of melanoma in the limbs can be treated by extracorporeal circulation, perfusing the limb with high concentration of IL-2/LAK. Preliminary results with this type of therapy from our Institute are encouraging in terms of low toxicity and high response rate (RR). In a similar setting, Lejeune's group has recently reported a 90% RR by administering Melphalan combined with IFN-γ and TNF-α in melanoma or osteosarcoma patients, although this treatment is accompanied by high toxicity (personal communication). These and other clinical studies indicate that loco-regional therapy with cytokines with or without LAK warrants further evaluation.

Adoptive immunotherapy with TIL appears more effective than LAK in advanced melanoma patients resulting in a RR of approximately 40% (Rosenberg et al., 1988), whereas the results in kidney cancer are disappointing (Bukowski et al., 1991). The reason for such a discrepancy might be ascribed to the different immunogenicity of the two types of neoplasms, since tumor-specific TIL are easily found in melanoma but not in kidney cancer (Balch et al., 1990). An effort should be made, therefore, to improve the selection of tumor-specific T cells from patients with cancer. The molecular analysis of TCR of tumor-specific lymphocytes may facilitate selection and propagation of such cells, if oligoclonality of TCR will be confirmed among tumor-specific CTL (Nitta et al., 1991; Sensi et al., 1991).

Finally, the somatic gene therapy of cancer appears to offer new, although still uncertain, perspectives. By introducing cytokine genes (e.g. TNF or IL-2) into tumor-specific T cells and improving their direct or indirect killing activity, and by rendering tumor cells more immunogenic through gene transfer and expression of cytokines (IL-2, IL-4, IL-6, IL-7, IFN-γ, G-CSF, GM-CSF), one should be able either to better stimulate T cells to be infused into the patients or to vaccinate such patients with potent immunogenic cells. The discovery of genes encoding tumor-specific antigens of human melanomas (Van den Bruggen et al., 1991) will most likely allow oncologists to vaccinate patients with normal or neoplastic cells engineered to express appropriate amounts of MHC, adhesion molecules, cytokines and tumor-specific antigens. Although the obstacles are still formidable, we all hope that immunotherapy will finally be able to cure a sizeable fraction of tumors resistant to other standard therapies. As in the past, Ron Herberman will certainly assist us with his wisdom and intelligence to critically evaluate and design future approaches to immunotherapy of cancer.

References

1. Balch CM, Riley LB, Bae YJ, Sameron MA, Platsoucas CD, Von Eschenbach A, Itoh K. Patterns of human tumor-infiltrating lymphocytes in 120 human cancers. Arch Surg, 1990; 125:200-205.
2. Basse P, Herberman RB, Nannmark U, Johansson BR, Hokland M, Wasserman K, Goldfarb RH. Accumulation of adoptively transferred adherent, lymphokine activated killer cells in murine metastasis. J Exp Med, 1991; 174: 479-498.
3. Bukowski RM, Sharfman W, Murty S, Rayman P, Tubbs R, Alexander J., et al. Clinical results and characterization of tumor-infiltrating lymphocytes with or without recombinant interleukin 2 in human metastatic renal cell carcinoma. Cancer Res, 1991; 51:4199-4205.
4. Cohen DJ, Lotze MT, Roberts JR, Rosenberg SA, Jaffe EB. The immunopathology of sequential tumor biopsies in patients treated with interleukin-2. Am J Pathol, 1987; 129:208-216.
5. Lienard D, Ewalenko P, Delmotte JJ, Renard N, Lejeune FJ. High-dose recombinant tumor necrosis factor α in combination with interferon γ and Melphalan in isolation perfusion of the limbs for melanoma and sarcoma. J Clin Oncol, 1992; 10:52-60.
6. Negrier S, Philip T, Stoter G, Fossa SD, Janssen S, Incom A, et al. Interleukin-2 with or without LAK cells in metastatic renal cell carcinoma: a report of a European multicenter study. Eur J Cancer clin Oncol, 1989; 25 (S3):21-28.
7. Nitta T, Sorto K, Okumura K, Steinman L. An analysis of T-cell receptor variable-region genes in tumor infiltrating lymphocytes within malignant tumors. Int J Cancer, 1991; 49:545-550.
8. Parmiani G. An explanation of the variable clinical response to interleukin 2 and LAK cells. Immunol. Today, 1990; 11:113-115.
9. Rodolfo M, Salvi C, Bassi C, Parmiani G. Adoptive immunotherapy of a mouse colon carcinoma with recombinant interleukin-2 alone or combined with lymphokine-activated killer cells or tumor-murine T lymphocytes. Cancer Immunol Immunother, 1990; 31:28-36.

10. Rosenberg SA, Packard BS, Aeberesold PM, Solomon D, Topalian SL, Toy ST, et al. Use of tumor-infiltrating lymphocytes and interleukin-2 in the immunotherapy of patients with metastatic melanoma. New Expl J Med, 1988; 319:1676-1680.
11. Sensi ML, Castelli C, Anichini A, Grossberger D, Mazzocchi A, Mortarini R, Parmiani G. Two autologous melanoma-specific and MHC-restricted human T cell clones with identical intra-tumor reactivity do not share the same TCR Vα and Vβ gene families. Melanoma Res, 1991; 1:261-271.
12. Van den Bruggen P, Traversari C, Chomez P, Lurquin C, De Plaen E, Van den Eynde B, Knuth A, Boon T. A gene encoding an antigen recognized by cytolytic T lymphocytes on human melanoma. Science, 1991; 254:1643-1647.

19

New Perspectives in Immunotherapy of Leukemia

Eva Lotzová[1]
The University of Texas
M.D. Anderson Cancer Center
Houston, Texas

I REEXAMINATION OF IMMUNOTHERAPY WITH INTERLEUKIN-2 ACTIVATED LYMPHOCYTES

The demonstration that interleukin-2 (IL-2) and IL-2- activated peripheral blood lymphocytes (lymphokine-activated killer (LAK) cells are effective in reducing numbers of established metastases in experimental animal systems have recently paved new avenues to cancer immunotherapy (1-3). Although IL-2 as a single agent or combined with other cytokines continues to be investigated for its optimal antitumor efficacy in some cancers, the enthusiasm for adoptive therapy with LAK cells has been tempered. This stems primarily from the observation that despite the predictions based on animal models, in human clinical trials, adoptively transferred IL-2 activated lymphocytes were only marginally more therapeutic than IL-2 alone in treatment of solid human cancers (2-4). However, this view may need revision in light of more current research data and more extensive experience with IL-2 as a single agent in cancer treatment.

One of the main concerns about treating patients with IL-2 alone relates to the high dose required for optimal activation of oncolytic lymphocytes. As extrapolated from in vitro data, the doses of IL-2 necessary to achieve optimal activation of oncolytic lymphocytes directed against fresh tumor cells (5-7) are too high to administer in vivo, because of the severe toxicity associated with systemic delivery of such high IL2 doses. On the other hand, other data demonstrate that treatment with clinically-tolerable doses of IL-2 has resulted in the in vivo generation of lymphocytes with oncolytic activity against tissue culture cell lines, but not against fresh tumor cells (8,9). Additional in vitro activation of these in vivo generated lymphocytes with higher doses of IL-2 was

[1] Recipient of the Florence Maude Thomas Research Professorship

necessary to induce oncolysis of fresh tumors. It has to be acknowledged that in vivo IL-2-activated lymphocytes may mediate antitumor activity by a pathway other than direct cytotoxicity, in particular, by production of cytokines with direct or indirect effects against tumor cells (3,10,11). However, it appears that for optimal cytokine production, activation with relatively high doses of IL-2 is also required.

Another concern regarding the efficacy and possible benefit of therapy with IL-2 relates to the immunocompetence of cancer patients. Some of these patients have impairments of the immune system due to previously administered chemotherapy or tumor-associated inhibitory factors, and this may be unable to respond to IL-2. These examples indicate that the in vivo manipulation of the immune system may not be uniformly feasible in all patients with cancer and that treatment with IL-2-activated lymphocytes and lower IL-2 doses could represent a therapeutic alternative. Before this therapy is considered, however, modifications are needed to optimize its therapeutic benefit.

One of the prerequisites for improving adoptive immunotherapy with activated lymphocytes is a better understanding of the tumor-lymphocyte relationship and subsequent selection of lymphocytes with the strongest antitumor activity. In this context, it is essential to analyze positive and negative aspects of the conventional LAK therapy and to more accurately define characteristics of effector cells involved in LAK activity.

In contrast to some presumptions, LAK is neither a new lymphocyte population nor a unique T lymphocyte subset (12,13); instead, LAK is a functional activity of IL-2-activated lymphocytes able to lyse highly-resistant tumor cell lines and fresh tumor cells in a short-term cytotoxicity assay. Accordingly, the designation "LAK cell" is inappropriate, since it specifies the cell rather than its function. Studies have unequivocally demonstrated that LAK activity is mediated primarily by natural killer (NK) cells; the major histocompatibility complex (MHC)-nonrestricted T cells were also implicated in LAK activity, although to a lesser degree (12,13). For consistency in terminology and for accuracy, LAK should be quoted generally as "LAK activity" and only after a definition of the particular NK- or T-cell oncolytic subset against a tumor target, this designation could be qualified as "NK-LAK" or "T-LAK".

Despite the demonstration that LAK activity is mediated primarily by NK cells, LAK cultures prepared by conventional methods and used for cancer treatment, contain mostly T cells and only a small proportion of NK cells. It is possible that this imbalance between highly cytotoxic (NK cells) versus poorly oncolytic or nononcolytic (T cells) populations could explain marginal effectiveness of LAK therapy in cancer treatment. It would be important to determine whether utilization of highly enriched NK cell populations could improve the efficacy of adoptive therapy. Conceivably, lower numbers of lymphocytes and lower doses of IL-2 could suffice if highly oncolytic lymphocytes are utilized for therapy. In addition to their direct cytotoxic effects mediated via LAK activity, NK cells may facilitate antitumor effects through cytokines they have been shown to produce (14).

The interest in LAK therapy and its improvement diminished further after it was reported that lymphocytes residing within the tumors (tumor-infiltrating lymphocytes, TIL) were significantly more efficacious after activation with IL-2 in vitro in treatment tumors than peripheral blood lymphocytes (PBL) with LAK activity (15). These reports, however, should be interpreted with caution. First, this evaluation of the effectiveness is based on results obtained from animal studies (15), and there is no clear evidence in man that points to the benefit of TIL over LAK therapy. Moreover, TIL therapy has been used primarily in patients with malignant melanoma or renal cell carcinoma (16,17), and the efficacy of this therapy in other cancers remains to be determined. Also, TIL therapy was designed to induce specific cytotoxic T cells (CTL) mediating lysis of autologous tumor cells. This aim has not been accomplished, since TIL obtained from most human tumors and cultured in the presence of IL2- are cytotoxic against autologous as well as against an array of allogeneic tumor cells (7,18,19). The one notable exception are studies with malignant melanoma, where a proportion of cultured TIL often displays specific autologous tumor lysis (20). Actually, similarly to LAK, IL-2-activated TIL cultures obtained from most tumors are composed of a variety of cell types, including $CD3^-$, $CD56^-$ NK cells and $CD3^+$. $CD56^{+/-}$ T cell subsets. In general, T cells predominate in TIL cultures, although antitumor cytotoxicity can be mediated by both NK and T cells (7,18,19). Which population actually mediates predominant antitumor activity may depend on the tumor type. Consequently, careful analysis of tumor sensitivity to NK or T cells, together with evaluation of the susceptibility of the tumors to cytokines produced by either of these populations, may be important in designing improved or more specific therapy with TIL.

Another concern regarding therapy with TIL relates to the limited availability of tumor specimens for propagation and activation of TIL in vitro and the failure of some TIL (but not autologous PBL) to respond to IL-2 (18). Taken together, the available data provide no direct evidence demonstrating the superiority of TIL therapy and suggest instead that this interesting approach might not be generally applicable or may need significant modifications.

II NEW ASPECTS OF LEUKEMIA THERAPY

Our earlier studies demonstrated that naive (unstimulated) as well as IL-2-activated NK cells mediated lysis of acute myelogenous leukemia (AML) blasts and inhibited clonogenic activity of more primitive leukemia progenitors (5,21). In contrast, MHC-nonrestricted T cells did not lyse or inhibit AML blasts. These observations indicated the sensitivity of AML blasts to NK cells and the potential of these cells in adoptive therapy of leukemia. Based on this possibility, we have initiated investigations of the pathways leading to optimal NK cell activation, enrichment, and growth.

Although various methods are available for NK cell enrichment, two methods achieve high NK cell purity without excessive manipulation of lymphocytes. The first rapidly generates NK cells by adherence to plastic, shortly after activation of PBL with IL-2. Because of their abilities to adhere to and to mediate IL-2-dependent oncolytic function, NK cells generated by this method were designated adherent lymphokine-activated killer cells (A-LAK) (6,22). It has been obvious from the beginning that A-LAK cells were highly oncolytic and surpassed conventional IL-2- activated lymphocytes in lysis of fresh tumor cells in vitro as well as in cell growth. Even AML blasts, which are resistant to lysis by LAK, were sensitive to lysis by A-LAK (22). This suggests that A-LAK, containing mainly NK cells, may be highly efficacious in leukemia therapy. However, therapeutic application of A-LAK cells may be limited. Since A-LAK cells have been studied predominantly in normal donors, and those data are not universally applicable to cancer patients, we investigated generation of A-LAK cells from PBL of AML patients in remission. Results of these studies demonstrated that A-LAK cultures were not always rich in NK cells. Whereas NK cells predominated in A-LAK cultures of some patients, T cells predominated in others (22). A-LAK cultures with a majority of NK cells were always oncolytically superior. Additionally, removal of NK cells (but not T cells) from A-LAK cultures resulted in a substantial impairment of antileukemia oncolysis. These data, together with the observation that A-LAK cultures from some AML patients did not proliferate optimally/or at all, indicate that it may be possible to generate oncolytic NK-A-LAK for adoptive therapy from all patients. However, according to the recent report it may be possible to enhance NK cell purity and proliferation in A-LAK cultures by adding mitogen-stimulated PBL, especially those with the CD4^{+} phenotype (23).

A second approach for obtaining highly purified NK cells is through collection of peripheral blood or bone marrow from patients receiving continuous IL-2 infusion. As we have shown previously, this therapy is not only well tolerated, but results in induction of growth of NK cells in PBL (>70% of all lymphocytes) and bone marrow (>60% of all lymphocytes) (24). These NK cells are cytotoxic for certain tumor cell lines, but require additional activation with higher doses of IL-2 for oncolysis of AML blasts. This activation and subsequent propagation of high numbers of NK cells for therapeutic use can be achieved in IL-2 cultures in vitro.

III SIGNALS INVOLVED IN INDUCTION OF ONCOLYTIC CELLS AGAINST RESISTANT LEUKEMIA

The IL-2 activated lymphocytes acquire lytic activity against most tumor cell lines that are resistant to lysis by nonactivated lymphocytes. The degree of sensitivity of fresh tumor cells to IL-2-activated lymphocytes may vary considerably, with some tumors displaying resistance to lysis. Furthermore, high lymphocyte-to-tumor cell ratios are required for lysis of most fresh tumor cells. Although NK-A-LAK cells are highly oncolytic, some AML blasts have exhibited resistance to this population. Therefore, we investigated possible approaches to overcoming the leukemia resistance.

Focusing on the potentiation of the cytotoxic mechanism, we found that leukemia resistance to lysis could be diminished after simultaneous activation of PBL with irradiated AML blasts and IL-2 (25). This approach resulted in reproducible induction of antileukemia-directed oncolytic cells from PBL of both normal donors and AML patients. Autologous as well as allogeneic AML blasts (with respect to prospective AML targets) were effective in triggering lysis of leukemia targets. To delineate the type of PBL (NK or T) responsive to AML/IL-2 signal, purified NK or T cells were cultured with AML blasts and IL-2. The leukemia-directed cytotoxicity of purified NK cells and T cells was significantly facilitated or induced by activation with AML and IL-2 (25), and growth of both of these populations was enhanced. Thus, it was possible to trigger the NK as well as T cell arm of oncolytic response against leukemia, and propagate these cells in vitro this may represent a new approach to efficient adoptive therapy. Since allogeneic AML blasts effectively induce cytotoxic cells, they may be practical in situations when the autologous AML blasts are either unavailable or are suppressive for autologous cytotoxic lymphocytes.

As we reported previously, bone marrow tissue can be used as an alternative source of highly oncolytic lymphocytes in vitro (26). As with PBL, the oncolytic effect of bone marrow lymphocytes against fresh leukemia can be enhanced by addition of AML blasts to IL-2 cultures (27). During these studies we also explored the efficacy of various cell lines in the generation of oncolytic cells from bone marrow and showed that it may be possible to induce antileukemia-directed NK cells or T cells, depending on the tumor used for stimulation. Although AML blasts and the Daudi and BSM cell lines all facilitated antileukemia activity of IL-2 activated bone marrow cells, cultures stimulated with Daudi, contained predominately NK cells, while those stimulated with BSM line and AML blasts were composed of T cells (28). The highest oncolysis of AML blasts was induced by stimulation of bone marrow cells with autologous AML blasts.

Fluorescence-activated cell sorting showed that NK cells were the major antileukemia effectors in Daudi-stimulated cultures, while both T cells and NK cells could mediate cytotoxic activity in AML-stimulated cultures (28). Interestingly, cultures activated with any of these stimulator cells retained high and stable levels of lytic activity for several weeks and promoted higher proliferation of lymphocytes than did cultures stimulated with IL-2 alone. Thus, by using different stimulator cells, it may be possible to preferentially induce activation of a particular lymphocyte subset and promote its growth.

IV CONCLUSIONS

Human adult nonlymphocytic leukemia is still a serious clinical problem, despite progress in cancer chemotherapy. Chemotherapy-induced remissions are usually transient, and chemotherapy alone rarely cures the disease. The rate of disease recurrence is lower following allogeneic bone marrow transplantation, which has been used as an alternative treatment for AML. However, because of histocompatibility restraints, allogeneic bone marrow transplantation may not be available to majority of AML patients, necessitating the development of new therapeutic modalities for this disease.

Although immunotherapeutic trials with IL-2 and IL-2- activated lymphocytes have been mainly used in patients with solid cancers, these approaches appear to be applicable to treatment of leukemia. However, therapeutic benefits of adoptive immuno therapy and IL-2 in this disease remain to be determined. At least two pieces of evidence indicate that lymphocytes, especially NK cells, can mediate antileukemia effect. First, the benefit of bone marrow transplantation in the control of leukemia recurrence appears to be due to the effect of the immuno-competent cells (present in the bone marrow graft) against leulemic cells resistent to chemotherapy or radiation therapy. The immune effector cells involved in this phenomenon are likely NK cells, although a role for various T cell subsets has not been excluded. Second, our data and those of others demonstrate the susceptibility of leukemic blasts to lysis by NK cells, and under certain condition, by T cells. This information indicates that application of oncolytic lymphocytes to the treatment of leukemia is feasible and may be beneficial.

Acknowledgement

The work cited from this laboratory was supported by Grant CA 39632 from the National Cancer Institute and by the Pawelek fund.

References

1. Lafraniere, R. and Rosenberg, S.A. Successful immunotherapy of murine experimental hepatic metastases with lymphokine-activated killer cells and recombinant interleukin-2. Cancer Res. 1985; 45:3735-41.
2. Rosenberg, S.A., Lotze, M.T., Yang, J.C., Aebersold, P.M., Linehan, W.M., Seipp, C.A. and White, D.E. Experience with the use of high dose interleukin-2 in the treatment of 652 patients with cancer. Ann. Surg. 1989; 210:474-84.
3. Rosenberg, S.A. The immunotherapy and gene therapy of cancer. J. Clin. Oncol. 1992; 10:180-99.
4. Bradley, E.C., Louie, A.C., Paradise, C.M., Carlin, D.A., Bleyl, K.L., Groves, E.S. and Rudolph, A.R. Antitumor response in patients with metastatic renal cell carcinoma is dependent upon regimen intensity. Proc. ASCO 1989; 8:133 (abstract).
5. Lotzová, E., Savary, C.A. and Herberman, R.B. Induction of NK cell activity against fresh human leukemia in culture with interleukin-2. J. Immunol. 1987; 138:2718-27.
6. Melder, R.J., Whiteside, T.L., Vujanovic, N.L., Hiserodt, J.C. and Herberman, R.B. Human adherent lymphokine-activated killer (A-LAK) cells: A new approach to generating antitumor effectors for adoptive immunotherapy. Cancer Res. 1988; 48:3461-9.

7. Itoh, K. and Balch, C.M. Cell-mediated cytotoxicity against fresh solid tumor cells. Regulation by soluble mediators. In: Lotzová, E., Herberman, R.B., (Eds.) Interleukin-2 and killer cells in cancer. Boca Raton, FL: CRC Press. 1989: 65-87.
8. Lotzová, E., Savary, C.A., Schachner, J.R., Huh, J.O. and McCredie, K. Generation of cytotoxic NK cells in peripheral blood and bone marrow of patients with acute myelogenous leukemia after continuous infusion with recombinant interleukin-2 Am. J. Hematol. 1991; 37:88-99.
9. Hank, J.A., Kohler, P.C., Weil-Hillman, G., Rosenthal, N., Moore, K.H., Storer, B., Minkoff, D., Bradshaw, J., Bechhofer, R. and Sondel, P.M. In vivo induction of the lymphokine-activated killer phenomenon: Interleukin 2-dependent human non-major histocompatibility complex-restricted cytotoxicity generated in vivo during administration of human recombinant interleukin-2. Cancer Res. 1988; 48:1965-71.
10. Ortaldo, J.R., Ranson, J.R., Sayers, T.J. and Herberman, R.B. Analysis of cytostatic/cytotoxic lymphokines: Relationship of natural killer cytotoxic factor to recombinant lymphotoxin, recombinant tumor necrosis factor and leukoregulin. J. Immunol. 1986; 137:2857-63.
11. Stotter, H., Wiebke, E.A., Tomita, S., Belldegrun, A., Topalian, S., Rosenberg, S.A. and Lotze, M.T. Cytokines alter target cell susceptibility to lysis. II. Evaluation of tumor infiltrating lymphocytes. J. Immunol. 1989; 142:1767-73.
12. Lotzová, E. and Herberman, R.B. Reassessment of LAK phenomenology: A Review. Nat. Immun. Cell Growth Regul. 1987; 6:109-15.
13. Herberman, R.B., Hiserodt, J., Vujanovic, N., Balch, C., Lotzová, E., Bolhuis, R., Golub, S., Lanier, L.L., Phillips, J.H., Riccardi, C., Ritz, J., Santoni, A., Schmidt, R.E. and Uchida, A. Lymphokine-activated killer cell activity. Immunol. Today 1987; 8:178-81.
14. Scala, G., Djeu, J.Y., Allavena, P., Kasahara, T., Ortaldo, J.R,, Herberman, R.B. and Oppenheim, J.J. Cytokine secretion and noncytotoxic functions of human large granular lymphocytes. In: Lotzová, E., Herberman, R.B. (Eds.). Immunobiology of Natural Killer Cells. Volume 2. Boca Raton, FL: CRC Press. 1986;133-44.
15. Rosenberg, S.A., Spiess, P. and Lafreniere, R. A new approach to adoptive immunotherapy of cancer with tumor infiltrating lymphocytes. Science 1986; 233:1318-21.
16. Rosenberg, S.A., Packard, B.S., Aebersold, P.M., Solomon, D., Topalian, S.L., Toy, S.T., Simon, P., Lotze, M.T., Yang, J.C. and Seipp, C.A. Use of tumor-infiltrating lymphocytes and interleukin-2 in the immunotherapy of patients with metastatic melanoma. Special Report. N. Engl. J. Med. 1988; 319:1676-80.
17. Topalian, S.L., Solomon, D., Avis, F.P., Chang, A.E., Freerksen, D.L., Linegan, W.M., Lotze, M.T., Robertson, C.N., Seipp, C.A., Simon, C.G. and Rosenberg, S.A. Immunotherapy of patients with advanced cancer using tumor-infiltrating lymphocytes and recombinant interleukin-2: A Pilot Study. J. Clin. Oncol. 1988; 6:839-53.

18. Lotzová, E. and Savary, C.A. Growth kinetics, function and characterization of lymphocytes infiltrating ovarian tumors. In Interleukin-2 and Killer Cells in Cancer. Lotzová, E., Herberman, R.B. (Eds.) Boca Raton, FL: CRC Press, Inc. 1990: 153-61.
19. Heo, D.S., Whiteside, T.L., Johnson, J.T., Chen, K., Barnes, E.L. and Herberman, R.B. Long term interleukin-2-dependent growth and cytotoxic activity of tumor-infiltrating lymphocytes (TIL) from human squamous cell carcinomas of the head and neck. Cancer Res. 1987; 47:6353-62.
20. Topalian, S.L., Muul, L.M., Solomon, D. and Rosenberg, S.A. Expansion of human tumor infiltrating lymphocytes for use in immunotherapy-trials. J. Immunol. Methods 1987; 102:127-41.
21. Lotzová, E., Savary, C.A. and Herberman, R.B. Inhibition of clonogenic growth of fresh leukemia cells by unstimulated and IL-2 stimulated NK cells of normal donors. Leuk. Res. 1987; 11:1059-66.
22. Lotzová, E., Savary, C.A., Totpal, K., Schachner, J., Lichtiger, B., McCredie, K.B. and Freireich, E.J. Highly oncolytic adherent lymphocytes: therapeutic relevance for leukemia. Leuk. Res. 1991; 15:245-54.
23. Rabinowich, H., Sedlmayr, P., Herberman, R.B. and Whiteside, T.L. Increased proliferation, lytic activity, and purity of human natural killer cells cocultured with mitogen-activated feeder cells. Cell. Immunol. 1991; 135:454-70.
24. Lotzová, E., Savary, C.A., Schachner, J.R., Huh, J.O. and McCredie, K. Generation of cytotoxic NK cells in peripheral blood and bone marrow of patients with acute myelogenous leukemia after continuous infusion with recombinant interleukin-2. Am. J. Hematol. 1991; 37:88-99.
25. Lotzová, E., Savary, C.A. and Cristoforoni, P.M. Triggers involved in NK cells and T lymphocytes oncolysis. In: Lotzová, E., Herberman, R.B. (Eds.) NK Cell Mediated Cytotoxicity: Receptors, Signalling and Mechanisms. CRC Press, Inc., Boca Raton, FL: CRC Press. 1992; 423-29.
26. Lotzová, E. and Savary, C.A. Generation of NK cell activity from human bone marrow. J. Immunol. 1987; 139:279-84.
27. Fuchshuber, P.R. and Lotzová, E. Feeder cells enhance oncolytic and proliferative activity of long-term human bone marrow interleukin-2 cultures and induce different lymphocyte subsets. Cancer Immunol. Immunother. 1991; 33:15-20.
28. Fuchshuber, P.R., Lotzová, E. and Savary, C.A. Generation of MHC-nonrestricted and restricted oncolytic subsets from human bone marrow. Cell. Immunol. 1992; 139:30-43.

20

The Role of Interferons in the Therapy of Melanoma

John M. Kirkwood
Pittsburgh Cancer Institute
and University of Pittsburgh School of Medicine
Pittsburgh, Pennsylvania

I INTRODUCTION

The medical therapy of melanoma has not improved significantly for the past 30 years, despite the experimental application of a multitude of new cytotoxic drugs. Dacarbazine is the single cytotoxic drug approved for the indication of melanoma by the US FDA. Dacarbazine has consistently achieved response rates in the range of 20%, but has not yet demonstrated improved survival in patients with either inoperable metastatic melanoma, or patients with resectable disease at high risk of recurrence. Thus, biological and immunological therapies for melanoma have received increasing emphasis, which dating from the first half of this century and the exploration of a range of microbial immunostimulants that are still relevant, and the more recent pursuit of recombinant DNA-produced or highly purified cytokines. A considerable body of evidence suggests that melanoma may be susceptible to a variety of mediators of the host immune system. The clinical occurrence of spontaneous regression and paraneoplastic syndromes of depigmentation associated with prolonged survival of melanoma have represented a topic of investigation by our group for more than a decade (1-3). These clinical observations, together with the frequent histological appearance of host lymphocytes in the bed of the primary tumor, and correlation with prognosis of the primary tumor (4) and the demonstration of humoral and cellular immunity to melanoma in certain patients with melanoma in multiple laboratories, suggest that the immune response may be harnessed to improve the therapy of human melanoma. (5-8)

II INTERFERON ALPHA

The clinical experience with the interferons (IFNs) in melanoma is the topic I have been asked to discuss at this Symposium. Interferon alpha is the most extensively studied of biological response modifiers produced by recombinant DNA technology to date. Interferon alpha exerts multiple, pleiotropic effects that may be divided into categories that are: (a) directly mediated against the tumor and (b) indirectly mediated through the host immune system, and may be subcategorized according to their efferent and afferent impact. (9-11) Which of these activities is principally responsible for the clinical benefits reported from therapy of melanoma with IFN remains uncertain.

The anti-proliferative activity of interferon alpha is believed to be a consequence of its activities mediated through specific receptors, resulting in the induction of interferon response elements at cytoplasmic and nuclear levels that are beyond the scope of this discussion (12,13). The down-regulation of protein synthesis in IFN-treated tumor cells is associated with the appearance of novel oligo 2',5' adenylate synthetase studied in some detail, both in host effector and tumor cell populations. (14,15) To date, neither the induction of this enzyme, nor the inhibition of tumor cell proliferation by IFN, have shown useful correlation with therapeutic effects of the IFNs in clinical trials against melanoma.

Interferon alpha modulates the expression of tumor-associated antigens as well as MHC Class I tumor cell surface antigens, and these effects may account for enhanced host immune response, through afferent, efferent, or composite mechanisms of antitumor action (16-22).

Clinical trials of IFN alfa-2 of several industrially produced types have demonstrated response rates of 16-22% on average, approximating the response rates reported for dacarbazine, the single chemotherapeutic agent approved for the indication of metastatic melanoma. The relatively low frequency of objective clinical response observed among melanoma and other solid tumors treated with IFNs as single agents, or in combinations, has posed an obstacle to efforts to tease out primary mechanisms of antitumor activity. Despite the considerable evidence for activation of natural killer (NK) cell function by IFN in vitro, clinical studies have not established a meaningful correlation between NK activation and antitumor activity of IFN in subjects with a range of neoplasms including melanoma. The modulation of potentially tumor-specific host immune T cell and B cell mechanisms in vitro is even more difficult to approach in the context of clinical trials of the IFNs against melanoma. To date, no compelling evidence is available to establish a causal relation between any immunomodulatory effects and the therapeutic benefits of IFN alpha in patients with melanoma (23).

The preclinical studies of Belardelli et al. (24,25) and Gresser et al. (26) provide suggestive evidence that IFN achieves its antitumor effects in vivo through the host (immune) system, and not through direct tumor inhibition. These investigators have demonstrated successful treatment of murine tumors that are resistant to the direct in vitro effects of IFN. Clinical studies of IFN alpha have demonstrated a correlation of clinical response to the bulk and sites of metastatic tumor in patients with melanoma; patients with limited tumor burdens, and disease limited to cutaneous, soft tissue, lymphatic, and/or pulmonary sites show the highest response rates (23,27,28). Few clinical trials have attempted to define the optimal dose/route/schedule for IFN alpha, using rigorous randomized designs and accruing large enough numbers of subjects to yield meaningful data. Those trials conducted to date in solid tumors have suggested that doses approximating maximum tolerable levels may achieve higher response rates (29,30). The analysis of correlations between dosage regimens, clinical antitumor responses, and immunomodulatory effects of IFN alpha upon the circulating pools of blood T cells in treated subjects (rises in the CD4 "T helper" phenotypic population in particular, and falls in the CD8 "T suppressor/cytotoxic" subpopulations) spur further interest in the specific host immunoregulatory effects of IFN, in the treatment of solid tumors (30).

The wide range of candidate mechanisms represents one of the difficulties encountered in the analysis of potential mechanisms of action. Few studies have attempted to assess the non-MHC-restricted (NK-, and monocyte-mediated) non tumor-specific and MHC-restricted tumor-specific (T cell) and antibody effector mechanisms simultaneously so as to allow assessment of their relative correlation to clinical outcomes. Studies that we recently completed (31,32), encompassing a particularly complete range of immunomodulatory functions and phenotypic parameters, suggested that IFN-mediated effects upon the distribution of circulating T cells correlate best with antitumor effects of a sequenced combination of IFNs gamma and alpha in metastatic renal cell carcinoma (31,32). Given the indirect evidence for the role of T cells in murine experimental (33,34) as well as human clinical trials, the T cell is now the focus of analysis in current trials.

III RECOMBINANT INTERFERON ALFA-2

The advent of highly purified recombinant human interferon alpha permitted careful investigations of a range of dosage regimens and durations of treatment using dosages higher than previously possible for metastatic melanoma (35-44). Initial phase I/II trials of interferon alfa-2b conducted by our group while at Yale University enrolled 23 melanoma patients for treatment with interferon alfa-2b at dosages of 3-100 M units intramuscularly or intravenously daily for four weeks. Four objective responses were noted, including two complete and two partial responses. Experience in our subsequent studies, as well as those of other groups using high dosages of IFN alfa-2, showed that complete responses may be rendered disease-free for many years, suggesting the induction of durable host immunity to tumor (40).

Dosages from 12 Mu/M^2 to 50 Mu/M^2, as well as 50 Mu/M^2 plus cimetidine, gave comparable results at the Mayo Clinic with recombinant interferon alfa-2a in 96 patients with metastatic melanoma (45). A large experience has been reported with recombinant interferon alfa-2b given at 10 Mu/M^2 administered subcutaneously three times a week, with overall response rates and durable complete response rates comparable to higher dose IV, and IM experiences (42). These are summarized in Table 1.

Table 1
Recombinant Interferon Alfa-2b in Metastatic Melanoma

Author/Year	Dose/ Schedule	Entered/ Evaluable	CR/PR	(%)	Duration (Mos.)*
Kirkwood 1985(40	10-100/M^2/QD	23/23	3/2	22	36
Dorval 1986(41)	10/M^2/TIW	22/22	2/4	27	4.6
Robinson 1986(42)	30/M^{2}TIW	51/40	4/6	25	6
Sertoli 1989(43)	10/M^2/TIW	21/21	0/3	14	12.5
TOTAL		117/106	9/15		22.6

*Duration expressed as median duration of response

While these and other trials indicate no clear dose-response curve, dose schedule appears to be important in the therapy of melanoma and other solid tumors. Daily or alternate-daily (QOD) treatment without interruption are associated with higher response rates. Patients with minimal disease, including lymph node and subcutaneous sites, tend to be responsive to interferon alpha; these patients are more responsive to therapy with cytotoxic drugs as well. Patients with lung and liver metastases were noted to respond to interferon therapy, and neither performance status nor prior chemotherapy appears to influence the likelihood of response to interferon alpha (44). Attempts to improve upon these responses have led to incorporation of other agents into trials with interferon alpha and to evaluation of alternative routes and schedules of administration.

IV COMBINED MODALITY BIOLOGIC/CHEMOTHERAPEUTIC APPROACHES TO MELANOMA

Trials of combinations of interferon alpha with chemotherapy have employed dacarbazine (DTIC), low dose cyclophosphamide, and difluormethylornithine (DFMO) (36,45,46). Dacarbazine is the single FDA-approved cytotoxic drug available for the indication of melanoma. Combinations of interferon alpha and dacarbazine have yielded conflicting results. Australian (44) and South African (47) results are encouraging, while one of the largest multicenter studies of these agents in the US and Europe (48) failed to demonstrate evidence of synergism and was aborted before full accrual could be completed. A major difference between the negative and positive trials was the dosage and schedule of therapy employed, where the positive trial employed daily IV therapy for several weeks with IFN alfa-2 before chemotherapy, while the negative trial utilized concurrent administration of IFN at a lower dosage subcutaneously on an alternate daily schedule. Additionally, the species of IFN alfa-2 differed among these trials. The Eastern Cooperative Oncology Group is testing dacarbazine combinations with IFN alfa-2 and with tamoxifen in a 2 x 2 design that will definitively assess the role of these preliminarily encouraging combined modality results (36,49). Combinations with cyclophosphamide and DFMO have been less rewarding to date.

A. Combined Biological Therapy with Interleukin-2 and Other Biologics

Combinations of interferon alpha and other cytokines such as interleukin-2, tumor necrosis factor, and interferon gamma have been pursued on the basis of known immunomodulatory pathways of these cytokines. The use of combined modality therapy with interferon alpha has been of recent interest, based on the suggestion that interleukin-2 mediated effects on the host response tumor might be augmented if interferon could induce tumor cell surface antigen expression to higher levels. Early trials of interferon alfa-2 in combination with interleukin-2 suggested increased responses in melanoma and renal cell carcinoma (50-52), but toxicity of the combination was significant, and the level of ≈25% clinical activity for this regimen may not represent synergism, but subadditive activity. Given the observation that interferon response rates do not appear to be compromised by prior patient therapy, one may require synergism rather than additive effects to argue for combinations unless a survival advantage is evident.

Combinations of interferon alpha with cytotoxic antibodies were reported to improve <u>in vivo</u> localization of radiolabelled murine monoclonal antibody 96.5 administered systemically to melanoma patients in conjunction with interferon alpha (partially purified leukocyte HuIFN alpha) (53). Improved antibody localization to tumor may result both from augmented tumor cell surface antigen expression and from protracted host clearance of the monoclonal antibody. Controlled studies will be required to confirm and develop these observations, and to establish whether a diagnostic or therapeutic advantage is available from this approach. The role of cytotoxic and hormonal agents, as well as radiotherapy in combination with alpha interferon, require further exploration in melanoma.

V ADJUVANT USAGE OF INTERFERONS IN HIGH-RISK RESECTED MELANOMA

As previously noted, moderate or high doses of interferon alfa-2 doses appear to be associated with the highest frequencies of clinical response in metastatic melanoma, and clinical response appears to vary inversely with the size of tumor masses. This evidence has prompted the study of interferon alfa-2b as an adjuvant to surgery, given at maximum tolerable dosage for a period of one year to patients with resected regional lymph node metastasis or high risk deep stage I melanoma in the Eastern Cooperative Oncology Group (EST 1684, JM Kirkwood, P.I.). The North Central Oncology Group has also tested the efficacy of moderately high doses of interferon alfa-2a, but for shorter periods of only three months (E. Creagan, PI). These adjuvant protocols of investigation have reached their projected accrual requirements and early results may be anticipated within the next year or two. The World Health Organization has more recently undertaken an investigation of much lower dosages of interferon alfa-2a in a protocol of therapy that is intended to span at least 2 years, compared to an observation control.

The ECOG recently initiated a study of interferon alfa-2b to extend the encouraging preliminary trial results available from EST 1684, the controlled trial of IFN alfa-2b at maximum tolerable dosage administered for 1 year. The new trial includes an arm which duplicates the treatment arm of EST 1684, with IFN alfa-2b administered for 1 year at maximal tolerable dosage. This new trial tests a lower dosage of 3Mu/d selected as an immunobiologically active dosage which is the highest dosage compatible with long-term (at least 2 year) therapy, in a trial design that preserves a control arm of careful observation. (Table 2).

The evidence of antitumor activity which may be emerging in both the advanced disease setting, with IFN-dacarbazine combinations, and in the adjuvant disease setting, where several ongoing trials may have relapse preventative and survival prolonging outcomes, now afford us with an opportunity to evaluate the mechanism of IFN alfa-2. Preliminary indications point toward immunomodulatory activity upon the T cell, which could be envisioned as the outcome of effects upon tumor cell surface histocompatibility markers, tumor-associated antigens, or alternatively upon the host T cell response mechanism apart from tumor cell antigen induction. These studies have, in fact, recently been initiated, and will require close interdisciplinary collaboration of surgical, medical, and immunological laboratory investigators to obtain the tumor samples available at lymphadenectomy for patients who are to be candidates for the adjuvant trial. Only by analyzing the impact of IFN upon host immune response against a patient's autologous tumor cells harvested for lymph node metastasis at surgery may we determine the effects of this therapy upon specific immune response to melanoma.

Table 2
Interferon Applications Under Current Investigation: Postoperative Adjuvant Therapy

Cooperative Group	Agent	Eligible Subjects	Determine Role of:
ECOG II	IFN alfa-2b	Stage I (deep)	Intensive-dose/prolonged Stage IV --> SC regimen
	IFN alfa-2b	Stage I (deep) Stage II	Intensive-dose vs. Low-dose prolonged SC
NCCTG	IFN alfa-2a	Stage I (deep) Stage II	Moderate-dose/limited IM regimen
WHO	IFN alfa-2a	Stage II	Low-dose Prolonged SC regimen
SWOG	IFN gamma	Stage I* Stage II	Optimal immunological active dose; limited course

* >1.50 mm intermediate depth (cf deep >4.0 mm depth)

VI CONCLUSION

Reports indicate that interferon alpha as a single agent induces clinical remissions comparable in frequency and quality with the best known single chemotherapeutic agents for melanoma to date. Additionally, a number of complete remissions following therapy with interferon alpha have shown remarkable durability. The exact mechanisms of interferon-induced tumor regression in humans have yet to be demonstrated unequivocally. A greater understanding of these mechanisms may result in the adoption of more rationally designed and effective treatment designs. Large randomized controlled phase III trials of maximally tolerable dosages of interferon alpha are underway in the Intergroup Cooperative Trials mechanism. These adjuvant trials will establish whether interferon alpha as a single agent is able to reduce or retard relapse among subjects with high relapse risk following resection of deep primary melanoma lesions or regional lymph node metastasis.

The potential activation of several effector mechanisms in IFN treated subjects is under study in the context of the Intergroup Phase III adjuvant IFN alfa-2 trial (EST-1690/INT-2055), where the relevance of these mechanisms to disease course, as well as to IFN alfa-2 treatment dosage, is to be studied (EST 2690 Laboratory Corollary Trial, JM Kirkwood, PI). This trial in the adjuvant setting may provide a firmer foundation for the subsequent development of IFN alpha in a range of solid tumors.

References

1. Wagoner, M.D., Albert, D.M., Lerner, A.B., Kirkwood, J., Forget, B.M. and Nordlund, J.J. New observations on vitiligo and ocular disease. Am. J. Ophthalmol. 1983; 96:16.
2. Nordlund, J.J., Kirkwood, J.M. and Forget, B.M. Vitiligo in patients with metastatic melanoma: A good prognostic sign. J. Am. Acad. Dermatol. 1983; 9(5):689-696.
3. Kirkwood, J.M., Nordlund, J.J., Lerner, A.B., et al. Favorable prognosis of melanoma associated with hypopigmentation (HYP) in a randomized adjuvant trial comparing DTIC-BCG (DB) vs monobenzyl ether of hydroquininone (HQ) vs null (NL) treatment. Proc. Am. Soc. Clin. Oncol. 1985; 4:149-#C-581 (Abstract).
4. van Vreeswijk, H., Ruiter, D.J., Brocker, E.B., Welvaart, K. and Ferrone, S. Differential expression of HLA-DR, DQ, and DP antigens in primary and metastatic melanoma. J. Invest. Dermatol. 1988; 90(5):755-760.
5. Kirkwood, J.M. and Ernstoff, M.S. Potential applications of the interferons in oncology: Lessons drawn from studies of human melanoma. Semin. Oncol. 1986; 13(3):48-56.
6. Vlock, D.R., DerSimonian, R. and Kirkwood, J.M. Prognostic role of antibody reactivity to melanoma. J. Clin. Invest. 1986; 77(4):1116-1121.
7. Knuth, A., Danowski, B., Oettgen, H.F. and Old, L.J. T-cell mediated cytotoxicity against autologous malignant melanoma: analysis with interleukin 2-dependent T-cell cultures. Proc. Natl. Acad. Sci. USA 1984; 81:3511-3515.
8. Anichini, A., Fossati, G. and Parmiani, G. Clonal analysis of cytotoxic T-lymphocyte response to autologous human metastatic melanoma. Int. J. Cancer 1985; 35:683-689.
9. Stewart, T. The interferon system, New York:New York Academic Press, 1979.
10. Gupta, V., Singh, S.V., Ahmad, H., Medh, R.D. and Awasthi, Y.C. Glutathione and glutathione S-transferases in a human plasma cell line resistant to melphalan. Biochem. Pharmacol. 1989; 38(12):1993-2000.

11. Herberman, R.B. Natural cell-mediated immunity against tumors, New York:Academic Press, 1980.

12. Levy, D.E., Kessler, D.S., Pine, R. and Darnell, J.E.,Jr. Cytoplasmic activation of ISGF3, the positive receptor of interferon-α-stimulated transcription, reconstituted in vitro. Genes Dev. 1989; 3:1362-1371.

13. Schindler, C., Shuai, K., Prezioso, V.R., Darnell, Jr.,J.E. Interferon-dependent tyrosine phosphorylation of a latent cytoplasmic transcription factor. Science. 1992; 257:809-813.

14. Williams, B.R.G. Mechanisms of interferon actions. In: Sikora (Ed.) Interferon in cancer. 1983; New York: Plenum Press.

15. St.Laurent, G., Yoshie, U. and Floyd-Smith, G. Interferon action: Two (2'5') A (A)n synthetase specified by distinct mRNAs in Erhlich ascites tumor cells treated with interferon. Cell. 1983; 33:95.

16. Murray, J.L., Stuckey, S.E., Pillow, J.K., Rosenblum, M.G. and Gutterman, J.U. Differential in vitro effects of recombinant α-interferon and recombinant gamma-interferon alone or in combination on the expression of melanoma-associated surface antigens. J. Biol. Response Modif. 1988; 7:152-161.

17. Giacomini, P., Aguzzi, A., Pestka, S., Fisher, P.B. and Ferrone, S. Modulation by recombinant DNA leukocyte (α) and fibroblast (β) interferons of the expression and shedding of HLA- and tumor-associated antigens by human melanoma cells. J. Immunol. 1984; 133(3):1649-1655.

18. Greiner, J.W., Schlom, J., Pestka, S., et al. Modulation of tumor associated antigen expression and shedding by recombinant human leukocyte and fibroblast interferons. Pharmac. Ther. 1985; 31:209-236.

19. Giacomini, P., Aguzzi, A. and Ferrone, S. Differential susceptibility to modulation by recombinant immune interferon of HLA-DR and -DQ antigens synthesized by melanoma COLO 38 cells. Hybridoma 1986; 5(4):277-288.

20. Giacomini, P., Aguzzi, A., Tecce, R., Fisher, P.B. and Ferrone, S. A third polypeptide associated with heavy and light chain subunits of class I HLA antigens in immune interferon-treated human melanoma cells. Eur. J. Immunol. 1985; 15:946-951.

21. Graham, G.M., Guarini, L., Moulton, T.A., et al. Potentiation of growth suppression and modulation of the antigenic phenotype in human melanoma cells by the combination of recombinant human fibroblast and immune interferons. Cancer Immunol. Immunother. 1991; 32:382-390.

22. Giacomini, P., Imberti, L., Aguzzi, A., Fisher, P.B., Trinchieri, G. and Ferrone, S. Immunochemical analysis of the modulation of human melanoma-associated antigens by DNA recombinant immune interferon. J. Immunol. 1985; 135(4):2887-2894.

23. Kirkwood, J.M. and Ernstoff, M.S. Cutaneous melanoma. In: DeVita, V.T.Jr., Hellman, S., Rosenberg, S.A. (Eds.) Biologic Therapy of Cancer. 1991; Philadelphia: J.B. Lippincott Co.: p. 311-333.

24. Belardelli, F., Gresser, I., Maury, C. and Maunoury, M.T. Antitumor effects of interferon in mice injected with interferon-sensitive and interferon-resistant Friend leukemia cells. I. Int. J. Cancer 1982; 30:813-820.

25. Belardelli, F., Gresser, I., Maury, C. and Maunoury, M.T. Antitumor effects of interferon in mice injected with interferon-sensitive and interferon-resistant Friend leukemia cells. II. Role of host mechanisms. Int. J. Cancer 1982; 30:821-825.

26. Gresser, I., Maury, C., Carnaud, C., DeMaeyer, D., Maunoury, M.T. and Belardelli, F. Anti-tumor effects of interferon in mice injected with interferon-sensitive and interferon-resistant friend erythroleukemia cells. VIII. Role of the immune system in the inhibition of visceral metastases. Int. J. Cancer 1990; 46(3):468-474.

27. Kirkwood, J.M. and Ernstoff, M.S. The role of interferons in the therapy of melanoma. In: Hersey, P. (Ed.) Biological Agents in the Treatment of Cancer: Proceedings of the International Conference held in Newcastle, September 4-7, 1990. 1990; Newcastle, NSW: Goverment Printing: p. 303-320.

28. Kirkwood, J.M. and Ernstoff, M.S. Role of interferon in the therapy of melanoma. J. Invest. Dermatol. 1990; 95:180S-185S.

29. Kirkwood, J.M., Harris, J.E., Vera, R., et al. A randomized study of low and high doses of leukocyte alpha-interferon in metastatic renal cell carcinoma: The American Cancer Society Collaborative Trial. Cancer Res. 11985; 45(2):863-871.

30. Silver, H.K., Connors, J.M., Kong, S., Karim, K.A. and Spinelli, J.J. Survival, response and immune effects in a prospectively randomized study of dose strategy for alpha-N1 interferon. Br. J. Cancer 1988; 58(6):783-787.

31. Ernstoff, M.S., Gooding, W., Nair, S., et al. Immunological effects of treatment with sequential administration of recombinant interferon gamma and alpha in patients with metastatic renal cell carcinoma during a phase I trial. Cancer Res. 1992; 52:851-856.

32. Ernstoff, M.S., Nair, S., Bahnson, R.B., et al. Sequential administration of recombinant DNA-produced interferons: A phase IA trial of combination rIFN gamma and rIFN alpha in patients with metastatic renal cell carcinoma. J. Clin. Oncol. 1990; 8:1637-1649.

33. Sayers, T.J., Wiltrout, T.A., McCormick, K., Husted, C. and Wiltrout, R.H. Antitumor effects of alpha-interferon and gamma-interferon on a murine renal cancer (Renca) in vitro and in vivo. Cancer Res. 1990; 50:5414-5420.

34. Markovic, S.N. and Murasko, D.M. Role of natural killer and T cells in interferon-induced inhibition of spontaneous metastases of the B16-F10L murine melanoma. Cancer Res. 1991; 51:1124-1128.
35. Kirkwood, J.M. and Ernstoff, M.S. Melanoma: therapeutic options with recombinant interferons. Semin. Oncol. 1985; 12(4-5):7-12.
36. Kirkwood, J.M., Ernstoff, M.S., Guiliano, A., et al. Clinical trials of interferon alfa-2B (Intron A, alpha-IFN) in melanoma: review of phase I, II, and III studies. Proc. XIV Int. Cancer Cong.,Budapest August:1986.
37. Coates, A., Rallings, M., Hersey, P. and Swanson, C. Phase II study of recombinant alpha 2-interferon in advanced malignant melanoma. J. Interferon Res. 1986; 6:1-4.
38. Elsasser-Beile, U. and Drews, H. Interferon in the treatment of malignant melanoma. Results of clinical studies. Fortschr. Med. 1987; 105(21):401.
39. Steiner, A., Wolf, C. and Pehamberger, H. Comparison of the effects of three different treatment regimens of recombinant interferons (r-IFN alpha, r-IFN gamma, and r-IFN-alpha + cimetidine) in disseminated malignant melanoma. J. Cancer Res. Clin. Oncol. 1987; 113:459-465.
40. Kirkwood, J.M., Ernstoff, M.S., Davis, C.A., Reiss, M., Ferraresi, R. and Rudnick, S.A. Comparison of intramuscular and intravenous recombinant alpha-2 interferon in melanoma and other cancers. Ann. Intern. Med. 1985; 103(1):32-36.
41. Dorval, T., Palangie, T., Jouve, M., et al. Clinical phase II trial of recombinant DNA interferon (interferon alfa 2b) in patients with metastatic malignant melanoma. Cancer 1986; 58(2):215-218.
42. Robinson, W.A., Mughal, T.I., Thomas, M.R., Johnson, M. and Spiegel, R.J. Treatment of metastatic malignant melanoma with recombinant interferon alpha-2. Immunobiol. 1986; 172:275-282.
43. Sertoli, M.R., Bernengo, M.G., Ardizzoni, A., et al. Phase II trial of recombinant alpha-2b interferon in the treatment of metastatic skin melanoma. Oncology. 1989; 46(2):96-98.
44. McLeod, G.R.C., Thomson, D.B. and Hersey, P. Recombinant interferon alfa-2a in advanced malignant melanoma: A phase I-II study in combination with DTIC. Int J Cancer 1987; Suppl.1:31-35.
45. Croghan, M.K., Booth, A. and Meyskens, F.L.,Jr. A phase I trial of recombinant interferon-α and α-difluoromethylornithine in metastatic melanoma. J. Biol. Response Modif. 1988; 7(4):409-415.
46. Abdi, E.A., McPherson, T.A. and Tan, Y.H. Combination of fibroblast interferon (HuIFN beta), carboxamide (DTIC), and cimetidine for advanced malignant melanoma. J. Biol. Response. 1986; 5:423-428.

47. Vorobiof, D.A., Sarli, R. and Falkson, G. Combination chemotherapy with dacarbazine and vindesine in the treatment of metastatic malignant melanoma. Cancer Treat. Rep. 1986; 70:927-928.

48. Kirkwood, J.M., Ernstoff, M.S., Guiliano, A., et al. Interferon alpha 2 and dacarbazine (DTIC) in melanoma. J. Natl. Cancer Inst. 1990; 80(12):1062-1063.

49. Albert, D.M., Niffenegger, A.S. and Willson, J.K.V. Treatment of metastatic uveal melanoma: Review and recommendations. Surv. Ophthalmol. 1992; 36(6):429-438.

50. Rosenberg, S.A., Lotze, M.T., Yang, J.C., et al. Combination therapy with interleukin-2 and alpha-interferon for the treatment of patients with advanced cancer. J. Clin. Oncol. 1989; 7(12):1863-1874.

51. Talpaz, M., Lee, K., Papadopoulos, N., Plager, C., Benjamin, R. and Gutterman, J. Concomitant administration of recombinant human interleukin-2 and recombinant interferon alpha-2a in cancer patients: A phase I study. Proc. Amer. Soc. Clin. Oncol. 1989; 8:289.

52. Stoter, G., Fossa, S.D., Rugarli, C., et al. Metastatic renal cell cancer treated with low-dose interleukin-2. A phase II multicentre study. Cancer Treat. Rev. 1989; 16:111-113.

53. Rosenblum, M.G., Lamki, L.M., Murray, J.L., Carlo, D.J. and Gutterman, J.U. Interferon-induced changes in pharmacokinetics and tumor uptake of ^{111}In-labeled antimelanoma antibody 96.5 in melanoma patients. J. Natl. Cancer Inst. 1988; 80:160-165.

21

Therapy with Interleukin-2 and Tumor-Derived Activated Lymphocytes

Robert K. Oldham
Biological Therapy Institute
Franklin, Tennessee

I INTRODUCTION

The heterogeneity of human cancer is a major impediment to successful cancer therapy (1). Attempts to manage heterogeneity therapeutically using both conventional (2) and biological (3) methodologies have met with limited success. Heterogeneity is based on cell cycle, cell differentiation, tissue or organ distribution, and results in differential cell surface antigen expression (4-8). This variability supports the use of a population of cytolytic effector cells called lymphokine-activated killer (LAK) cells (9-12). Although some reactivity against normal cells has been reported (12,13), the bulk of cytolytic activity is seen against autologous or allogeneic tumor cells. More recently, methods have become available to generate tumor derived cytotoxic T lymphocytes (TDAC - tumor-derived activated cells). These cells, also called TIL (tumor infiltrating lymphocytes) because of their specificity, are less likely to destroy normal host tissue and show more specificity for each patient's cancer (14-16).

In this paper, an overview will be presented on the production of activated cells, including LAK, TDAC, and helper T cells from cancer patients' peripheral blood, lymph node, or tumors.

II LYMPHOKINE ACTIVATED KILLER (LAK) CELLS

When peripheral blood mononuclear cells (PBMC) are cultured in vitro for short periods of time (1-3 days) with interleukin-2 (rIL-2), dramatic morphological, biochemical, functional, and phenotypic changes occur. Close examination of an activated lymphocyte reveals a broad end, ruffled and organelle free, and a tapered end with a uropod containing a large population of granules (17). Biochemical changes in differentiating

LAK cells include both an increase in the number of granules as well as an increase in granule associated serine esterases (18,19). PBMCs contain a subset of effector cells termed natural killer (NK) cells with the capacity to lyse the tumor cell line, K562 (20,21). Little or no lytic activity is detectable using unstimulated NK cells against the Daudi cell line or most other fresh or cultured tumor cell targets. In the presence of rIL-2, cytolytic activity is generated against Daudi and other tumor lines but not normal cells. Recent experiments in our laboratory have shown that a brief pulse in rIL-2 (15 min-1 hr) is all that is necessary for the delivery of the appropriate signal causing LAK precursors to develop cytotoxic potential (22). Maximum cytolytic potential was only achieved using PBMCs from cancer patients primed in vivo with rIL-2. Thus, LAK precursors are pre-activated in vivo with rIL-2 or other lymphokines generated during the priming process. This method of LAK-cell generation is now utilized in clinical trials (23).

Phenotypic changes in LAK populations during culture in the presence of IL-2, as determined by flow cytometric analysis, are characterized by increased expression of common T-cell markers (Table 1). Over time, the percentage of lymphocytes expressing CD4 decreases, while the percentage of $CD8^+$ and $NKH\text{-}1^+$ cells increases. These changes are associated with DNA synthesis by the $NKH\text{-}1^+$ cells (24). There is also an increase in the percentage of $NKH\text{-}1^+$ cells co-expressing either the T cell activation antigen TA1 or the Fc receptor antigen Leu 11A (CD16). Similar phenotypic changes have been reported by others (25,26).

A. Clinical Trials with rIL-2/LAK

Because of their potent cytolytic activity against tumor cells, the ease of generation from peripheral blood, and the lack of requirement for antigen stimulation, LAK cells have been utilized clinically by a number of groups (27-30). In our initial clinical studies, cancer patients received 18×10^6 IU/m^2/day of rIL-2 by continuous infusion for 5 days. After a 24-48 hour rest and four daily leukapheresis procedures, the mononuclear cells were sent to the laboratory and cultured in rIL-2 for 3-5 days. The LAK cells were then harvested, washed, and reinfused into the cancer patients along with additional rIL-2 (29).

Mononuclear cells removed during leukapheresis procedures were not significantly cytolytic toward tumor cells, even though the cancer patients received rIL-2 by continuous infusion for 5 days. The cells became highly cytolytic after further culture in vitro in high concentrations of rIL-2 (1000-3000 IU/3×10^6 cells) (Table 2). In brief, we developed an automated closed system to produce LAK cells in 1 or 3 liter Fenwal PL732 plastic platelet storage bags (31-33), with cell harvesting performed by a continuous flow harvest system developed on the CS3000. Such an automated system allows for efficient and safe handling of the large numbers of cells obtained in the leukapheresis procedure. By utilizing an automated closed system for handling the cells, we reduced the incidence of bacterial and fungal contamination of the LAK cell product. The mean levels of endotoxin as determined by the limulus ameobocyte lysate test (LAL) has been 0.19 ± 0.10 endotoxin U/ml (n = 125), giving no endotoxin-related clinical toxicities.

Table 1
Flow cytometric analysis of leukapheresis products obtained from cancer patients and after in vitro culture in the presence of recombinant interleukin 2

Patient no.	Percent of cells expressing:					
	CD4[a]	CD8	NKH-1	TA1	NKH-1/ TA1	NKH-1/ Leu11-A
1						
Before culture	15	27	34	7	3	12
After culture[b]	9	53	76	54	40	35
2						
Before culture	15	26	35	17	1	1
After culture	13	64	48	60	34	19
3						
Before culture	52	31	5	1	2	ND
After culture	32	36	20	23	14	ND

[a]T cell antigens studied were: CD4=helper/inducer T; CD8=cytotoxic/suppressor T; NKH-1 = Natural Killer; TA1≈T cell activation marker; Leu11A=Fc receptor
[b]Mononuclear cells were cultured for 3 days in the presence of 1000 U/ml rIL-2

Rosenberg et al. (27,28) activated LAK cells in RPMI-1640 containing 2% human AB serum, a very expensive and potentially dangerous reagent (34,35). We demonstrated that autologous plasma obtained from sequential leukapheresis procedures could be used to generate LAK cells and, later, that human serum was not required for the generation of LAK cells (36). Although LAK activity could be generated in RPMI-1640 without serum, more consistent results were recently obtained using commercially available serum-free formulations such as AIMV (Gibco, BRL, Grand Island, New York) (37). The routine use of serum-free, defined medium provided consistency to the laboratory LAK cell procedures, reduced costs, and eliminated safety and availability problems associated with the use of human serum.

We recently completed an expanded analysis of rIL-2/LAK therapy in 222 patients with advanced cancer (30). The likelihood of response was correlated with: 1) histology (19% for melanoma, 13% for renal, 0% for colon, 10% for other histologies), 2) visceral metastases (30% with no visceral metastases versus 7% with visceral involvement) and

3) rebound lymphocytosis (23% with rebound ≥ 6000 lymphocytes/mm^3 versus 9% with rebound ≤ 6000). Eleven of 33 patients (33%) with soft-tissue disease only and lymphocytosis > 6000 responded to treatment, compared to 1 of 72 patients (1%) with visceral disease and lymphocytosis < 6000. The duration of response appeared to be favorably impacted by recycling of rIL-2, but long-term progression-free survival was generally confined to renal cancer patients and, less frequently, to melanoma patients with favorable prognostic features (30,38).

Table 2
Cytolytic activity of leukapheresis[a] products obtained from rIL-2 treated cancer patient before and after in vitro culture in high concentration of rIL-2[b]

	% Cytotoxicity[c] against K562 (NK target)							
	Effector to target cell ratio:							
	100:1	50:1	25:1	12:1	6:1	3:1	1.5:1	0.75:1
Before culture	13	14	11	9	7	2	0	1
After culture	83	85	84	84	8	47	35	21
	% Cytotoxicity against Daudi (LAK target)							
Before culture	8	7	6	5	3	1	0	2
After culture	93	96	84	81	76	66	50	28

[a]Mononuclear cells obtained 36h after termination of in vivo rIL-2 infusion.
[b]Mononuclear cells cultured in 1000 U/ml rIL-2 for 3 days.
[c]Measured in a 4h ^{51}Cr release assay. The data are presented as percent specific lysis.

B. Clinical trials with rIL-2/pulse LAK

Recently, we have determined that extensive laboratory cell culturing may not be necessary for the generation of cytolytic LAK cells (22). Instead, a brief 15 minute to 1 hour pulse of the PBMC with a high concentration of IL-2 (3000 IU/ml) can deliver the required signal for LAK cell differentiation (Table 3). These pulsed LAK cells are now in clinical trials (23). The entire pulsing procedure has been modified for use in the CS3000, thus eliminating the requirement for a laboratory. Alternatively, if the LAK pulse procedure is ineffective or requires further manipulation, we have also developed culture methodology for the continuous in vitro generation of large numbers of LAK cells from a single leukapheresis procedure using bioreactor technology.

Table 3

Effect of rIL-2 pulse on PMBC obtained from cancer patients[a]

Patient no.[b]	Lytic units/10^7 effector cells against Daudi targets				
	Procedure	Control[d] (0 U/ml rIL-2)	Nonpulsed[e] (1000 U/ml rIL-2)	Pulsed[f] (0 U/ml rIL-2)	Pulse/strip[g] (0 U/ml rIL-2)
1	C1	9	6282	3701	--
2	C1	46	5755	6196	8127
3	C1	54	5337	> 7000	9437
4	C6	25	7350	5799	--
5	C1	37	> 7000	> 7000	--
6	C1	50	5512	> 7000	> 7000
7	C1	88	4379	> 7000	--
8	C1	35	> 7000	7155	
9	C5	23	8129	6228	5448
	Mean ± SD[h]		6347±1125	6352±1038	

[a]Cytotoxicity was measured in a 4h ^{51}Cr release assay. Lytic units defined as the number of effector cells per 10^7 cells capable of causing 33% lysis of 2.5×10^3 tumor cell targets.

[b]All patients were undergoing rIL-2/LAK cell therapy (3).

[c]Procedures refer to PBMC obtained via cytapheresis at different times after the cessation of rIL-2 infusion: C1, 24h; C5, 36h; C6, 60h.

[d]Cells were cultured for 3 days in the absence of rIL-2.

[e]Cells were cultures for 3 days in the presence of 1000 U/ml rIL-2.

[f]Cells were pulsed with rIL-2 then cultured for 3 days in the absence of rIL-2.

[g]Cells were pulsed, stripped of rIL-2 then cultured for 3 days in the absence of rIL-2.

[h]$p<0.05$, Student's t test.

III TUMOR-DERIVED ACTIVATED CELLS (TDAC)

Another source of potent cytolytic effector cells is a cancer patient's tumor biopsy (15, 39-43). Originally called TIL (tumor infiltrating lymphocytes) by Rosenberg, this methodology relies on IL-2 to expand lymphocytes which have infiltrated the tumor bed or are derived from draining lymph nodes (14). TDAC offer the expansion of T lymphocytes, either helper or cytotoxic, which are potentially specific for the patients own tumor-associated antigens. These TDAC have limited reactivity to allogeneic tumor cells and none to normal cells, thus representing a much more individualized approach to cancer biotherapy. It has been suggested that T cells may be more potent in vivo than LAK cell populations (43). Finally, it has also been suggested that TDAC cells, once expanded, possess a memory component for the patients' tumor antigen. Thus, after reinfusion, these cells, which are presumed to be long lived, might be capable of responding to the tumor antigens in the future, if the tumor reappears.

Our TDAC methodology is divided into a number of stages (15). The initiation stage begins when the tumor arrives in the laboratory and is prepared for cell culture (Figure 1). The tumor is mechanically minced and placed in cell culture containing 3000 IU of rIL-2/ml. The induction phase follows, when stimulation of antigen-specific T cells generally occurs. We speculate that rIL-2, T lymphocytes, tumor cells, and antigen-presenting cells (APC) interact to select autotumor-specific T lymphocytes. The lymphocytes are then fed with fresh media and rIL-2 on a weekly basis and reseeded into culture at 5-10^5 cells/ml. After the first week of culture, it is important to supplement the TDAC cultures with supernatant derived from LAK cultures. This provides a source of other lymphokines and growth factors necessary for expansion of the TDAC cultures (Figure 2).

Following induction, the TDAC cultures enter the expansion phase, at the end of which the cultures are harvested, utilizing our automated system, as described above for LAK (31). It is important to periodically restimulate the TDAC cultures with antigen in order to maintain the specificity which presumably developed during the induction phase. Restimulation is accomplished using autologous tumor antigen (15), sometimes with exposure of the lymphocytes to plastic-bound anti-CD3 (OKT3) antibody (44). As illustrated in Table 4, lymphocytes from three or four tumor biopsies tested were shown to continue to proliferate when exposed to plastic-bound OKT3.

TDAC cultures, as opposed to LAK, may need to be in culture for up to 3 months in order to reach a therapeutic dose (approximately 1.0×10^{11} cells). In our initial studies where TDAC were grown in plastic bags, over 200 liters of media and up to 200 Fenwal PL732 culture bags (3 liter size) were used. Our goal to reduce the time, complexity and expense in generating the therapeutic dose of TDAC may ultimately be realized via the bioreactor technology we currently employ.

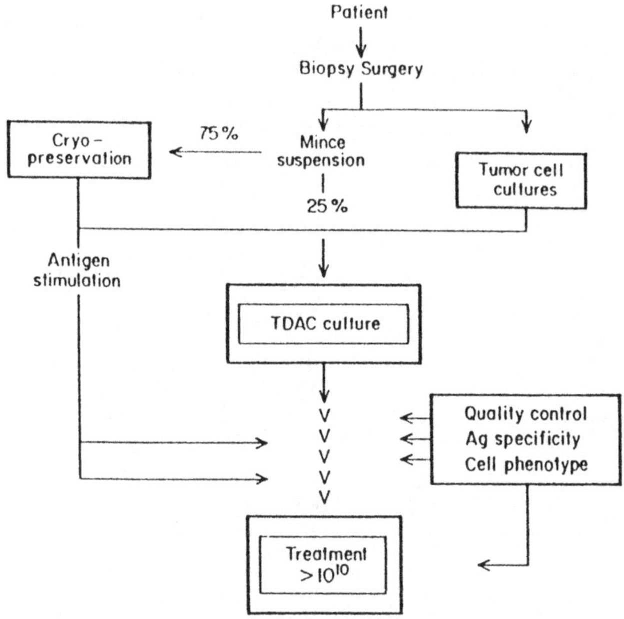

Fig. 1 Schematic diagram of the procedure used to generate TDAC for therapeutic use

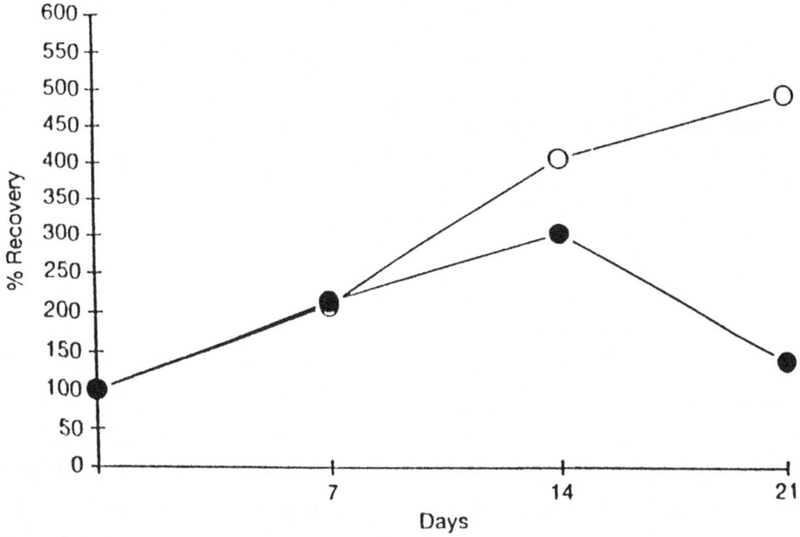

Fig. 2 The effect of LAK conditioned medium on the expansion of TDAC. The growth rates of TDAC initiated in the presence -0- or absence -0- of 20 percent (v/v) LAK conditioned medium and additionally supplemented 1 wk after culture initiation were observed for the first 3 weeks of culture

Table 4
Recovery of lymphocytes cultured from tumor biopsies in the presence of interleukin 2 (rIL-2) or rIL-2 and monoclonal antibody OKT3

Expt no.	Weeks in culture	Total no. x10^6 lymphocytes recovered	
		Culture condition[a]	
		rIL-2	rIL-2 + OKT3
	1	ND	ND
1	2	0.8	8.6
(Testicular carcinoma)	3	1.8	32.0
	4	2.4	138.0
2	1	1.0	4.2
(Breast carcinoma)	2	6.6	38.2
	3	12.8	168.0
	4	13.8	135.0
3	1	6.8	7.1
(Melanoma)	2	14.6	33.1
	3	21.0	148.0
	4	19.0	156.0
4	1	4.8	1.9
(Endometrial sarcoma)	2	1.2	1.6
	3	0.8	1.0
	4	1.1	0.9

[a]Tumor biopsies, following enzymatic digestion, were placed in culture either in the presence of rIL-2 alone (3000 IU/ml) or rIL-2 and plastic-bound OKT3.

A. TDAC Biology

We have examined the outgrowth of lymphocytes from tumors obtained from 142 patients with a success rate of 75% (Table 5). Lymphocytes from melanoma grew most consistently while those from renal cell carcinoma fared less well. Interestingly, our overall success rate increased to 92% when patients had been treated with LAK/rIL-2 before the tumor was removed. The in vivo rIL-2 infusions seemed to recruit additional lymphocytes into the tumor bed and prime the patients TDAC (45).

Table 5
Expansion of TDAC from various tumor types

Tumor Type	Number of patients	% positive growth[a]
Melanoma	41	85
Renal carcinoma	18	50
Ovarian carcinoma	15	80
Colon carcinoma	18	67
Breast carcinoma	10	100
Lung carcinoma	12	83
Miscellaneous[b]	28	75
Total	142	Mean %77

[a]Cultures were growing when terminated and contained 1.0×10^9 cells or were predicted to grow to that level.

[b]Miscellaneous tumors included but not limited to: esophageal carcinoma, gastric carcinoma, hepatoma, lymphoma, mesothelioma, neuroepithelial carcinoma, pancreatic carcinoma, prostatic carcinoma, and sarcoma.

In general, the early growth of lymphoid cells from tumor is characterized by cells possessing LAK activity. These cells lyse K562, Daudi, and autologous tumor as well as allogeneic tumor cells of varying histologies. As time passes, specific activity against autologous tumor cells is observed with a loss in LAK activity (Figure 3). In the final TDAC products, we have observed cultures which are predominantly $CD4^+$, some which are predominantly $CD8^+$, and those which are a mixture of the two. However, killing has only been observed in these cultures containing predominantly $CD8^+$ T cells. The specificity that is observed is induced and maintained by the periodic addition of autologous tumor cells. Our OKT3 methodology was successful in expanding lymphocytes to therapeutic doses which retained similar specificity to T cells stimulated with antigen. As can be seen in Tables 6 and 7, OKT3-stimulated TDAC can retain the specificity of T cells stimulated with antigen.

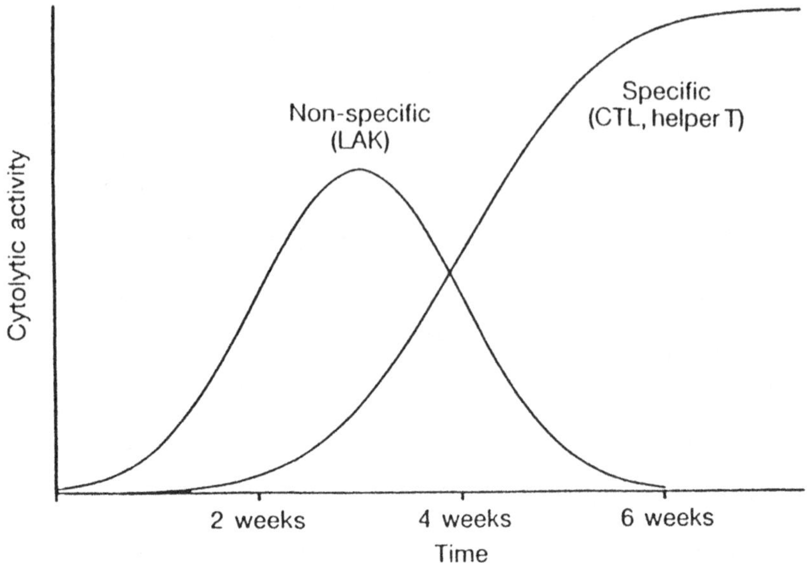

Fig. 3 Generalized representation of the patterns of cytotoxicity observed in TDAC cultures derived from tumor biopsies. The first 4 weeks are generally characterized by cytolytic activity against the NK target K562, the LAK target Daudi, and both autologous and allogeneic tumor cell lines. Specific activity (both helper and cytolytic) generally develops around week 4. After this point, nonspecific activity diminishes

B. Clinical trials with rIL-2/TDAC

Melanoma metastases were harvested from 82 patients for the purpose of growing and expanding TDAC (45). Results for twenty-one treated patients are summarized in Table 6. Tumor tissue cell suspensions were incubated with rIL-2 followed by repeated exposure to tumor antigen with or without OKT3 monoclonal antibody. Initial growth success was achieved in 56 of 82 cultures (72%). Efforts were made to expand 26 of these 56 cultures for therapy; 23 of 26 early cultures (88%) were successfully expanded for in vivo therapy. It took a mean of 78.5 $\pm$ 25.4 days to grow sufficient TDAC for treatment. Therapy included cyclophosphamide (1 g/m^2) on day 1, followed by a 96-hour continuous infusion of rIL-2 (18 IU/m^2/d) on days 2 to 5, and approximately 10^{11} (mean 1.49 $\pm$ 0.93x10^{11}) TDAC on day 2. Patients who responded received monthly rIL-2 as a 96-hour infusion. Median patient age was 45 years of age. Sixty-seven percent of the patients were men. Performance status was 0 to 1 in 77 percent of patients. Thirty-four percent of the patients had liver metastases. Response rate for 21 patients was 24 percent (95 percent confidence interval, 10 percent to 49 percent). One complete response was achieved with cells 98 percent $CD4^+$; four partial responses were achieved with cells 80 percent, 94 percent, 98 percent, and 98 percent $CD8^+$, respectively. Four of eight patients who received TDAC, which had never been stimulated with OKT3, had tumor response. TDAC therapy with rIL-2 is technically difficult, costly, and effective for only a minority of patients. Overall, clinical results were not clearly superior to those obtained with other rIL-2 regimens (46).

Table 6
Summary of rIL-2/TDAC results in 21 patients with melanoma

Patient no.	Response	Survival (mo)	No. of grams tumor cultured	OKT3 used to stimulate TIL
1	CR	20.1	2.3	-
2	PR	9.8	3.4	+
3	PR	29.6	8.0	-
4	PR	19.9	2.2	-
5	PR	11.6+	1.4	-
6	MR	6.1	1.1	+
7	SD	19.2	**	-
8	SD	15.6	1.7	+
9	SD	5.8	2.0	-
10	SD	5.6	2.0	+
11	SD	4.2	0.6	+
12	SD	6.3	**	+
13	PD	11.3+	2.2	+
14	PD	5.0	0.5	+
15	PD	3.4	1.0	+
16	PD	2.6	0.2	-
17	PD	1.8	**	+
18	PD	1.4	2.0	+
19	PD	0.9	55.0	+
20	PD	1.6	2.0	+
21	PD	4.0	**	+

ND: not done; CR: complete response; PR: partial response; SD: stable disease; PD: progressive disease; MR: mixed response

*Cells staining positive for monoclonal antibodies against CD4 and CD8, respectively.

Table 6 (continued)

T4/T8* x 10^{11}	TIL killing of tumor cell targets	Days in culture	No. of cells
98.0		91	2.0
02/98	ND	#	2.0
20/80	+	56	0.6
02/98	+	42	2.9
06/94	ND	#	2.0
60/45	ND	94	0.7
**	**	**	**
20/80	ND	132	2.5
02/98	-	#	2.3
28/78	ND	102	1.2
20/80	ND	#	0.9
**	**	**	**
17/81	ND	70	3.1
11/84	ND	81	2.0
02/98	-	49	2.7
98/02	ND	70	1.8
**	**	**	**
19/81	+	63	0.1
**	ND	44	1.4
51/48	**	**	**
**	**	**	**

#Culture time extended because patient was medically unable to be treated at time cells were initially prepared.
**No data available.

In patients with non-melanoma cancers (summarized in Table 7), metastases from patients with solid tumors were harvested from 196 patients for the purpose of growth TDAC. Cells were prepared from autologous tumor cultures by incubation with rIL-2 followed by repeated exposure to tumor antigen and/or anti-CD3 monoclonal antibody. Initial growth success was achieved in 66%; 45 of 56 (80%) of these early cultures were subsequently expanded for in vivo therapy. It took a mean of 69.4 $\pm$ 24.0 days to grow TDAC for treatment. Thirty-eight patients were treated with cyclophosphamide (1 g/m^2) on day one followed by a 96-hour continuous infusion of rIL-2 (18 x 10^6 IU/m^2/day) on days 2-5 and approximately 10^{11} TDAC on day 2. Patients subsequently received monthly rIL-2 as a 96-hour constant infusion if their cancers were stable or regressing. Median age was 51 years; 58% were male. Performance status was 0-1 in 64%, 29%

had lung metastases; 34% had liver metastases. Responses were seen only in 1/38 patients (3%); a partial response in a patient with lymphoma. Forty-two percent were stable 90 days post-treatment, the rest were progressive or inevaluable. We conclude that a treatment plan for rIL-2/TDAC is technically difficult, costly, and not practical under these conditions for these tumor types. Clinical results to date are not clearly different than those obtained with other rIL-2 regimens (47).

Table 7
Summary of rIL-2/TDAC in non-melanoma patients

Initials	Response	Survival (MO)	No. grams in culture	Total no. grams
Colon				
WG	STABLE	3.6	2.0	23.3
GH	STABLE	14.6+	3.2	7.0
GL	STABLE	3.1	20.0	88.0
HM	STABLE	5.3	0.5	63.3
RO	PROG	0.5	5.0	18.5
LP	STABLE	7.4	2.0	5.0
JR	STABLE	8.7	0.7	1.9
MS	PROG	0.4	NA	NA
PS	PROG	10.3+	NA	NA
Renal				
RB	STABLE	0.7	0.5	30.3
JD	PROG	12.7	4.0	14.6
FF	PROG	5.3	2.0	99.0
EH	PROG	2.8	2.0	19.2
DK	PROG	15.1+	1.0	4.3
MK	PROG	3.6	7.5	7.5
LL	STABLE	8.6+	2.0	20.0
HM	STABLE	8.4+	2.0	13.0
DM	PROG	13.9	1.5	4.5
Lung				
RB	PROG	3.5+	3.0	12.3
JC	STABLE	1.5+	2.0	13.3
FD	PROG	0.2	1.4	3.6
DG	PROG	1.3	0.5	1.4
SG	PROG	0.3		
ML	PROG	2.0+	0.8	2.0
TZ	PROG	1.1	1.0	1.0

Table 7. Summary of rIL-2/TDAC in non-melanoma patients (continued)

Initials	Response	Survival (MO)	No. grams in culture	Total no. grams
Sarcoma				
LA	STABLE	11.5+	40.0	171.9
LC	STABLE	8.2+	43.0	45.7
HN	PROG	4.1	2.4	12.0
Ovarian				
SF	INEVAL	9.4	4.9	5.0
AK	PROG	3.0	0.8	0.8
Gastric				
EK	PROG	1.0	0.5	1.8
Lymphoma				
JL	PR	15.4+	2.0	33.0
Fibrohistocytoma				
LE	STABLE	3.1+	2.0	13.2
Unknown				
VS	STABLE	1.9+	4.0	16.0
Breast				
CT	PROG	3.9	2.0	5.5
Pancreas				
RC	STABLE	15.2+	0.7	2.4
Bladder				
WM	STABLE	0.3+	NA	NA
Esophagus				
GW	PROG	0.5	NA	NA

NOTE: Abbreviations: NA, not available; ND, not done; PROG, progressive; INEVAL, inevaluable; PR, partial response.

Table 7. (Part 2)

T4/T8	Killing autologous	Days in culture	No. cells ($x10^{11}$)
20/88	ND	84	0.4
70/30	-	35	0.9
84/10	-	56	0.8
27/73	ND	a	1.0
20/80	-	56	2.5
50/62	ND	119	0.9
03/93	+	77	1.9
88/18	ND	63	0.3
97/02	-	84	0.7
05/95	ND	56	1.2
04/96	+	35	0.9
11/76	ND	73	0.8
49/24	-	92	0.9
10/95	ND	70	1.2
98/00	ND	32	0.9
99/01	ND	56	1.2
99/01	ND	63	0.9
93/07	+	56	0.4
04/95	ND	123	2.4
42/32			
99/01	-	a	1.0
01/99	+	56	1.3
94/06	-	84	0.7
04/95	ND	56	1.0
01/99	ND	a	0.1
34/72	ND	38	1.0
23/74			
28/50	ND	99	1.8
60/40	ND	63	1.3
87/10	ND	85	1.0
90/02	-	106	0.6
98/01			
30/75	ND	42	2.3
75/20	ND	a	1.0
36/78	-	70	0.9
30/70	-	a	1.0

IV CONCLUSIONS

Given the accumulated data from nearly 20 years of immunotherapy, it is now clear that T lymphocytes can mediate significant antitumor responses in man. A limiting factor has always been the ability to expand T cells, whether they were derived from the peripheral blood, lymph node, spleen, or tumor. With rIL-2, a T cell growth factor, it is now feasible to expand T cells from any of these sources, some of which have demonstrated clinical efficacy as antitumor effectors in murine models and in man.

Concomitant with the demonstration that rIL-2 can provide the growth stimulus for T cells, the LAK cell technology came to the clinic, demonstrating that peripheral blood cells (NK cells), which are functionally distinct from T cells, could be cultured, expanded, and reinfused to patients with significant clinical effects. Although the most promising results have been in patients with melanoma and renal cancer, antitumor effects have been seen in patients with a wide variety of cancers, even those with bulky tumors (38).

Recent improvement in LAK generation allows for a brief pulse of rIL-2 in vitro after in vivo rIL-2 priming, leading to cytolytically active, lymphokine releasing cells. Studies are underway to determine the clinical efficacy of pulse LAK cells (23).

In tumor biopsy specimens, lymphocytes have been recognized as a component of mononuclear cell infiltrates for many years. As in the mouse, the major limiting factors have been the ability to culture large numbers of these infiltrating cells and the limited understanding of the tumor antigens involved in T cell stimulation. The TDAC technology makes it clear that the technical problems of T cell expansion are now being solved. The restimulation by antigen provides the ongoing stimulus needed to maintain selective killing of tumor cells. Various factors in the medium that support and enhance growth and T cell activation are being defined. Thus, the components are now available to develop a broad attack on advanced cancer using this laboratory-based technology of peripheral blood and tumor-derived activated cell stimulation, expansion, and therapy (48,49).

These studies of adoptive cellular biotherapy have confirmed that expanded and activated cell populations from the cancer research laboratory can provide a method by which clinicians can effectively treat advanced cancer (48). Although the original studies with rIL-2/LAK were time consuming and expensive, we have undertaken a new generation of LAK studies using a method of producing LAK cells at the bedside which is relatively efficient and inexpensive. Should these LAK cells prove effective, they can easily be generated in any community hospital. Furthermore, our current studies are focused on generating TDAC cells through bioreactor technology with the use of small, portable bioreactors from Cellco (Gaithersburg, Maryland). We anticipate being able to produce multiple doses of TDAC cells to extend our earlier studies using rIL-2 with a single dose of these cells. If a larger number of these cells proves more therapeutically effective, we now have a method to produce these cells in relatively large numbers at a cost that would be consistent with other forms of cancer therapy (50). In summary, cancer biotherapy continues to expand and the use of rIL-2 with activated cell populations continues to be of interest with regard to new opportunities for patients with advanced cancer (49).

References

1. Aukerman, S.L. and Fidler, I.J. The heterogeneous nature of metastatic neoplasms: relevance to biotherapy. In: Principles of Cancer Biotherapy - Second Edition. (ed. RK Oldham). Marcel Dekker, New York 1991; 23-54.
2. DeVita, V.T. Progress in cancer management: Keynote Address. Cancer 1983; 51:2401-9.
3. Oldham, R.K. Biologicals and biological response modifiers: The 4th modality of cancer treatment. Cancer Treatment Rep. 1984; 68:221-232.
4. Albino, A.P., Lloyd, K.O., Houghton, A.N., Oettgen, H.F. and Old, L.J. Heterogeneity in surface antigen and glycoprotein expression of cell lines from different melanoma metastases of the same patient. J. Exp. Med. 1981; 154:1764-1778.
5. Fogel, M., Gorelik, E., Segal, S. and Feldman, M. Differences in cell surface antigens of tumor metastases and those of the local tumor. J. Natl. Cancer Ins. 1979; 62:585-588.
6. Natali, P.G., Cavaliere, R., Bigotti, A., et al. Antigenic heterogeneity of surgically removed primary and autologous metastatic human melanoma lesions. J. Immunol. 1983; 130:1462-1466.
7. Roht, J.A., Restrep, C., Scuderi, P., Baldwin, R.W., Reichert, C.M. and Hosoi, S. Analysis of antigenic expression by primary and autologous human metastatic sarcoma using murine monoclonal antibodies. Cancer Res. 1984; 44:5320-5325.
8. Liao, S.K., Meranda, C., Avner, B.P., Romano, T., Husseini, S., Kimbro, B. and Oldham, R.K. Immunohistochemical phenotyping of human solid tumors with monoclonal antibodies in devising biotherapeutic strategies. Cancer Immunol. 1989; 28:77-86.
9. Grimm, E.A., Mazumder, A., Zhang, H.Z. and Rosenberg, S.A. Lymphokine activated killer cell phenomenon: lysis of natural killer-resistant fresh solid tumor cells by interleukin-2 activated autologous peripheral blood lymphocytes. J. Exp. Med. 1982; 155:1823-1827.
10. Grimm, E.A., Ramsey, K.M., Mazumder, A., Wilson, D.J., Djeu, J.Y. and Rosenberg, S.A. Lymphokine activated killer cell phenomenon. II. Precursor phenotype is seriologically distinct from peripheral T lymphocytes, memory cytotoxic thymus derived lymphocytes and natural killer cells. J. Exp. Med. 1983; 157:884-893.
11. Rosenstein, M., Yrou, I., Kaufman, Y. and Rosenberg, S.A. Lymphokine activated killer cells: lysis of fresh syngeneic natural killer resistant murine tumor cells by lymphocytes cultured in interleukin-2. Cancer Res. 1984; 44:1946-1953.

12. Sondel, P.M., Hank, J.A., Kohler, P.C., Chen, B.P., Minkoff, D. and Molenda, J.A. Destruction of human lymphocytes by interleukin-2 activated cytotoxic cells. J. Immunol. 1986; 137:502-509.
13. Moltenberg, A.M., Miejer-Paapa, M.E., Daha, M.R. and Paul, L.E. Endothelial cell lysis induced by lymphokine activated human peripheral blood mononuclear cells. Eur. J. Immunol. 1986; 17:1783-1789.
14. Rosenberg, S.A., Packard, B.S., Aerbersold, P.M., et al. Use of tumor infiltrating lymphocytes and interleukin-2 in the immunotherapy of patients with metastatic melanoma: a preliminary report. New Engl. J. Med. 1988; 319:1676-1680.
15. Maleckar, J.R., Fridell, C.S., Sfferuzza, A., Thurman, G.B., Lewko, W.M., West, W.H., Oldham, R.K. and Yannelli, J.R. Activation and expansion of tumor derived activated cells for therapeutic use. J. Natl. Cancer Inst. 1989; 81:1655-1660.
16. Oldham, R.K., Maleckar, J., Fridell, C., Lewko, W.M., West, W. and Yannelli, J.R. Tumor derived activated cells: preliminary laboratory and clinical results. Clin. Chem. 1989; 35(8):1576-1580.
17. Yannelli, J.R., Sullivan, J.A., Mandell, G.L. and Engelhard, V.H. Reorientation and fusion of cytotoxic T lymphocyte granules after interaction with target cells as determined by high resolution cinemicrography. J. Immunol. 1986; 136:377-382.
18. Hameed, A., Lowrey, D.M., Lichtenheld, M. and Podack, E.R. Characterization of three serine esterases isolated from human IL-2 activated killer cells. J. Immunol. 1988; 141:3142-3147.
19. Manyak, C.L., Norton, G.P., Fabe, C.G., Bleackley, R.C., Gershenfled, H.K., Weissman, I.L., Kuman, V., Sigal, N.H., and Koo, G.C. IL-2 induces expression of serine probase enzymes and genes in natural killer and nonspecific T killer cells. J. Immunol. 1989; 142:3707-3713.
20. Oldham, R.K. Natural killer cells: history and significance. J. Biol. Response Modifiers. 1982; 1:217-231.
21. Oldham, R.K. Natural killer cells: history, relevance and clinical applications. In: Natural immunity and cell growth regulation. (Eds): Lotzova, ES Karger AG, Switzerland 1990; pp. 297-312.
22. Horton, S.A. and Yannelli, J.R. Generation of lymphokine activated killer cells following a brief exposure to high dose interleukin-2. Cancer Res. 1989; 50:1686-1692.
23. Oldham, R.K., Lewis, M., Yannelli, J. and West, W.H. Treatment of advanced cancer with rIL-2 and pulsed mononuclear cells: preliminary results of phase I study. Program proceedings of the twenty-sixth annual meeting of the American Society of Clinical Oncology 1990; 9:194.
24. Yannelli, J.R., Desch, C., Shults, K., Houston, J. and Stelzer, G. Characterization of human lymphokine activated killer cells (LAK cells): cytotoxicity and cell surface phenotype. Fed. Proc. 1987; 46-483.

25. Phillips, H.J., Gembo, B.T., Myers, W.W., Rayner, A.A. and Lanier, L.L. In vivo and in vitro activation of natural killer cells in advanced cancer patients undergoing combined recombinant interleukin-2 and LAK cell therapy. J. Clin. Oncol. 1987; 15:1933-1940.
26. Tilden, A.B., Itoh, K. and Balch, C.M. Human lymphokine activated killer cells (LAK): identification of two types of effector cells. J. Immunol. 1987; 138:1068-1072.
27. Rosenberg, S.A., Lotze, M.T., Muul, L.M., Leitman, S., Chang, A.E., Ettinghausen, S.E., Matory, Y.Y., Shiloni, J.M., Vetto, J.T., Seipp, C.A., Simpson, C. and Reichert, C.M. Observations on the systemic administration of autologous lymphokine-activated killer cells and recombinant interleukin-2 to patients with metastatic cancer. N. Engl. J. Med. 1985; 313:1485-1492.
28. Rosenberg, S.A., Lotze, M.T., Muul, L.M., Leitman, S., Chang, A.E., Avis, F.P., Leitman, S., Leneman, M., Robertson, C.N., Lee, R.E., Rubin, J.I., Seipp, C.A., Simpson, C.G. and White, D.E. A progress report on the treatment of 157 patients with advanced cancer using lymphokine activated killer cells and interleukin-2 or high dose interleukin-2 alone. N. Engl. J. Med. 1987; 316:889-897.
29. West, W.H., Tauer, K.W., Yannelli, J.R., Orr, D.W., Thurman, G.B. and Oldham, R.K. Constant infusion recombinant interleukin-2 in adoptive immunotherapy of advanced cancer. N. Engl. J. Med. 1987; 316:898-905.
30. Dillman, R.O., Oldham, R.K., Tauer, K.W., Orr, D.W., Barth, N.M., Blumenschein, G., Arnold, J., Birch, R. and West, W.H. Continuous interleukin-2 and lymphokine-activated killer cells for advanced cancer: a National Biotherapy Study Group trial. J. Clin. Oncol. 1991; 9:1233-1240.
31. Yannelli, J.R., Thurman, G.B., Mrowca-Bastin, A., Pennington, C.S., West, W.H. and Oldham, R.K. Enhancement of human lymphokine-activated killer cell cytolysis and a method for increasing lymphokine activated killer cell yields to cancer patients. Cancer Res. 1988; 48:5696-5700.
32. Muul, C.M., Director, E.P., Hyatt, C. and Rosenberg, S.A. Large scale production of human lymphokine activated killer cells for use in adoptive immunotherapy. J. Immunol. Methods 1986; 88:265-270.
33. Yannelli, J.R., Thurman, G.B., Dickerson, S.G., Mrowca, A., Sharp, E. and Oldham, R.K. An improved method for the generation of human lymphokine activated killer cells. J. Immunol. Methods 1987; 100:137-145.
34. Yannelli, J.R., Dickerson, S.G., Sharp, E., Thurman, G.B., West, W.H. and Oldham, R.K. An automate procedure for the generation and harvesting of human lymphokine activated killer (LAK) cells. Proc of Amer Assoc for Cancer Res. 1987; 28:371.
35. Parkinson, D.R., Snydman, D.R., Weisfure, I., Werner, B., Grahma, D., Will, M., Marcus, S. and Boldi, D. Doroshow JHM, Ryner A, Glover A, Fisher R. Hepatitis A (HAV) infection occurring following IL-2/LAK cell therapy. Proc. Amer. Soc. Clin. Oncol. 1987; 6:234.

36. Yannelli, J.R., Mrowca-Bastin, A. and Jadus, M. Letter to the Editor. Cytotechnology 1988; 1:183.
37. Jadus, M.R., Thurman, G.B., Mrowca-Bastin, A. and Yannelli, J.R. The generation of human lymphokine activated killer cells in various serum free media. J. Immunol. Methods 1988; 190:169-174.
38. Dillman, R., Church, C., Oldham, R., West, W., Barth, N., et al. (In Press) Journal of the American Medical Association 1992.
39. Itoh, K., Tilden, A.B. and Balch, C.M. Interleukin-2 activation of cytotoxic T lymphocytes infiltrating into human metastatic melanomas. Cancer Res. 1986; 46:3011-3016.
40. Kradin, R.L., Boyle, C.A., Preffer, F.I., Callahan, R.J., Barlaikovach, R.J., Strauss, H.W., Dubinett, S. and Kurnick, J.T. Tumor derived interleukin-2 dependent lymphocytes in adoptive immunotherapy of lung cancer. Cancer Immunol. Immunother. 1987; 24:76-85.
41. Muul, L.M., Spiess, P.J., Director, E.P. and Rosenberg, S.A. Identification of specific cytolytic immune response against autologous tumor in humans bearing malignant melanoma. J. Immunol. 1987; 138:989-993.
42. Topalian, S.L., Muul, L.M., Soloman, D. and Rosenberg, S.A. Expansion of human tumor infiltrating lymphocytes for use in immunotherapy trial. J. Immunol. Methods. 1987; 102:127-132.
43. Rosenberg, S.A., Spiess, P. and Lafreniere, R. New approach to the adoptive immunotherapy of cancer with tumor-infiltrating lymphocytes. Science. 1986; 233:1318-1320.
44. Yannelli, J.R., Krumpacker, D.B., Good, R.W., Friddell, C.D., Poston, R., Horton, S., Maleckar, J.R. and Oldham, R.K. Use of anti-CD3 monoclonal antibodies in the generation of effector cells from human solid tumors for use in cancer biotherapy. J. Immunol. Methods. 1990; 10-20.
45. Oldham, R.K., Maleckar, J., West, W.H. and Yannelli, J.R. IL-2 and cellular therapy LAK and TDAC. In: Cytokines in hemopiuesus. Oncology and AIDS Symposium, (ed. Freund et al.) Springer-Verlag, West Germany, 1989.
46. Dillman, R.O., Oldham, R.K., Barth, N.M., Cohen, R.J., Minor, D.R., Birch, R., Yannelli, J.R., Meleckar, J.R., Sferruzza, A., Arnold, J. and West, W.H. Continuous interleukin-2 and tumor-infiltrating lymphocytes as treatment of advanced melanoma. Cancer. 1991; 68:1-8.
47. Oldham, R.K., Dillman, R.O., Yannelli, J.R., Barth, N.M., Maleckar, J.R., Sferuzza, A., Cohen, R.J., Minor, D.R., Spitler, L., Birch, R. and West, W.H. Continuous infusion interleukin-2 and tumor-derived activated cells as treatment of advanced solid tumors: A National Biotherapy Study Group trial. Mol. Biother. 1991; 3:68-71.
48. Stevenson, H.C. Adoptive Cellular Immunotherapy of Cancer. Marcel Dekker, New York 1989.

49. Oldham, R.K. Principles of Cancer Biotherapy. Marcel Dekker, New York, 2nd Edition 1991.

50. Oldham, R.K. Cancer cures: by the people, for the people at what cost. Mol. Biother. 1990; 2(1):2-3.

22

Combination Cytokine Therapy in Cancer

Marc S. Ernstoff
Dartmouth-Hitchcock Medical Center
Hanover, New Hampshire

I INTRODUCTION

Many cytokines have been characterized and genetically cloned over the past decade allowing for the production of highly purified proteins (1,2). The availability of these proteins has permitted the evaluation of cytokine function in vitro and in vivo as well as the clinical application of cytokines in human disease (3,4,5). Cytokines regulate many cellular functions. A very complex cytokine network has been described that involves cell growth and differentiation, immune function, vascular permeability, coagulation, wound healing, fibrosis, and other functions committed to body homeostasis (Figure 1) (6,7,8).

Cytokines act locally by binding to specific receptors on the cell membrane (9). The pleiotropic activity and intricate interactions of the cytokines form a system containing positive and negative feedback signals and redundancy which function to maintain homeostasis. Attempts to influence the immune system or the cellular functions of tumor cells with cytokines in the treatment of human cancer are limited by feedback mechanisms which bring the immune system and cellular function back toward equilibrium following provocation. Given the activity of these mechanisms, it seems unlikely that the use of a single cytokine in either physiologic or pharmacologic dosages would be therapeutically successful against metastatic cancer. However, when a cytokine functions as a principal regulator of growth or differentiation of a particular cell, a single cytokine may successfully control the human malignancy derived from that cell. Such is the case of interferon alpha in the treatment of hairy cell leukemia (10).

To improve our ability to manipulate the immune system for therapeutic advantage we must understand the anti-tumor mechanisms of action. Even when these mechanisms are described optimal combinations of cytokines or drugs, dosage, schedule, route and sequence of administration require investigation in the clinical setting.

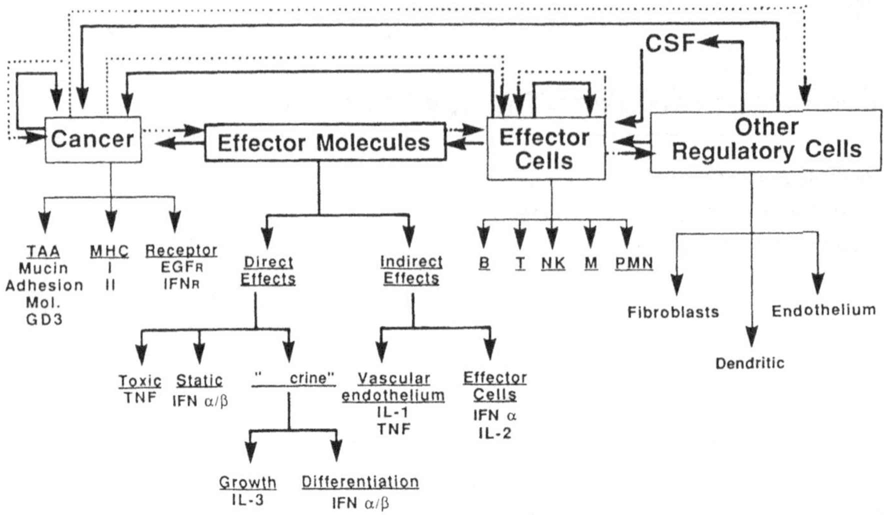

Fig. 1 Cytokine network.

II RATIONALE OF COMBINATION CYTORINE THERAPY

Let us consider the rationale for combination cytokine therapy which may be used in the clinical setting. One way to stimulate the effector arm of the immune system is to introduce cytokines in such a manner as to recapitulate the physiologic pathways. The use of colony stimulating factors to raise the level of circulating neutrophils in leukopenic patients is an example of this approach (11,12).

Another premise is to correct any deficiency states which may exist in patients with cancer by replacing the missing cytokine(s) or cells. Patients with metastatic melanoma or renal cell carcinoma have been shown to have either a qualitative or quantitative deficiency in LAK cells. Therapy with high dose IL 2 and adoptive transfer of LAK cells has had limited success in the treatment of these patients (13,14,15).

The third approach would be to use cytokines in a pharmacologic fashion. The use of interferon gamma in patients with chronic granulomatous disease, even though these patients are not deficient in interferon gamma, has improved their overall ability to ward off serious infections (16).

Numerous laboratory experiments have demonstrated the synergy of combining cytokines to stimulate the natural killer cell population, T cell populations, and other immune functions (17,18,19,20). Examples of these experiments include: the observations of synergy between IL-2 and IL-4 on both LAK cell and T cell growth and function, and the synergy between IFN gamma and tumor necrosis factor on T-cell and macrophage function.

Human studies have found specific deficiencies of cytokine production which are responsible for a impaired response to specific stimuli. The impaired response has been corrected by reintroducing the necessary cytokines. Specifically, T-cell response to BCG and M. lepra is poor in patients with lepromatous leprosy. The T-cell response is improved by the introduction of interferon gamma, IL-2 and IL-4 (21).

Putting these concepts into the clinical setting faces a number of hurdles. Patient compliance will depend upon schedule, route of administration, length of therapy and toxicities of the cytokines employed. Determination of optimal dose, schedule and sequence will hinge on: 1. identification of toxicities and the principal mechanism(s) of anti-tumor activity, 2. monitoring and ameliorating unwanted toxicities, and 3. monitoring and modulating appropriate immune function in vivo. Modulation of the immune system will also be influenced by host heterogeneity and tumor heterogeneity. The use of cytokines in the treatment of human malignancies is in its infancy, and observations from clinical trials should be viewed with an eye toward gaining insight into anti-tumor mechanisms and improvement of anti-tumor activity by modulating treatment combinations, schedule, route and dose.

With these thoughts in mind, we have undertaken a trial evaluating the sequential administration of interferon gamma and alpha in patients with metastatic renal cell carcinoma.

III COMBINATION INTERFERON (IFN) GAMMA AND ALPHA THERAPY IN THE TREATMENT OF METASTATIC RENAL CELL CARCINOMA

This study investigated the clinical and immunologic effects of sequentially administered recombinant interferon gamma and recombinant interferon alpha in 36 patients with metastatic renal cell carcinoma (22). The schedule of interferon administration was based on clinical observation in single agent trials in patients with metastatic renal cell carcinoma. The sequence was based on the observation that IFN alpha receptor reaches maximum upregulation six hours following exposure to IFN gamma (23). Recombinant IFN alpha was given by subcutaneous injection daily for 70 days at dosages that varied from 2.5 to 20 million units / m2 across four cohorts of nine patients. Subsets of patients within each IFN alpha dosage tier received either 30, 300, 1000 ug/m2 IFN gamma. Recombinant IFN gamma was administered intravenously for five days every third week, six hours prior to administration of IFN alpha.

Patients with histologically documented metastatic RCC were treated in this trial. History and physical examination were obtained prior to therapy, and every 2-4 weeks thereafter or as medically indicated. Patients underwent staging evaluation for determination of tumor measurements at weeks 5 and 10 of the first cycle of treatment and every 6 to 8 weeks among patients receiving maintenance treatment. All patients judged to have stable disease or an objective response at the completion of 70 days of therapy were continued on the same schedule of treatment for a maximum of one year or until progression of disease. Recombinant IFN alpha-2a (Hoffman LaRoche, Inc., Nutley N.J.) and IFN gamma (Biogen, Inc., Cambridge, Mass) were supplied through the

Biological Response Modifier Program as part of the National Cancer Institute Master Agreement Program (NCI-CM-37613-18, MAO #7).

Natural killer cell activity, T-cell phenotype (CD4, CD8, CD56, CD16, CD4/HLA-DR, CD8/HLA-DR, CD56/HLA-DR) and 2'5' oligoadenylate synthetase (2'5'As) were measured prior to therapy, during therapy and following completion of treatment; complete immune data were available in 28/36 patients (24). For immunologic monitoring, blood samples were drawn into preservative-free heparinized syringes between 8:00 and 11:00 am. Peripheral blood mononuclear cells (PBMNC) were separated by centrifugation on Ficoll-Hypaque gradients, washed and counted in a trypan blue dye. Natural Killer Activity (NK) was measured in a standard short-term (4 hr) cytotoxicity assay using cultured K562 cells as targets (25). T-cell populations were evaluated by two-color flow cytometry performed on a FACScan. 2'5' oligoadenylate synthetase (2'5'As) were determined by the method previously described by Merritt et al (26). The data were combined to create five time points for statistical analyses. Time effect, treatment effect, effect of initial dose level and relationship between clinical response and potential explanatory factors were evaluated.

Objective responses to IFN alpha and gamma were noted in eight of 30 (27%) evaluable patients. All patients with a complete or partial response had developed an objective response by day 70 of treatment. Two patients experienced complete remission and have remained in complete remission 46 and 48+ months. An additional six patients have shown partial responses for 4-32+ months. Two patients demonstrating a partial remission continued to show slow regression of pulmonary and liver lesions after termination of therapy. Dose-limiting toxicities included constitutional symptoms, leukopenia, nephrotic syndrome with acute renal failure, hypotension associated with death, and congestive heart failure, and were more often related to the IFN alpha dose level than to IFN gamma dose level. Maximum tolerated dose was 10 million u/m2 IFN alpha and 1000 micrograms/m2 IFN gamma. The results of this study suggest that toxicities associated with combination IFN therapy can be reduced by administering these agents sequentially as opposed to simultaneously.

Statistical analysis of all immune parameters was performed for the entire group, by individual patient, by dosage, by time and by clinical response. Overall significant depressions in NK activity, and in percent of circulating CD56, CD16, and CD8+ cells were noted. Significant increases in 2'5'As and in the percent of circulating CD4 cells were also noted. An association between the magnitude of change in percent CD16 + cells and 2'5'As and dosage of IFN gamma and IFN alpha, respectively, was observed. A decrease in percent circulating CD8+ cells was observed among patients with objective clinical response (partial and complete).

Sequentially administered IFN gamma and IFN alpha can modulate immunologic parameters in vivo in patients with metastatic RCC. A fall in percent circulating CD8+ cell was associated with response and suggests that this sequence of IFN alpha and IFN gamma might influence T-cell mediated anti-tumor activity.

IV CONCLUSION

Combination cytokine therapy for human malignancies is now possible with the production and availability of the interferons (alpha,beta, gamma), interleukins (1,2,3,4,6), colony stimulating factors (G, GM), tumor necrosis factors, and others will be forthcoming soon. The concept of combination cytokine therapy is well founded in numerous in vitro and animal models. Clinical progress in the application of these agents will be slow due to the need to determine for each cytokine the maximum tolerated dose and the individual toxicity spectrum. Most likely, cytokines will be tested for single agent activity in standard phase II studies, at which point combination therapies can be evaluated. Combination studies will be limited and dictated by availability of agents from industrial and federal agencies. Sensible choices of the many permutations of combinations, sequences, doses, and routes require that clinical trials have a firm foundation in rationale from laboratory studies. Further understanding of cell biology, immunology, and body homeostasis will open the opportunity to test new concepts in cancer treatment, one of which may lead to significant therapeutic improvement for our patients.

References

1. Welte, K., Wang, C.Y., Mertelsmann, R., et al. Purification of human interleukin-2 to apparent homogenicity and its molecular heterogenecity. J. Exp. Med. 1983 156:454-4.
2. Tanaguchi, T.,Matsui, H., Fugita, T., et al. Structure and expression of a cloned cDNA for interleukin-2. Nature. 1983 302:305-10.
3. Punnonen, J. and Vilganen, M.K. Interferon alpha and interferon gamma as regulators of human adult and newborn B cell function. Int. J. Immunopharmacol. 1989 11:717-23.
4. Talmadge, J.E., Black, P.L., Tribble, H., et. al. Preclinical approaches to the treatment of metastatic disease: therapeutic properties of rH TNF, rM IFN gamma, and rH IL-2. Drugs Esp. Clin. Res. 1987 13:327-37.
5. Ernstoff, M.S., Fusi, S., and Kirkwood, J.M. Parameters of interferon action: II. Immunological effects of recombinant leukocyte interferon (IFN alpha-2) in phase I-II trials. J. Biol. Response Mod. 1983 2:540-7.
6. Lorre, K., van Damme, J., and Ceuppens, J.L. A bidirectional regulatory network involving IL 2 and IL 4 in the alternative CD2 pathway of T cell activation. Eur. J. Immunol. 1990 20:1569-75.
7. Hawrylowicz, C.M., and Unanue, E.R. Regulation of antigen presentation I. IFN gamma induces antigen presenting properties on B cells. J. Immunol. 1988 141:4083-8.
8. Schultz, R.M. Interleukin 1 and interferon gamma: cytokines that provide reciprocal regulation of macrophage and T cell function. Toxicol. Pathol. 1987 15:333-7.

9. Robb, R. J., and Greene, W.C. Direct demonstration of the identity of the T cell growth factor binding protein and the Tac antigen. J. Exp. Med. 1983 158:1332-7.
10. Heslop, H.E., Bianchi, A.C. Cordingley, F.T. et al. Effects of interferon alpha on autocrine growth factor loops in B lymphoproliferative disorders. J. Exp. Med. 1990 172:1729 34.
11. Gabrilove, J.L., Jakubowski, A., Scher, H., et al. Effects of granulocyte colony -stimulating factor on neutropenia and associated morbidity due to chemotherapy for transitional cell carcinoma of the urothelium. N. Engl. J. Med. 1988 318:1414-22.
12. Nemunaitis, J., Buckner, C.D., Appelbaum, F.R., et al. Phase I/II trial of recombinant human granulocyte-macrophage colony-stimulating factor following allogeneic bone marrow transplantation. Blood. 1991 77:2065-71.
13. Herberman, R.B., and Ortaldo, J.R. Natural killer cells: Their role in defenses against disease. Science. 1981 214:24-30.
14. Grimm, E.A., Mazumder, A., Zhang, H.Z., and Rosenberg, S.A. Lymphokine activated killer cell phenomenon. Lysis of natural killer-resistant fresh solid tumour cells by interleukin 2-activated autologous peripheral blood lymphocytes. J. Exp. Med. 1982 155:1823-41.
15. Parkinson, D.R., Fisher, R.I., Rayner, A.A., et al. Therapy for renal cell carcinoma with interleukin-2 and lymphokine activated killer cells: phase II experience with a hybrid bolus and continuous infusion of interleukin-2 regimen. J. Clin. Oncol/ 1990 8:1630-6.
16. A controlled trial of interferon gamma to prevent infection in chronic granulomatous disease. The international Chronic Granulomatous Disease Cooperative Study Group. N. Engl. J. Med. 1991 324:509-16.
17. Ley, D.E., Lew, D.J., Decker, T., Kessler, D.S., and Darnell, J.E. Jr., Synergistic interaction between interferon alpha and interferon gamma through induced synthesis of one subunit of the transcription factor ISGF3. EMBO J. 1990 9:1105-11.
18. Dett, C.A., Gatanaga, M., Ininns, E.K. et al. Enhancement of lymphokine activated T killer cell tumor necrosis factor receptor mRNA transcription, tumor necrosis factor receptor membrane expression, and tumor necrosis factor/lymphotoxin release by IL-l beta, IL-4, and IL-6 in vitro. J. Immunol. 1991 146:1522-6.
19. Pace, J.L. Synergistic interactions between IFN gamma and IFN beta in priming murine macrophages for tumor cell killing. J. Leukoc. Biol. 1988 44:514-20.
20. Nathan, C.F., Prendergast, T.J., Wiebe, M.E., Stanley, E.R., Platzer, E., Remold, H.G., Welte, K., B.Y. Rubin, and H.W. Murray. Activation of human macrophages. Comparison of other cytokines with interferon gamma. J. Exp. Med. 1984 160:600-5.
21. Ottenhoff, T.H., Wondimu, A., and Reddy, N.N. A comparative study on the effects of rIL-4, rIL-2, rIFN gamma, and rTNF alpha on specific T-cell non-responsiveness to mycobacterial antigens in lepromatous leprosy patients in vitro. Scand. J. Immunol. 1990 31:553-65.

22. Ernstoff, M.S., Nair, S., Bahnson, R.R. et al. A phase IA trial of sequential administration of recombinant DNA produced interferons: Combination recombinant interferon gamma and recombinant interferon alfa in patients with metastatic renal cell carcinoma. J. Clin. Oncol. 1990 8:1637-49.
23. Hannigan, G.E., E.N. Fish, and Williams, B.R. Modulation of human interferon alpha receptor expression by human interferon gamma. J. Biol. Chem. 1984 259:8084-6.
24. Ernstoff, M.S., Gooding, W., Nair, S., et al. The Immunological Effects of Treatment with Sequential Administration of Recombinant Interferons gamma and alpha in Patients with Metastatic Renal Cell Carcinoma, A Phase Ib Study. submitted.
25. Pross, H.S., Baines, M.G., Rubin, P., P. Shragge, and M.S. Patterson. Spontaneous human lymphocyte-mediated cytotoxicity against tumor target cells. XI. The quantitation of natural killer cell activity. J. Clin.Immunol. 1981 1:51-63.
26. Merritt, J.A., E.C. Borden, and A. Ball. Measurement of2'5'-oligoadenylate synthetase in patients receivinginterferon-alpha. J. Interferon Res. 1985 5:195-198.

23

Anti-Idiotype Antibodies: Novel Therapeutic Approach to Cancer Therapy

Kenneth A. Foon
Malaya Bhattacharya-Chatterjee
University of Kentucky
Lexington, Kentucky

I INTRODUCTION

Immunoglobulin (Ig) molecules possess variable regions specific for antigen recognition. The variable region is encoded by V_H, D and J_H genes for the heavy chains and V_L and J_L chains for the light chain (1). The variable region contains determinants known as idiotypes (Ids), which are themselves immunogenic. Antibodies can be made to many structures in the variable region associated with the light chain, heavy chain, or a combination of both chains (2,3). Early studies by Oudin and Michel (3) and Kunkel and co-workers (2) indicated that an Id was unique to a small set of antibody molecules. However, the idiotypic determinants may show a continuum of specificity from more or less private to semi-public (4,5), e.g., if different antibodies are coded by the same V_H gene segment, a shared or semi-public Id may be found. The Id is often defined by the antibody made against it known as anti-Id antibody.

In 1973 Jan Lindemann (6) and in 1974 Niels Jerne (7) proposed theories which describe the immune system of interacting antibodies and lymphocytes. According to this original network hypothesis, the Id-anti-Id interactions regulate the immune response of a host to a given antigen (Ag). Both Ids and anti-Ids have been used to manipulate cellular and humoral immunity. Ids are distinguished by their topographical location on the immunoglobulin structures and are classified by their physical relation to the Ag binding site of the antibody (Ab1). If the target for an anti-Id antibody (Ab2) is an idiotope that does not interfere with Ag binding, it is designated the α type (8). If the target idiotope is close to the binding site of Ab1 so that it interferes with antigen binding, it is called the γ type (9). If the Ab2 binds to the antigen binding site (paratope) of Ab1, then this Ab2 is referred to as β type (8,10). Anti-Id antibodies of the β type express the internal image of the Ag for Ab1. The internal image Ag may be seen as stereochemical copies of nominal Ag. They can induce specific immune responses similar to responses induced by nominal Ag. Immunization with Ab2β can lead to the generation of anti-anti-Id antibodies (Ab3) that recognize the corresponding original antigen identified by the Ab1. Because of this Ab1-like reactivity, the Ab3 is also called Ab1' to indicate that it might differ in its other idiotopes from Ab1. This cyclic nature of complimentary binding sites and idiotopes is the basis for the approach to Id vaccines (11).

The idea of anti-Ids as vaccines against infectious disease has been derived from the successful preparation and characterization of anti-Id antibodies able to mimic bacterial and viral antigens. When appropriately manipulated, anti-Ids can serve as effective inducers of T- and B-cell immunity to pathogens.

II ADVANTAGES OF ANTI-IDIOTYPE ANTIBODIES OVER CONVENTIONAL VACCINES

The approach to produce vaccines is entering a critical state of transition. The small but serious risk factors associated with certain essential vaccines have become unacceptable and have, in one instance, produced a shortage of the vaccine against pertussis. This has created a pressing need for safer vaccines in situations where large segments of the population require vaccination.

In recent years, emphasis has been directed towards the use of synthetic peptides containing the antigenic or immunogenic determinants of viral surface coat proteins (12). In effect, this approach presents to the immune system only a fragment of the nominal antigen, thereby alleviating the difficulties associated with antigen isolation and purification or the immunization with attenuated and live viruses.

The network hypothesis of Niels Jerne (7) offers still another elegant concept for developing vaccines which is not based on the conventional approach of using the nominal antigenic material. These so-called anti-Id vaccines or internal Ag vaccines take advantage of the fact that the repertoire of external or nominal antigens is mimicked by Id structures on immunoglobulins and possibly on receptors and products of T cells as well. Thus, with this approach, Id-based vaccines do not contain nominal Ag nor its fragments. This excludes the possibility that Id vaccines would have the same undesired side effects which are associated with conventional vaccines.

Besides the increased safety of Id vaccines, these new kinds of Ags have other practical, economical and biological advantages over conventional vaccines. Id vaccines do not depend on the availability of large amounts of pure Ag, which often is a limiting economical factor in vaccine production. By virtue of their being proteins, Id vaccines can be easily manipulated; they can be coupled to potent immunogenic carriers to become T-cell-dependent antigens which can receive full T help. Eventually, it might be possible to produce fully synthetic Id vaccines using essential sequence information obtained from Id hybridoma Ags.

T-dependent protein vaccines can become a decisive factor in situations where the responding immune system is immature or suppressed. From experimental studies on animals we know that the response to T-cell dependent Ags matures earlier than the T-independent response to carbohydrate Ags, and that often a genetically or acquired abnormal immune system responds better to T-dependent Ags than to T-independent Ags.

Finally, data exist showing that an acquired state of tolerance to one Ag form can be broken by using a different molecular form of the same antigenic moiety. This could become an important consideration in a broader context such as in the immunotherapy of cancer patients, who may be often immunodeficient or tolerant against their own tumor. In this review, we will discuss various examples where anti-Id antibodies are used in cancer therapy.

III PRECLINICAL AND CLINICAL TRIALS WITH ANTI-ID VACCINES

Active immunization with tumor-specific Id vaccines has been shown to inhibit the growth of tumor in animal models (13-15). A series of studies (16-19) on the effect of anti-Id therapy in a mouse leukemia model L1210 in DBA/2 mice has been described, which has provided us with basic information on B- and T-cell-induced responses using the anti-Id approach. These investigators generated a number of anti-Id hybridomas against mAb to the L1210 tumor. These anti-Id mAbs have been shown to induce tumor-specific DTH, inhibition of tumor growth, CTL antibodies and T helper cells in this system. These findings are very promising since they demonstrate a cross-reaction of nominal Ag and internal image Ag for a tumor-associated Ag system at the T- and B-cell level. In a recent study, 100% cure of established tumors was achieved in DBA/2 mice by combining anti-Id vaccines with cyclophosphamide, whereas 50% cure rate was obtained with anti-Id therapy alone (20). Similar findings have also been obtained when cyclophosphamide (100 mg/kg), administered in combination with Id vaccines to mice bearing 10-day-old, 1-2 cm diameter subcutaneous B-cell lymphoma (38C13), results in a dramatic survival benefit (21). An anti-Id antibody was used to induce immunity to SV-40 transformed cells (22). Mice vaccinated with this anti-Id demonstrated prolonged survival after tumor transfer. The role of Id interactions in regulating the immune response of mice to chemically induced, syngeneic sarcomas has been recently studied (23). Treatment with anti-Id mAb of mice with the established sarcomas (MCA-490 and MCA-1511) had significant antitumor activity.

Anti-Id responses have been implicated in the induction of anti-tumor immunity to colorectal cancer (24). Clinical trials in human colorectal patients (25) with a polyclonal anti-Id raised against the mAb 17-1A that recognizes a colon cancer-associated Ag have been initiated. Thirty patients with advanced colorectal cancer were immunized with alum-precipitated goat Ab2 in doses between 0.5 and 4 mg per injection. All patients developed Ab3 with binding specificities on the surfaces of culture tumor cells similar to the specificity of Ab1. Furthermore, the Ab3 competed with Ab1 for binding to colorectal cancer cells. Fractions of Ab3-containing sera, obtained after elution of the serum immunoglobulins from colorectal cancer cells, bound to purified tumor Ag and inhibited binding of Ab2 to Ab1. Although the aim of this trial was to evaluate Ab3 responses in patients with advanced colorectal cancer, objective clinical improvement was observed in 13 of the 30 treated patients (25). In another recent study it was demonstrated that intradermal injection of 2 mg of anti-Id mAb MK2-23, which mimics a high molecular weight human melanoma antigen, elicited anti-tumor antibody responses in the host (26). In agreement with the data obtained in animal model systems, conjugation of anti-Id MK2-23 to a carrier molecule (KLH) and mixing with an adjuvant (BCG) enhanced the immunogenicity of mAb MK2-23. Antibodies reacting with melanoma cells Colo 38 have been detected in 73% of the patients immunized with mAb MK2-23 conjugated to KLH and mixed with BCG and in only 16% of the patients immunized with the unconjugated mAb MK2-23. Repeated injections of murine anti-Id mAb were not associated with any side effects. Furthermore, reductions in the size of metastatic lesions were observed in at least six of the immunized patients (26).

We have generated monoclonal Id cascades for two different human tumor-associated antigens. The first cascade originated from a T-cell leukemia/lymphoma (27,28) and the other from carcinoembryonic antigen (CEA) (29). In both cascades we have produced TAA-mimicking monoclonal anti-Ids (Ab2) and monoclonal anti-anti-Ids (Ab3) which bind to the original TAA. We have initiated our first clinical trial for patients with cutaneous T-cell lymphoma. Two patients have been treated and anti-anti-idiotypic responses have been demonstrated.

To the best of our knowledge, 3H1 is the only potential monoclonal anti-Id reagent available for active immunization therapy of CEA-positive cancer patients. The primary questions to be resolved are whether this anti-Id reagent can evoke an Ab3 as well as cellular immune response in patients; whether any Ab3 so derived behaves as an Ab1-like antibody and whether this Ab3 and/or cellular immunity can mediate a potential antitumor effect. There is evidence that certain Ab2 can induce an immune response in "silent clones" of lymphocytes which are not activated by mere exposure to nominal antigen (30,31). Active immunization of patients with Ab2β might generate an Ab3 (Ab1') response which will hopefully be able to recruit human effector cells or complement to destroy the tumor by antibody-dependent, cell-mediated cytotoxicity (ADCC) or complement-mediated cytotoxicity (CMC). In addition, anti-Id antibodies could induce various T-cell activities such as stimulation of DTH and cytotoxic T cells specific for the tumor. Syngeneic monoclonal anti-Id mAbs have been successfully utilized as specific immunogens to elicit both T- and B-cell-mediated immunity in experimental L1210/GZL tumor system and viral systems (32).

In order to break tolerance to TAA, patients might have to be immunized with Ab2β after conjugation with a potential immunogenic carrier and mixed with an appropriate adjuvant. This further alteration of the molecular environment for expression of the idiotope antigens was indeed required in the work on the generation of an idiotope vaccine for pneumococcal infection (33). Without coupling the idiotope Ag to a carrier, the internal image was ineffective in producing a serum titer or in providing immunity against infection. Other studies have demonstrated the immunoprophylactic effects of Id vaccination against a murine B-cell lymphoma (15,21). These studies also revealed that conjugation of the Id protein to an immunogenic carrier protein, such as KLH, and administration with an adjuvant, such as the muramyl dipeptide analog (MDP-derivative) were necessary for optimal antitumor responses. Although the mouse-specific Fc portion may function as a carrier determinant, we plan to use $F(ab')_2$ fragments of 3H1 coupled to KLH and mixed with BCG.

Allergic reactions to murine antibodies is a concern. Previous experience with murine antibodies in humans has not shown serious anaphylactic type reactions, although low-grade fever, nausea, diarrhea and urticaria are reported. These reactions have not led to discontinuation of therapy, however.

Potential heterologous and homologous anti-Id vaccines may be replaced by synthetic peptide antigens. It is possible to obtain the three-dimensional shape of idiotope internal internal antigens using computer modeling on the sequence of internal antigen hybridomas (34). The molecular modeling studies will indicate whether the internal idiotope antigen is expressed as a linear sequence or a complex determinant involving non-contiguous sequence contributions to the molecular architecture. After obtaining sufficient information as to the most likely peptide regions which might depict the antigenic structure, it should be possible to synthesize the relevant peptides which can mimic the structure and couple them to new synthetic adjuvants to increase the relative level of immunogenicity.

Heterogeneous expression and modulation of target Ags represent other potential problems to immunotherapy and modulation with anti-Id vaccine preparations. However, these problems are not insurmountable and they may be minimized by several complementary approaches. Heterogeneity of TAA expression may be addressed by utilizing cocktails of anti-Id vaccine preparations directed against multiple target antigens collectively expressed by the vast majority of tumor cells. With respect to antigenic modulations by Ab3 induced in response to Ab2β, it is known that modulation rates for different antigens may vary considerably. Hence, TAAs with slow modulation characteristics should be chosen as targets for anti-Id active immunotherapy. Furthermore, recent studies (35,36) have indicated that some biological response modifiers such as interferon are capable of inducing the expression of TAAs in several tumor systems. Hence, in addition to providing general stimulation to the immune system, the administration of biological response modifiers known to increase the expression of relevant TAAs may further enhance the efficacy of anti-Id tumor vaccine application.

The anti-Id approach needs to be compared to other tumor therapies, both established and experimental. A realistic assessment of anti-Id therapy predicts that complete remissions in patients with advanced disease will be unlikely. However, evidence exists that partial remission and responses can be achieved with anti-Ids. It is apparent that successful therapy of established tumors will depend on a multi-modality approach. Since the immune response has a limited capacity, immunotherapeutic manipulations will probably be most effective against small tumor masses. Consequently, immunotherapy with Id vaccine would be most attractive in the adjuvant setting. Compared to chemotherapy of lymphokine therapy, the Id vaccine approach is considerably less toxic.

IV ANTI-ID APPROACH TO B-CELL TUMORS

Another therapeutic approach with anti-Id mAbs involves the use of anti-Id mAbs in B-cell malignancies. In this setting, anti-Id antibodies have as their target a tumor-specific antigen, the Id of the cell surface immunoglobulin present on B cells. Indeed, this antigen is the closest we have come to identifying a tumor-specific Ag in man. This specificity is based on the fact that individual B cells are committed to the synthesis of only one immunoglobulin species with a unique variable region structure (Id). Moreover, since B-cell lymphomas and leukemias are clonal in nature, members of the malignant clone should express the same immunoglobulin molecule, and hence the same Id. This feature thus represents a marker by which these tumor cells can be distinguished from normal cells of the host. These facts also imply that an individual patient's tumor cell Id will be different from that of other patients, hence anti-Id antibodies must be "tailor-made" for the individual patient. Because of the highly specific nature of these antibodies, treatment with these antibodies has yielded important results regarding the ultimate potential of mab therapy.

The largest experience reported with anti-Id therapy is the work of Levy and co-workers. Their first attempt at this therapy was in a patient originally diagnosed as having a malignant lymphoma of the nodular, poorly differentiated, lymphocytic type (follicular small cleaved cell lymphoma) (37). At the time of treatment, the patient had evidence of rapidly progressive systemic disease symptoms which were resistant to chemotherapy and interferon. Following eight continuous six-hour intravenous infusions spaced over the period of one month, the patient entered a complete clinical remission for six years (R. Levy, personal communication). The mechanisms accounting for this dramatic response are not clear. Because it was noted that the patient's antitumor response continued after the period of passive antibody administration, evidence of an anti-anti-Id antibody response by the patient himself was investigated, but none was detected. It is still possible that indirect mechanisms could have been involved. Since the immune system may be regulated in part by networks of interactions between Ids and anti-ids, the administered anti-Id could have triggered these types of networks of interactions which led to an antiproliferative response against the patient's tumor.

Additional patients have been treated with individually tailored anti-Id antibodies of varying antibody subclasses (38). Some patients have been treated with more than one antibody (differing in isotype or epitope specificity) during the course of an individual treatment period. Significant tumor responses have been demonstrated in 50% of the patients with limited complete responses. Several important lessons have been learned from these studies. It was found that >1 gram of anti-Id mAb could be infused safely as a single dose, provided the level of circulating free antigen (Id) was low or non-detectable and if no immune response by the host against the infused mouse protein (human antimouse antibodies) was present. The presence of both serum Id and human antimouse antibodies was correlated with acute toxicity during infusions consisting of fever, rigors, dyspnea, arthralgias, and headache, with thrombocytopenia occurring less commonly. This was presumably due to immune complex formation. The presence of significant levels of serum Id was found to clearly be a barrier to antibody penetration to tumor sites and thus to a clinical response. Plasmapheresis was shown to transiently reduce serum Id levels but not to a degree sufficient to eliminate this barrier. The effect of the presence of an antimouse response by the host was similar in that tissue penetration and clinical response were prevented by these antibodies. About one-third of patients developed this response within a two-week period after the initial infusion. This thus appears to be a less frequent phenomenon in B-cell lymphoma patients than in patients with solid tumors and T-cell lymphomas.

Another means by which patients' tumors could evade the therapeutic effects of anti-Id antibodies was by the emergence of Id variants within tumors during treatment (39). This phenomenon was recognized when tumors from two patients lost reactivity with the anti-Id antibody generated against the respective original tumors during treatment. Subsequent studies have shown that this loss of reactivity was not due to antigenic modulation. Comparison of immunoglobulin gene rearrangements by Southern blot analysis in pre-therapy and post-therapy tumors taken from each patient revealed identical rearrangements in each case. This strongly suggests that all cell populations studied were part of a single monoclonal lymphoma in each patient. In one of these cases, the anti-Id antibody was known to react with only the heavy chain variable region of the surface IgM protein of the pre-therapy tumor and not with γ light chain regions. Eight independent heavy chain variable region isolates from tumors prior to and after treatment were subjected to nucleotide sequence analysis (40). Extensive point mutations were demonstrated in all isolates and no two sequences were identical. A clustering of mutations encoding for amino acid changes was observed in the CDR2 region. Comparison of pre-therapy and post-therapy sequences implicated a single amino acid in CDR2 at position 54 as being important in determining reactivity with the anti-Id antibody. Three of the post-therapy sequences had a common substitution at that position, and a fourth post-therapy sequence had other substitutions in a neighboring position. Thus, clones with mutations in this region apparently escaped the antibody's strong negative selection pressure in vivo. Further analysis indicated that there was a significant bias against mutations resulting in amino acid changes in portions of the V region gene other than CDR2, even in the absence of any selection by antibody treatment. Thus the non-random clustering in CDR2 may have been due to endogenous selective forces interacting with tumor cell surface immunoglobulin. The generality of these concepts is now being explored in other patients' tumor samples. It is now believed that somatic mutation accounted for tumor escape in more than these two patients than the exception (41,42). This poses an additional problem for anti-Id antibody therapy in that more than one antibody may need to be developed for each individual patient so that Id variants within the tumor can be recognized.

Various factors have been studied to predict response to this therapy. Included among these are the isotype of the anti-Id antibody used, the density of cell surface Id, the epitope recognized by the anti-Id antibody, the affinity of anti-Id antibody for antigen, the relative ability of the anti-Id antibody to modulate surface antigen, the direct effect of antibody on tumor cell proliferation in vitro, and the degree of T-cell infiltration present in pre-therapy tumor specimens. None of these factors has been positively correlated with good clinical outcome, except the number of T cells present in pre-therapy tumor tissue (43,44). In the two best responding cases, the T cells actually outnumbered the tumor cells. The majority of these T cells were the helper/inducer phenotype (CD4). Whether the anti-Id antibodies given to these patients augmented an ongoing cell-mediated cytotoxic response by the host against the tumor is not clear. Certainly more observations on pre-therapy T-cell infiltration must be made before the actual significance and function of this finding become apparent. Another factor which has yet to be fully explored is the nature of somatic mutation in the immunoglobulin genes of the various tumors of patients undergoing treatment, as they may more fully define an endogenous host response that may regulate tumor growth and eventual response to therapy.

Some anti-Id mAbs made against an individual patients' tumor cells are recently shown to cross-react with more than one patient's tumor cells. A panel of 29 anti-shared Id (SId) mAbs that reacted with approximately one-third of cases of B-cell lymphoma has been identified (45). These anti-SId mAbs also detected rare normal B cells and minor components of serum Ig. Using this same panel of anti-SId mAbs, we extended previous observations to other B-cell malignancies such as chronic lymphocytic leukemia (CLL) and diffuse B-cell lymphoma. We demonstrated 25% cross-reactivity of CLL and 40% cross-reactivity with NHL (46). Thus far, it is too early to determine whether anti-SId mAbs will be as effective as "tailor-made" anti-Id mAbs.

There is also an active approach to anti-Id therapy with B-cell diseases that simulate anti-Id "internal image" tumor vaccines. In this case, purified immunoglobulin derived from patient's tumors was used as immunogen to generate an active immunity. In one study, eight of nine patients immunized with autologous tumor-derived immunoglobulin coupled to KLH and mixed with adjuvant demonstrated idiotype-specific immune responses, either humoral, cellular or both (47). Tumor regressions were also reported. We have been interested in a similar approach to active anti-Id therapy in B-cell diseases. We have generated a murine anti-Id mAb to an antibody designated I0I0 (48) that binds to a highly restricted antigen on malignant B cells. This antigen is not found on any normal cells, including lymphocytes and activated lymphocytes. It is, however, identified on over 75% of B-cell malignancies and represents an ideal target antigen for active immunotherapy.

V CONCLUSION

Immunization with anti-Id antibodies represents a novel new approach to active immunotherapy. Preclinical trials using internal image anti-Ids for infectious diseases and cancer have supported the "immune network theory." Whether this will have clinical applicability in man is currently under investigation.

Acknowledgements

This work was supported by United States Public Health Service Grants CA47860 and CA51434.

References

1. Tonegawa, S. Somatic generation of antibody diversity. Nature 1983, 302:575-581.
2. Kunkel, H.G., Mannik, M., Williams, R.C. Individual antigenic specificity of isolated antibodies. Science 1963, 140:1218-1219.
3. Oudin, J., Michel, M. A new allotype form of rabbit serum gammaglobulins, apparently associated with antibody function and specificity. C.R. Acad. Sci. Paris 1963, 257:805-808.
4. Williamson, A.R. The biological origin of antibody diversity. Annu. Rev. Biochem. 1976, 45:467.
5. Stevenson, G.T., Glennis, M.J. Surface immunoglobulin of B-lymphocytic tumours as a therapeutic target. Cancer Surv. 1985, 4:213-44.
6. Lindenmann, J. Speculations on idiotypes and homobodies. Ann. Immunol. (Paris) 1973, 124:171-184.
7. Jerne, N.K. Towards a network theory of the immune system. Ann. Immunol. (Paris) 1974, 125C:373-389.
8. Jerne, N.K., Roland, J., Cazenave, P.A. Recurrent idiotypes and internal images. EMBO J. 1982, 1:243-247.

9. Bona, C.A., Köhler, H. Anti-idiotypic antibodies and internal images. In: Monoclonal and Anti-Idiotypic Antibodies: Probes for Receptor Structure and Function, Venter, J.C., Fraser, C.M., Lindstrom, J. (Eds.) New York, Alan R. Liss, 1984, pp. 141-150.
10. Kohler, H. The immune network revisited. In: Idiotype in Biology and Medicine, Köhler, H., Urbain, J., Cazenave, P. (Eds.) New York, Academic Press, 1984, pp. 3-28.
11. Nisonoff, A., Lamoyi, E. Implications of the presence of an internal image of the antigen in anti-idiotypic antibodies: Possible application to vaccine production. Clin. Immunol. Immunopathol. 1981, 21:397.
12. Shinnick, T.M., Sutcliffe, J.G., Green, N., Lerner, R.A. Synthetic peptide immunogens as vaccines. Ann. Rev. Microbiol. 1983, 37:425-446.
13. Sugai, S., Palmer, D.W., Talal, N., Witz, I.P. Protective and cellular immune responses to idiotypic determinants on cells from a spontaneous lymphoma of NZB-NZW Fl mice. J. Exp. Med. 1974, 140:1547-1558.
14. Stevenson, F.K., Gordon, J. Immunization of idiotypic immunoglobulin protects against development of B lymphocytic leukemia, but emerging tumor cells can evade antibody attack by modulation. J. Immunol. 1983, 13:970.
15. Kaminski, M.S., Kitamura, K., Maloney, D.G., Levy, R. Idiotype vaccination against murine B-cell lymphoma. Inhibition of tumor immunity by free idiotype protein. J. Immunol. 1987, 138:1289-1296.
16. Raychaudhuri, S., Saeki, Y., Fuji H., Köhler, H. Tumor-specific idiotype vaccines. I. Generation and characterization of internal image tumor antigen. J. Immunol. 1986, 137:1743-1749.
17. Raychaudhuri, S., Saeki, Y., Chen, J.J., Iribe, H., Fuji, H., Köhler, H. Tumor-specific idiotype vaccines. II. Analysis of the tumor-related network response induced by the tumor and by internal image antigens (Ab2ß). J. Immunol. 1987, 139:271-278.
18. Raychaudhuri, S., Saeki, Y., Chen, J.J., Iribe, H., Fuji, H., Köhler, H. Tumor-specific idiotype vaccines. III. Induction of T helper cells by anti-idiotype and tumor cells. J. Immunol. 1987, 139:2096-2102.
19. Raychaudhuri, S., Saeki, Y., Chen, J.J., Köhler, H. Tumor-specific idiotype vaccine. IV. Analysis of the idiotypic network in tumor immunity. J. Immunol. 1978, 139:3902-3910.
20. Chen, J.J., Saeki, Y., Shi, L., Köhler, H. Syngernistic anti-tumor effects with combined "internal image" anti-idiotypes and chemotherapy. J. Immunol. 1989, 143:1053-1057.
21. Campbell, M.J., Esserman, L., Levy, R. Immunotherapy of established murine B-cell lymphoma. Combination of idiotype and cyclophosphamide. J. Immunol. 1988, 141:3227-3233.
22. Kennedy, R.C., Dreesman, G.R., Butel, J.S., Lanford, R.E. Suppression of in vivo tumor formation induced by simian virus 40-transformed cells in mice receiving anti-idiotypic antibodies. J. Exp. Med. 1985, 161:1432-1439.

23. Nelson, K.A., George, E., Swenson, C., Forstrom, J.W., Hellstrom, K.E. Immunotherapy of murine sarcomas with auto-anti-idiotypic monoclonal antibodies which bind to tumor-specific T cells. J. Immunol. 1987, 139:2110-2117.
24. Koprowski, H., Herlyn, D., Lubeck, M., DeFreitas, E., Sears, H.F. Human anti-idiotype antibodies in cancer patients: Is the modulation of the immune response beneficial for the patient? Proc. Natl. Acad. Sci. USA 1984, 81:216-219.
25. Herlyn, D., Wettendorf, M., Schmoll, E., Hopoulos, D., Schedel, I., Dreikhausen, U., Raab, R., Ross, A.H., Jaksche, H., Scriba, M., Koprowski, H. Anti-Id immunization of cancer patients: Modulation of immune response. Proc. Natl. Acad. Sci. USA 1987, 84:8055-8059.
26. Ferrone, S., Chen, Z.J., Yang, H., Yamada, M., Zheng, Y., Mittelman, A. Active specific immunotherapy with murine anti-idiotypic monoclonal antibodies which bear the internal image of the human high molecular weight melanoma-associated antigen (HMW-MAA). Proc. Annu. Meet. Am. Assoc. Cancer Res. 1990, 31:474.
27. Bhattacharya-Chatterjee, M., Pride, M.W., Seon, B.K., Köhler, H. Idiotype vaccines against human T cell acute lymphoblastic leukemia (T-ALL). I. Generation and characterization of biologically active monoclonal anti-idiotypes. J. Immunol. 1987, 139:1354-1360.
28. Bhattacharya-Chatterjee, M., Chatterjee, S.K., Vasile, S., Seon, B.K., Köhler, H. Id vaccines against human T cell leukemia. II. Generation and characterization of a monoclonal Id cascade (Ab1, Ab2 and Ab3). J. Immunol. 1988, 141:1398-1403.
29. Bhattacharya-Chatterjee, M., Murkerjee, S., Biddle, W., Foon, K.A., Köhler, H. Murine monoclonal anti-Id antibody as a potential network antigen for human carcinoembryonic antigen. J. Immunol. 1990, 145:2758-27654.
30. Bona, C.A., Herber-Katz, E., Paul, W.E. Idiotype-anti-idiotype regulation. I. Immunization with a levan-binding myeloma protein leads to the appearance of auto-anti-(anti-idiotype) antibodies and to the activation of silent clones. J. Exp. Med. 1981, 153:951-967.
31. Cazenave, P.A. Idiotypic-anti-idiotypic regulation of antibody synthesis in rabbits. Proc. Natl. Acad. Sci. USA 1977, 74:5122-5125.
32. Finberg, R.W., Ertl, H. The use of anti-idiotypic antibodies as vaccines against infectious agents. CRC Crit. Rev. Immunol. 1987, 7:269-284.
33. McNamara, M.K., Ward, R.E., Köhler, H. Monoclonal idiotype vaccine against Streptococcus pneumoniae infection. Science 1984, 226:1325-1326.
34. Kieber-Emmons, T., Ward, R.E., Raychaudhuri, S., Rein, R., Kohler, H. Rational design and application of idiotope vaccines. Int. Rev. Immunol. 1986, 1:1-26.
35. Carrel, S., Schmidt-Kessen, A., Giuffre, L. Recombinant interferon-gamma can induce the expression of HLA-DR and -DC on DR-negative melanoma cells and enhance the expression of HLA-ABC and tumor-associated antigens. Eur. J. Immunol. 1985, 15:118-123.

36. Liao, S.K., Kwong, P.C., Khosravi, M.,. Dent, P.B. Enhanced expression of melanoma-associated antigen monoclonal antigens and β_2-microglobulin on cultured human melanoma cells by interferon. J. Natl. Cancer Inst. 1982, 68:19-25.
37. Miller, R.A., Maloney, D.G., Warnke, R., Levy, R. Treatment of B-cell lymphoma with monoclonal anti-idiotype antibody. N. Engl. J. Med. 1982, 306:517-522.
38. Meeker, T.C., Lowder, J., Maloney, D.G., Miller, R.A., Thielemans, K., Warnke, R., Levy, R. A clinical trial of anti-idiotype therapy of B-cell malignancy. Blood 1985, 65:1349-1363.
39. Meeker, T.C., Lowder, J., Cleary, M.L., Stewart, S., Warnke, R., Sklar, J., Levy, R. Emergence of idiotype variants during treatment of B-cell lymphoma with anti-idiotype antibodies. N. Engl. J. Med. 1985, 312:1658-1665.
40. Cleary, M.L., Meeker, T.C., Levy, S., Lee, E., Trela, M. Sklar, J., Levy, R. Clustering of extensive somatic mutations in the variable region of an immunoglobulin heavy chain gene from a human B-cell lymphoma. Cell 1986, 44:97-106.
41. Raffeld, M., Neckers, L., Longo, D.L., Cossman, J. Spontaneous alteration of idiotype in a monoclonal B-cell lymphoma. Escape from detection by anti-idiotype. N. Engl. J. Med. 1985, 312:1653-1658.
42. Carroll, W.L., Lowder, J.N., Streifer, R., Warnke, R., Levy, S., Levy, R. Idiotype variant cell populations in patients with B-cell lymphoma. J. Exp. Med. 1986, 164:1566-1660.
43. Lowder, J.N., Meeker, T.C., Campbell, M., Garcia, C.F., Gralow, J., Miller, R.A., Warnke, R., Levy, R. Studies on B lymphoid tumors treated with monoclonal anti-idiotype antibodies: Correlation with clinical responses. Blood 1987, 69:199-210.
44. Garcia, C.F., Lowder, J., Meeker, T.C., Bindl, J., Levy, R., Warnke, R.A. Differences in "host infiltrates" among lymphoma patients treated with anti-idiotype antibodies: Correlation with treatment response. J. Immunol. 1985, 135:4252-4260.
45. Miller, R.A., Hart, S., Samoszuk, M., Coulter, C., Brown, S., Czerwinski, D., Kelkenberg, J., Royston, I., Levy, R. Shared idiotopes expressed by human B-cell lymphomas. N. Engl. J. Med. 1989, 321:851-857.
46. Chatterjee, M., Barcos, M., Han, T., Liu, X., Bernstein, Z., Foon, K.A. Shared idiotype expression by chronic lymphocytic leukemia and B-cell lymphoma. Blood 1990, 76:1825-1829.
47. Kwak, L.W., Campbell, M.J., Czerwinski, D., Levy, R. Active specific immunotherapy of B-cell lymphoma with purified tumor-derived immunoglobulin. Blood 1990, (Suppl 1) 76:211a.
48. Gingrich, R.D., Dahle, C.E., Hoskins, K.F., Senneff, M.J. Identification and characterization of a new surface membrane antigen found predominantly on malignant B lymphocytes. Blood 1990, 75(12): 2375-2387.

24

Immunity and Cancer Therapy: Present Status and Future Projections[1]

Enrico Mihich
Grace Cancer Drug Center
Roswell Park Cancer Institute
New York State Department of Health
Buffalo, NY

I INTRODUCTION

Cancer chemotherapy has achieved major successes during the past 40 years as a certain percentage of patients with certain types of cancer can now be brought into complete remission by the use of drugs alone or in combination with other modalities of treatment and are free of detectable disease five or more years after therapy. Major difficulties still remain to be overcome, however, before cancer therapeutics can provide curative treatments for many of the so-called common solid tumors. These difficulties are related to the fact that the drugs available to date do not have antitumor-specificity. In addition, they are not sufficiently selective against the tumor: therefore, even a minor degree of resistance at the tumor cell level cannot be overcome by dose increases without incurring unacceptable toxicity. While the search for new antitumor agents continues and is now focused on the increasing availability of new specific sites of intervention, particularly those related to unique tumor cell regulatory mechanisms, other modalities of treatments are being developed which may provide for greater selectivity of antitumor action. Prominent among these are modalities based on the exploitation of antitumor host defenses, whether through a drug-induced modification of regulatory mechanisms of immune responses, or through the administration of immune effectors and/or mediators, or through a modification of tumor cell populations. Indeed by its very nature immunotherapy may provide the kind of selectivity, perhaps even specificity, of antitumor action that has so far eluded chemotherapy.

---

[1] Based on the Keynote Address delivered on May 13, 1991 at the International Symposium "Tumor Immunology: Basic Mechanisms and Prospects for Therapy," Pittsburgh, PA, to honor Dr. Ronald Herberman on his 50th birthday.

Although the general concept that host defenses against tumor exist and might be augmented therapeutically preceded the beginning of modern chemotherapy, it is only in the late 1950's that results obtained in rigorous experimental tumor models proved the existence of specific antitumor immunity (1,2). Based on this information, and on data obtained in syngeneic transplantable tumor systems, much developmental work on immunotherapy has been carried out for many years, primarily in mice, with the expectation that also in humans effective antitumor immunity existed and could be activated and/or augmented. In the case of some human tumors, this assumption was supported by the identification of tumor associated antigens (TAA): even though these were found to be differentiation antigens and not tumor specific antigens in an absolute sense, their almost exclusive presence in malignant melanoma, some lymphomas and leukemias, sarcomas and carcinomas of pancreas, colorectum, breast, lung and ovaries, lent credence to that expectation. However, it soon became apparent that the presence of TAA on tumor cells or in circulating blood did not _per se_ indicate that effective host mechanisms directed against them could be exploited therapeutically.

At present much evidence supports the notion that tumor immunity in fact exists in humans and is expressed by both humoral and cellular effectors. For instance, _in vitro_, major histocompatibility complex (MHC) restricted proliferation of $CD4^+$ cells and the generation of cytotoxic T lymphocytes (CTL) with autologous tumor specificity have been demonstrated in the case of some tumors, e.g. malignant melanoma (3). _In vivo_, again in patients with melanoma, tumor infiltrating lymphocytes (TILs) have exhibited antitumor activity and tumor specificity when assayed _in vitro_ (4). Although lymphokine activated killer (LAK) cells do not exhibit the tumor specificity provided by T cells, they do represent a mechanism of host defense which could be exploited in humans (4). In terms of antibodies, specific binding to autochthonous tumor cells (5), generation of anti-idiotypes (6) and possibly specific cytotoxic activity have been demonstrated with various human tumor types (7). These and other examples indicated not only that tumor immunity exists in humans, but also that it may be exploited in therapeutics. While this conclusion provides encouragement towards further intensive pursuits in this direction, it is also fair to say that to this date the evidence for successful therapeutic exploitation of antitumor host defenses is still preliminary.

II MAJOR APPROACHES IN IMMUNE INTERVENTION

These can be somewhat artificially classified into approaches aimed at antitumor host response augmentation and approaches involving the transfer of antitumor humoral or cellular effectors; certain cytokines may be used as effectors or immunomodulators.

The use of tumor vaccines for immunoaugmentation was attempted already several decades ago before solid evidence of the existence of TAA had become available. At present this approach is being pursued with renewed intensity and is yielding some success, at least in the case of malignant melanoma (see below). Immunostimulation was also attempted in early years, using partially purified extraction preparations from natural products, most of which were shown to stimulate macrophage and other cell functions. An early example was that of bacillus Calmette-Guérin (BCG) (8,9). As greater knowledge is acquired about immune effector functions and their regulation, and as agents purified to homogeneity, chemically synthesized or made by genetic engineering become available, the therapeutic exploitation of immunoaugmentation can be studied with greater scientific rigor than in the past; attempts also are being made to identify the mode of action of the agents studied despite the fact that their pleiotropic effects on host responses make it difficult to identify the specific effector functions that may be primarily involved in the antitumor effects seen. Bone marrow stimulation, at present carried out with specific cytokines, also represents a type of immunostimulation which in most cases has the main objective of protecting or rescuing the bone marrow from tumor and/or therapy induced suppression (10).

The passive transfer of antibodies directed against tumor specificities has been attempted many years ago and has been greatly facilitated by the advent of hybridoma technology and the consequent availability of monoclonal antibodies. Nevertheless, the information obtained until recently with cytotoxic antibodies is mainly anecdotal in nature. More systematic trials are being carried out at present (11). The adoptive transfer of effector cells, obtained from the patient and expanded in culture, has been successful in initial therapeutic trials and promises to have increased applications in the future (4).

A. Tumor Vaccines

Recent trials of active immunization have been carried out in patients with malignant melanoma. In a study by Mastrangelo's group (12), of 40 patients with metastatic disease immunized with a vaccine prepared according to a modification of the method by Hanna and his group (13), 4 achieved a complete remission and 1 a partial remission with several additional patients showing responses of varying degrees; the 4 complete remissions were of substantial duration. In this study the patients had been treated with cyclophosphamide (300mg/m^2) 3 days before the first of the series of i.d. vaccinations. It is of interest that metastasis regression was associated with intra-tumor host cells infiltrates. Delayed type hypersensitivity (DTH) reactions to autologous melanoma cells were augmented in 60% of the patients, these increases were associated with tumor regression in 8 out of 10 patients and there was a linear relationship of DTH increases to disease-free interval (12). Other positive responses in patients with melanoma were obtained by Mitchell et al. (14) and by Morton et al. (15), and in patients with renal cell carcinoma by Sahasraludhe et al. (16) and McCune et al. (17). Prompted by these promising results current work is directed toward developing better vaccines and, hopefully, disease-specific or patient subset specific vaccines based on the existence of common tumor antigens.

B. Immunoaugmenting and Immunomodulating Agents

Bacterial products were the first to be tested within this group of agents, BCG being probably one of the most extensively studied (8,9). Significant antitumor effects were seen in patients with melanoma (18) and with superficial bladder tumors (19); in both cases in situ or regional administration were most effective. A direct antitumor effect was unlikely; indeed inoculation of BCG into one melanoma nodule led also to the regression of distant non-inoculated nodules. Although the mechanism of action of BCG has not been unequivocally clarified, it seems likely that activation of cells of the monocytic lineage is a relevant factor. This idea is corroborated also by the unique macrophage stimulating activity of MTP-PE (20), a synthetic analog of muramyl dipeptide, the smallest subunit of BCG that retains the adjuvant effect (21). Other bacterial products that have been tested clinically with uncertain success include C. parvum, Nocardia cell wall, streptococcal preparation OK432 (22) and FK565. Polysaccharide preparations such as Krestin, lentinan or glucans (22) have also been used in patients with a variety of neoplasias, especially in Japan, in conjunction with chemotherapeutic treatments. Although the majority of these preparations seems to provide non-specific immunoaugmentation, and are of some benefit to the patient in reducing the negative impact of opportunistic infections, rigorous evidence for antitumor effects mediated by an augmentation of host defense mechanisms is not available in most cases. In addition, modern oncologists are reluctant to use preparations not purified to homogeneity. It is conceivable that a synthetic preparation of a well characterized active principle may ultimately prove to have clinical utility. For example, Krestin has been shown to have augmenting effects on the mRNA of specific lymphokines (23) and this might lead to interesting applications after the active principles of this preparation have been completely characterized. Other curative products used in recent years as immunoaugmenting agents include the peptides tuftsin and bestatin (22).

Levamisol deserves particular mention among a group of miscellaneous immunoaugmenting agents studied to date. In fact, in suitable combination with 5-fluorouracil, in an adjuvant treatment, the agent provided for substantially improved therapeutic effects in patients with Dukes' stage C colon cancer (24). Whereas no evidence is available to indicate that this effect is related to the immunomodulating activity of levamisol, it is tempting to assume that this is the case. Should a role for immunomodulation be demonstrated in humans, it would greatly encourage exploring the use of other such agents in conjunction with optimal chemotherapeutic treatments of so-called solid tumors.

Several anticancer drugs have been found to have immunomodulating action expressed in specific effects on the immune system (25). For instance, 6-mercaptopurine affects antibodies regulation depending on dose, antigenic stimulus and time of administration relative to antigenic stimulation (26); cyclophosphamide at low doses inhibits precursors of T suppressor cells/function (27); Adriamycin positively affects cells of the macrophage-monocyte type and the regulation of T cells, and also induces stimulation of the production of specific cytokines (25,27), bleomycin inhibits T suppressor function and stimulates macrophages (28), and also induces the production of certain lymphokines (29); vincristine stimulates CTL function but not through an inhibition of T suppression whereas the closely related vinblastine augments antibody responses without affecting CTLs (30); cis-platinum, busulfan and melphalan each inhibit T cell suppressor function (25). Thus in recent years it has become apparent that several anticancer agents can exert immunomodulating effects; it is conceivable that these effects can be exploited, in addition to the well established cytotoxic effect of these drugs, in suitable therapeutic combination regimens. In this respect, it should be noted that in a syngeneic mouse tumor model system, Adriamycin in combination with interleukin 2 (IL-2) had curative effects that could not be obtained with either combinant alone (31). This effect was dependent on intact host defense functions, particularly those of $CD8^+$ cytotoxic T cells (32) and could be obtained also when a tumor resistant to Adriamycin was used, this suggesting that most of the therapeutic synergisms observed were effected through host mechanisms of antitumor defenses (31).

C. Antibodies

The passive transfer of antibodies directed against TAA is probably the longest studied modality among those based on the passive transfer of effectors of the immune system. Nevertheless, as viewed by Catane and Longo in 1988 (33), of the first 184 cancer patients given monoclonal antibodies (mAbs) at various centers in the world, only 3 achieved complete remission and 23 partial remission of their disease. Although in some cases the mAbs seem to exert direct antitumor cytotoxicity, in many cases it would appear that antibody dependent cellular cytotoxicity (ADCC) is the mechanism involved. Efforts have also been made to use mAbs conjugated with radioisotypes to effect localized radiation treatments (34) or with anticancer drugs for improved tumor localization (35). Several problems have become apparent which have not yet been satisfactorily solved; most of the mAbs studied are directed against weak TAA, they have low affinity for the TAA and limited penetrability into the tumor mass. The mouse mAbs used in many cases have limiting immunogeneity in humans. Heterogeneity of tumor cell population is reflected in TAA heterogeneity and this poses problems analogous to those encountered in chemotherapy for the same basic reason. When mAbs are used in conjugated form additional problems arise which are related to the need to optimize the stability of the conjugate, the need to use internalizing antibodies when selective uptake of drugs or toxins is sought, the occurrence of non-specific binding primarily to the RES. Some of these problems are currently being addressed with a measure of success (36). In this all too brief overview only two examples are mentioned because of their significance, namely the long-standing investigations by Levy's group of the use of anti-idiotype antibodies for the treatment of patients with B cell lymphomas (6) and the studies by Waldman's group of anti-TAC "hyperhumanized" antibodies for the treatment of IL-2 receptor expressing neoplasias (37).

Levy's group pioneered the development of anti-idiotype antibodies against epitopes on B cell lymphomas. Important principles have been established in this area, already starting from the bindings reported with a single patient in 1982 (6). It was found that treatment with the antibody caused progressive decrease in tumor volume which started to be evident only some 12 days into treatment when tumor epitopes circulating in blood were no longer measurable and free antibody titers became detectable. Complete remission was achieved in that patient after treatment had to be ended because the antibody had been used up; the mechanisms involved in this delayed effect are still obscure. This initial patient remained in complete remission for 6 years and, after recurrence and X-ray treatment, for 3 additional years until she died of a heart attack (Levy, personal communication). By now the experience gained by the Stanford group includes 55 patients with complete and partial responses in more than 50% of the cases. The complete responses lasted unmaintained in some cases for more than 4 years, with a median of 9 months as evaluated in 1991. Usually treatment lasted 3-4 weeks and positive responses occurred later, confirming the original observation of delayed effects. The responses were somewhat associated with increases of intratumor $CD4^+$ T cells and could not be ascribed to an ADCC mechanism. In this respect it might be of interest to set up combined treatments of antiidiotype antibodies with TILs. Treatments with these antibodies are rendered difficult by the need to develop individualized reagents. An antibody library is being built at Stanford based on the relatively recent recognition that about one-third of the patients exhibit cross-reactivities (6): the availability of antibodies ready for use in a substantial proportion of patients would represent a major improvement over the lengthy preparation of custom-made antibodies and should increase the number of patients who could be given this treatment. Still a suitable selection of patients with less than 50 μg/ml of idiotype protein in blood is necessary because of the requirement identified already in 1982, that circulating antigen be neutralized before antitumor effects can be seen. Numerous potentially exciting developments may stem from the excellent results obtained to date under Levy's leadership, including the design of combined treatment with drugs and/or cytokines which may overcome problems related to the existence of idiotype-negative cells within heterogeneous tumor cell populations.

Another approach pursued at present consists of raising antibodies against inducible epitopes expressed on tumor cells. A successful example is the development of anti-Tac peptide chimeric mouse mAbs. Anti-Tac mAb reacts with the ligand binding site of the p55 IL-2 receptor. The non-covalent linking of the p55 and p75 receptors form the high affinity IL-2 receptor which is induced physiologically. Tac antigen is expressed on several types of leukemic cells, such as adult T cell leukemia (ATCL), cutaneous T cell lymphoma, hairy cell leukemia, Hodgkin's disease, chronic and acute myelogenous leukemia (37). In initial investigations the chimeric mAb was found to be active against ATCL, but problems arose related to anti-mouse mAbs reactions. Through genetic engineering technology "hyper-humanized" mAbs were obtained which were "human" except for the complementarity-determining regions derived from the mouse mAbs. This hyperchimeric mAb had the same affinity for Tac as the parent chimeric anti-Tac mAb, but had no immunogenicity in humans. The chimeric and the hyperchimeric mAbs caused the same inhibition of T cell proliferative responses to tetanus antigen. Moreover, the hyperchimeric mAbs were uniquely capable of mediating an ADCC with IL-2-activated human mononuclear cells (37). The clinical experience gained to date with these hyper-humanized anti-Tac mAbs indicates that the preparation has significant therapeutic activity against ATCL.

D. Cytokines

Within the limits of this presentation, it is not possible to give justice to the rapidly expanding knowledge of the biological and therapeutic effects of the cytokines. The reader is referred to some excellent reviews on the subject (38,39). Here it is sufficient to say that cytokines as a group can be used in therapeutics essentially in three basic ways, namely as cytotoxic effectors, as cell regulators, and as carriers of toxic moieties; in each case their action is conditioned by the presence of specific receptors on the cells that are to be affected. Cytotoxic and regulatory functions may co-exist in the same molecule: both IL-1 and tumor necrosis factor (TNF) provide examples in this respect. In fact, IL-1 is cytotoxic for some types of neoplastic cells (40) in addition to having its well known functions in regulating T cell population through stimulation of thymocyte proliferation (38). TNF has direct and selective cytotoxic activity for tumor cells in culture and cytotoxic plus necrotizing effects _in vivo_ (42) but, at the same time, is a potent regulator of cells of the monocytic lineage including cytotoxic macrophages (42) as well as of T cell proliferation (43). While the full chemotherapeutic potential of the interferons (IFNs) is still being investigated, the effectiveness of IFNα in the treatment of malignant melanoma, renal cell carcinoma, lymphomas, Kaposi's sarcoma, hairy cell leukemia, mycosis fungoides, multiple myeloma, ovarian cancer and superficial bladder cancer has been recognized and ranges from very limited to highly significant depending on the disease (see specific sections in 39). Clinical effectiveness has also been demonstrated for IL-2 in such tumors as melanoma and renal cell carcinoma (44). Although TNF has activity against certain types of brain tumors after intracarotid administration, its toxicity by systemic administration limits its clinical utilization. This is in contrast to the selective antitumor effects seen in animal model systems (45). Notwithstanding the demonstrated immunomodulating effects of IFN, TNF and IL-2 in animals and in humans, unequivocal clinical evidence is not yet available which would demonstrate the causal relationship of an immunomodulating function to the anticancer effects observed with some tumors. In general, with cytokines as well as with immunoaugmenting agents as a group a greater augmentation of certain host functions has been observed in patients in whom these functions were below normal before treatment. For example, with IFNβ, increases in NK function were more significant in patients with low initial levels of this function (46).

Some cytokines, e.g., the colony stimulating factors (CSFs), have regulatory and proliferative activities in the bone marrow and are being investigated clinically in conjunction with intensive chemotherapy for their potential in counteracting bone marrow toxicity. However, after reduction of bone marrow toxicity, other limiting toxicities usually become evident with dose escalation. It is expected that the judicious usage of combinations of drugs with different limiting toxicities may result in a significant improvement of overall dose selection and consequently, increased potential for therapeutic effects.

E. Cell transfer

Again, within the context of this brief presentation, cell transfer approaches in immunotherapy cannot be discussed in any depth and the reader may wish to consult some excellent recent reviews on this subject (see specific sections in 39 and other chapters in this volume).

LAK cells are a mixture of cells derived from large granulocytes in peripheral blood and expanded in culture in the presence of IL-2 (4). The cellular composition of this mixture of phenotypically non-T cells is not yet completely clarified even though increasing evidence suggests that the relevant effectors are activated NK cells (47). More recently, attempts to proceed towards the identification of the effectors involved have led to the demonstration of the superiority of adherent LAK (A-LAK) cells in effecting antitumor function (48). The infusion of LAK cells into the patient of origin with supportive in vivo IL-2 has yielded significant therapeutic responses particularly in cases of malignant melanoma and renal cell carcinoma (4). Toxicity has also been observed and seemed to be the consequence of a "leaky vessel" syndrome causally related to the high doses of IL-2 apparently required in this regimen. It is important to define the relevant effector cells among the LAK cells and the minimum IL-2 requirements also in expectation that in the future these cells will almost certainly be used in combination with other biologicals and/or certain anticancer drugs. The clinical use of LAK cells has proven for the first time that autochthonous transfer of cells of the host defense system can exert antitumor action in patients after suitable amplification. Indeed this demonstration represents a milestone in the development of cell transfer based clinical immunotherapy and supports the concept that human tumors exhibit specificities that can be recognized by autochthonous effector cells.

The therapeutic utility of T cell transfers has been repeatedly demonstrated in the past in combination with certain anticancer drugs, in mice bearing experimental tumors (49). In humans, this approach has again been pioneered by Rosenberg and his group using TILs for transfer. These cells are obtained from tumor samples, expanded and activated (often with difficulty) in vitro and re-infused in the patient of origin where they appear to concentrate again with the context of the tumor (4,44). As reported by Rosenberg (4,44) about 50 patients with malignant melanoma were treated with autologous TILs in two different studies, the first including cyclophosphamide and short, high dose IL-2 administration and the second without cyclophosphamide and with some differences in TILs preparations. Overall, it seems that about 19 patients responded with tumor decreases of 50% or more, these responses being of somewhat short duration. TILs exhibit autologous tumor specificity and MHC restriction (4) and are 90% $CD3^+$ cells: there is no reason to believe that they are not T cells. Although the clinical therapeutic effects of TILs have been somewhat modest to date, the validity of the approach is getting rapidly established; at this time the biology of these systems should be carefully studied to improve and firm up this modality of treatment. Several questions may be asked in connection with the therapeutic use of TILs, for instance: the kinetics of redistribution of TILs in tumor and elsewhere upon re-infusion should be defined, optimal conditions need to be further determined for cell growth and activation in culture to reverse the functional depression of TILs which apparently develop in situ in the tumor of origin, the requirement of cyclophosphamide or other anti-suppressor treatments for optimal TIL activation should be defined, the possibility of replacing IL-2 with other cytokines or of reducing IL-2 doses by the interaction with other cytokines should be explored, the possibility of inducing TIL activation through exposure to tumor cells membrane perhaps presented on immunogenic matrices should be studied, individual tumors or tumor type determinants of TIL activity should be identified. At this time, there is little doubt that the transfer of T cells, as a basic procedure, will find an important place in cancer immunotherapy and will be used, most likely, in combination with other treatments.

It is not necessary to emphasize the importance of macrophages as cytotoxic effectors, as cells intimately involved in the regulation of the immune systems and as targets of numerous immunomodulating agents. Thus it is perhaps unfortunate that only two studies have been reported to date about the transfer of macrophages for therapeutic purposes (50,51). Because of their basic functions in the immune system it would seem desirable that more work be done in the area of macrophage transfer in humans despite the obvious difficulties involved.

Recently studies have been initiated by Rosenberg and his group to explore the possibility that TILs transfected with suitable cytokine genes may bring about intratumor cytotoxicity through "in situ" secretion of cytotoxic cytokines (4). Current studies are being carried out with the TNFα gene. This imaginative approach merges concepts of "genetic therapy" with immunotherapy approaches and opens up a possible fruitful area of future investigation. Questions are currently being answered about the regional and systemic immunobiological consequences of the TNF produced by transfected TILs, the pharmacokinetics of the secreted cytokine and the termination of the TNF function upon discontinuation of the required supportive IL-2 administration.

III FUTURE PROJECTIONS

Although it is obviously impossible to project future developments in immunotherapy with any measure of certainty, some reasonable assumptions can be made based on the present status of the field and on leads stemming from current knowledge.

An important and very difficult task ahead is represented by the need to define the optimal biological response modifier dose (OBRMD) whenever an agent with BRM action is used (52). Indeed the customary use of maximum tolerated doses (MTD) in cancer chemotherapy and the attempts to increase dose intensities towards achieving maximum chemotherapeutic effect (53) are not necessarily valid in the case of BRMs; the optimal dose for this type of agent is usually very different from the MTD. The determination of the OBRMD must be based on the measurement of relevant BRM actions. Because in many cases a BRM will modify the action of several types of immune functions, it is often difficult to identify the modification which is causatively relevant to the antitumor effect expected in humans. Moreover, each effect is likely to occur with a different dose/response and this further complicates the issue. An additional dimension is derived from the fact that the function of the immune system, and therefore, also the relative effects of BRMs on each of its components, can vary during tumor progression. Despite these unquestionable difficulties, it is very important that optimal immunotherapeutic treatments be designed and this can be done only with a knowledge of the mechanisms of immunomodulation involved and of the OBRMD at any one stage of tumor progression.

More and more emphasis is bound to be given to the design of ad hoc regimens tailored to the individual patient. While this approach is being given increasing attention in cancer chemotherapy based on increased knowledge of the mechanism of action of anticancer drugs and of the individual determinant of response in target tumors, it is already an intrinsic attribute in certain forms of cancer immunotherapy. Indeed the use of autochthonous vaccines, of antiidiotype antibodies, of cell transfer with or without cytokines, of suitable combinations with drugs, all operate through an individuality of action in any given patient.

The development of combination treatments is motivated not only by the biological individuality of tumors and the heterogeneity of their cell populations in terms of epitope profiles, but also by the well established fact that agents with different mechanisms of action can exert synergistic effects on tumors with reduced side effects. The use of mixtures of antibodies directed against epitopes on heterogeneous ovarian cancer cell populations (54), of cytokines such as IL-2 with drugs having cytoreductive and/or immunomodulating action such as Adriamycin (31), of antiidiotype antibodies with drugs affecting epitope negative cell subsets (see above), of LAK cells plus suitable mAbs (55), of IL-2 plus mAbs (56), the use of autologous bone marrow transplantation with IL-2 plus LAK cells (57), the reduction of the chemotherapy induced toxicity by CSFs or other cytokines (10); all these represent current examples of studies directed towards improving the utilization of BRMs through the design of combination treatments.

Another area that is likely to be emphasized in the future is the genetic engineering of effectors. As mentioned above, clinical studies are already being carried out with TILs transfected with the TNFα gene (4). Genes of other cytotoxic moieties, of adhesion molecules and/or recognition sites are likely to be transfected in suitable effector cells in attempts to establish in situ cytotoxic treatments or to increase the efficiency of the effector cells. Genetic engineering is also being applied to the development of hybrid and hyperhumanized antibodies (37); given the initial successes achieved it is likely that these approaches will be intensified.

Integrins and cell adhesion molecules in general are involved at some stages of the metastatic process as well as in the interactions among cells which determine immune or inflammatory responses (58,59). For instance, some integrins are involved in the monocyte dependent antigen induced T cell activation, in affecting CTL and/or T suppressor cell functions and the interactions of CTL and/or T suppressor cells with T helper cells (59). It is likely that intervention on an integrins function will be attempted to increase the efficiency of antitumor host mechanisms of defense or to reduce metastasis.

Most cytokines have pleiotropic effects. Work has begun in several laboratories aimed at fragmenting cytokine moieties in attempts to identify polypeptides which might selectively exhibit a given desired effect. Recently, fragments of IL-1 have been developed which retain immunostimulating and T helper cell activation properties but are deprived of pyrogenicity and of prostaglandins stimulating activities (60). This approach is also being pursued with TNF (61) and it is reasonable to predict that in the future it will be followed with increased intensity.

Modification of tumor cells has been already studied for a number of years in efforts to develop better vaccines. Thus the pioneering studies by Kobayashi indicated that xenogenization of tumor cells by infection with non-pathogenic viruses may lead to useful applications in this respect (62). Mutagenic gene amplification in melanoma cells with a drug like DTIC has been studied particularly by Bonmassar (63) and may also yield effective vaccines. Following the current utilization of BCG as an adjuvant in mixtures with tumor cell vaccines, it is possible that haptenic substitutions on tumor cells may be attempted in the future for similar purposes.

Tumor-induced suppression of the immune system and tumor escape mechanisms (64,65) represent important areas of investigations. That certain tumors suppress the immune system has been repeatedly observed in experimental models and in clinical situations. A reduction of this suppressive action would be very useful particularly in combination with treatments aimed at augmenting the host response against the tumor. Soluble tumor suppressor factors have been repeatedly reported to occur clinically and should be purified and characterized as possible targets of intervention. The development of T suppressor cells during tumor growth has been studied extensively in mouse models (66) and this cell function is being targeted for immunotherapy. The fact that the antitumor function of TILs is at times hard to reactivate during TILs expansion in culture seems consistent with the possibility that these cells are inactivated or "paralyzed" within the tumor context. Whether this represents a form of immunological tolerance or the result of the action of specific tumor suppressive factors should be clarified before rational attempts can be made to reactivate these cells in situ, a possible future immunotherapeutic approach. Should tumor suppressor factors play a role in inducing self-serving immunosuppression, it would seem possible that these factors require binding to a receptor for their action. If this is the case, protection of these receptors by biologically inactive substances may provide another approach aimed at counteracting tumor induced suppression of host defenses.

Attention is increasingly given to regional immunity as an area that needs careful investigation. Regional immunity is characterized by regional specificities of effector cells and of regulatory mechanisms which also involves the participation of cells such as epithelial cells which are not usually construed as being strictly a part of the immune system. In a sense, immunotherapy with TIL transfers represents a form of regional immunotherapy. In the liver, sinusoids lined by endothelial cells, NK and Kupffer's cells provide a first line of defense. A number of cytokines and factors are secreted by Kupffer's cells, and perhaps related to this, agents such as Indomethacin block suppressor factors in liver (67). It was shown that NK cells are major instruments for clearance of tumor cells from various organs (68,69). Moreover, adoptively transferred IL-2 activated NK (A-LAK or A-NK cells) accumulated within established metastases including, when administered locally through the hepatic vasculature, liver metastases (70,71). Undoubtedly clarification of mechanisms of regional defenses against tumor cells and subsequent attempts to augment them may provide new types of immunotherapeutic approaches aimed at inhibiting both primary and metastatic tumor.

In conclusion, it is fair to say that there are many opportunities now open for the pursuit of new leads which may substantially improve the effectiveness of immunotherapy and allow the rational design of multimodality regimens. Given the rapid rate at which basic information is being accrued and extended to clinical investigations, and the multiplicity of approaches that can be followed, one cannot help being optimistic for the future of immunotherapy, despite the caution that must be critically derived from the relative paucity of clinical therapeutic results achieved to date with this modality of treatment.

References

1. Prehn, R.T., Main, J.M. Immunity to methylcholanthrene-induced sarcomas. J. Natl. Cancer Inst. 1957; 18:769-778.
2. Klein, G., Sjogren, H.O., Klein, E., Hellstrom, K.E. Demonstration of resistance against methylcholanthrene-induced sarcomas in the primary autochthonous host. Cancer Res. 1960; 20:1561-1572.
3. Muul, L.M., Spiess, P.J., Director, E.P. and Rosenberg, S.A. Identification of specific cytolytic immune responses against autologous tumor in humans bearing malignant melanoma. J. Immunol. 1987; 138:989-995.
4. Rosenberg, S.A. Immunotherapy and gene therapy of cancer. J. Clin. Oncol. 1992; 10:180-199.
5. Carey, T.E., Takahashi, T., Resnick, L.A., Oettgen, H.F. and Old, L.J. Cell surface antigens of human malignant melanoma: Mixed hemadsorption assays for humoral immunity to cultured autologous melanoma cells. Proc. Natl. Acad. Sci. 1976; 73:3278-3282.
6. Levy, R. and Miller, R.A. Antibodies in cancer therapy: B-cell lymphomas, In: Biologic Therapy of Cancer, Devita, Jr., V.T., Hellman, S. and Rosenberg, S.A. (Eds.) Philadelphia, J.B. Lippincott, 1991; pp. 512-522.
7. Schlom, J. Antibodies in cancer therapy: Basic principles and applications. In: Biologic Therapy of Cancer, DeVita, Jr., V.T., Hellman, S. and Rosenberg, S.A. (Eds.), Philadelphia, J.B. Lippincott, 1991; pp. 464-481.
8. Oettgen, H.F. and Old, L.J. The history of cancer immunotherapy. In: Biologic Therapy of Cancer, DeVita, Jr., V.T., Hellman, S. and Rosenberg, S.A. (Eds.), Philadelphia, J.B. Lippincott, 1991; pp. 87-122.
9. Mastrangelo, M.J. and Berd, D. Immunotherapy with microbial products. In: Immunological Approaches to Cancer Therapeutics. Mihich, E. (Ed.), New York, John Wiley and Sons, 1982; pp. 75-106.
10. Gabrilove, J.L. Colony-stimulating factors: Clinical status. In: Biologic Therapy of Cancer, DeVita, Jr., V.T., Hellman, S. and Rosenberg, S.A. (Eds.), Philadelphia, J.B. Lippincott, 1991; pp. 445-463.
11. Houghton, A.N., Chapman, P.B., Bajorin, D.F. Antibodies in cancer therapy: Melanoma. In: Biologic Therapy of Cancer, DeVita, Jr., V.T., Hellman, S. and Rosenberg, S.A. (Eds.), Philadelphia, J.B. Lippincott, 1991; pp. 533-549.
12. Berd, D., Maguire, H.C., Jr., McCue, P. and Mastrangelo, M.J. Treatment of metastatic melanoma with an autologous tumor-cell vaccine: Clinical and immunologic results in 64 patients. J. Clin. Oncol. 1990; 8:1858-1867.
13. Hanna, M.G., Jr., Peters, L.C., Hoover, H.C., Jr. Immunotherapy by active specific immunization: Basic principles and preclinical studies. In: Biologic Therapy of Cancer, DeVita, Jr., V.T., Hellman, S. and Rosenberg, S.A. (Eds.), Philadelphia, J.B. Lippincott, 1991; pp. 651-669.

14. Mitchell, M.S., Kan-Mitchell, J., Kempf, R.A., Harel, W., Shau, H. and Lind, S. Active specific immunotherapy for melanoma: Phase I trial of allogeneic lysates and a novel adjuvant. Cancer Res. 1988; 48:5883-5893.
15. Morton, D.L., Nizze, A., Famatiga, E., Hoon, D.S.B., Gupta, R. and Irie, R. Clinical results of a trial of active specific immunotherapy with melanoma cell vaccine and immunomodulation in metastatic melanoma. Proc. Am. Assoc. Cancer Res. 1989; 30:383.
16. Sahasrabudhe, D.M., deKernion, J.B., Pontes, J.E., Ryan, D.M., O'Donnell, R.W., Marquis, D.M., Mudholkar, G.S. and McCune, C.S. Specific immunotherapy with suppressor function inhibition for metastatic renal carcinoma. J. Biol. Response Mod. 1986; 5:581-594.
17. McCune, C.S., Schapiro, D.V., Henshaw, E.C. Specific immunotherapy of advanced renal carcinoma: Evidence for the polyclonality of metastases. Cancer 1981; 47:1984-1987.
18. Morton, D.L., Hunt, K.K., Bauer, R.L., Lee, J.D. Immunotherapy by active immunization of the host using nonspecific agents: Clinical application using intralesional therapy. In: Biologic Therapy of Cancer, DeVita, Jr., V.T., Hellman, S. and Rosenberg, S.A. (Eds.) Philadelphia, J.B. Lippincott, 1991; pp. 627-642.
19. Herr, H.W. Immunotherapy by active immunization of host using nonspecific agents: Instillation therapy for bladder cancer. In: Biologic Therapy of Cancer, DeVita, Jr., V.T., Hellman, S. and Rosenberg, S.A. (Eds.), Philadelphia, J.B. Lippincott, 1991; pp. 643-650.
20. Fidler, I.J., Fogler, W.E., Tarscay, L., Schumann, G., Braum, D.G. and Schroit, A.J. Systemic activation of macrophages and treatment of cancer metastases by liposomes containing hydrophilic or lipophilic muramyl dipeptide. Adv. Immunopharmac. 1983; 2:235-241.
21. Lederer, E. and Chedid, L. Immunomodulation by synthetic muramyl peptides and trehalose diesters. In: Immunological Approaches to Cancer Therapeutics. Mihich, E. (Ed.), New York, John Wiley and Sons, 1982; pp. 107-136.
22. Mihich, E. Future perspectives for biological response modifiers: A viewpoint. Semin. Oncol. 1986; 13:234-254.
23. Hirose, K., Zachariae, C.O.C., Oppenheim, J.J. and Matsushima, K. Induction of gene expression and production of immunomodulating cytokines by PSK in human peripheral blood mononuclear cells. Lymphokine Res. 1990; 9:475-483.
24. Laurie, J.A., Moertel, C.G., Fleming, T.R. Surgical adjuvant therapy of large bowel carcinoma: An evaluation of levamisole and the combination of levamisole and 5-fluroruracil. A study of the North Central Cancer Treatment Group and Mayo Clinic. J. Clin. Oncol. 1989; 7:1447-1456.
25. Mihich, E. and Ehrke, M.J. Immunomodulation by anticancer drugs. In: Biologic Therapy of Cancer. DeVita, Jr., V.T., Hellman, S. and Rosenberg, S.A. (Eds.), Philadelphia, J.B. Lippincitt, 1991; pp. 776-786.

26. Schwartz, R., Eisner, A. and Dameshek, W. The effect of 6-mercaptopurine on primary and secondary immune responses. J. Clin. Investigation, 1959; 38:1394-1403.
27. Ehrke, M.J., Mihich, E., Berd, D. and Mastrangelo, M.J. Effects of anticancer drugs on the immune system in humans. Semin. Oncol. 1989; 230-253.
28. Morikawa, K., Hosokawa, M., Hamada, J., Sugaware, M., Kobayashi, H. Host-mediated therapeutic effects produced by appropriately times administration of bleomycin on a rat fibrosarcoma. Cancer Res. 1985; 45:1502-1507.
29. Abdul Hamied, T.A., Parker, D. and Turk, J.L. Potentiation of release of interleukin-2 by bleomycin. Immunopharmac. 1986; 12:127-134.
30. Ryoyama, K., Mace, K., Ehrke, M.J. and Mihich, E. The differential sensitivity of T-cell immune functions to vincristine and vinblastine . Intl. J. Immunopharmac. 1982; 4:187-194.
31. Ehrke, M.J., Ho, R.L.X. and Mihich, E. Modifications of antitumor immune effectors by Adriamycin. In: New Horizons of Tumor Immunotherapy, Torisu, M. and Yoshida, T. (Eds.), Amsterdam, Elsevier, 1989; pp. 467-475.
32. Ho, R.L.X., Maccubbin, D.L., Mihich, E. and Ehrke, M.J. Demonstration of the phenotype of effector cells critical to the effects of a chemoimmunotherapeutic protocol involving Adriamycin and interleukin 2. Proc. Am. Assoc. Cancer Res. 1991; 32:246.
33. Catane, R. and Longo, D.L. Monoclonal antibodies for cancer therapy. Isr. J. Med. Sci. 1988; 24:471-476.
34. Larson, S.M., Cheung, N-K.V., Leibel, S.A. Antibodies in cancer therapy: Radioisotope conjugates. In: Biologic Therapy of Cancer. DeVita, Jr., V.T., Hellman, S. and Rosenberg, S.A. (Eds.), Philadelphia, J.B. Lippincott, 1991; pp. 496-511.
35. Durrant, L.G., Robbins, R.A. and Baldwin, R.W. Flow cytometric screening of monoclonal antibodies for drug or toxin targeting to human cancer. J. Natl. Cancer Inst. 1988; 81:689-696.
36. Vitetta, E.S. and Thorpe, P.E. Antibodies in Cancer Therapy: Immunotoxins. In: Biologic Therapy of Cancer, DeVita, Jr., V.T., Hellman, S. and Rosenberg, S.A. (Eds.), Philadelphia, J.B. Lippincott, 1991; pp. 482-495.
37. Waldman, T.A. Antibodies in cancer therapy: T-cell leukemia/lymphoma, In: Biologic Therapy of Cancer, DeVita, Jr., V.T., Helmman, S. and Rosenberg, S.A. (Eds.), Philadelphia, J.B. Lippincott, 1991; pp. 523-532.
38. Thomson, A. (Ed.) The Cytokine Handbook, New York, Academic Press, 1991.
39. DeVita, Jr., V.T,Hellman, S. and Rosenberg, S.A. (eds.) Biologic Therapy of Cancer. Philadelphia, J.B. Lippincott, 1991.
40. Onozaki, K., Matsushima, K., Aggarwal, B.B. and Oppenheim, J.J. Human interleukin 1 is a cytocidal factor for several tumor cell lines. J. Immunol. 1985; 135:3962-3968.

41. Old, L.J. Tumor necrosis factor: Keynote address. In: Tumor Necrosis Factor Structure, Mechanism of Action, Role in Disease and Therapy, Bonavida, B. and Granger, G. (Eds.), Basel, S. Karger AG, 1990; pp. 1-30.
42. Hori, K., Ehrke, M.J., Mace, K., Maccubbin, D., Doyle, M.J., Otsuka, Y. and Mihich, E. Effect of recombinant human tumor necrosis factor on the induction of murine macrophage tumoricidal activity. Cancer Res. 1987; 47:2793-2798.
43. Ehrke, M.J., Ho, R.L.X. and Hori, K. Species-specific TNF induction of thymocyte proliferation. Cancer Immunol. Immunother. 1988; 27:103-108.
44. Rosenberg, S.A. Adoptive cellular therapy: Clinical studies. In: Biologic Therapy of Cancer, DeVita, Jr., V.T., Hellman, S. and Rosenberg, S.A. (Eds.), Philadelphia, J.B. Lippincott, 1991; pp. 214-236.
45. Spriggs, D.R. Tumor necrosis factor: Basic principles and preclinical studies. In: Biologic Therapy of Cancer, DeVita, Jr., V.T., Hellman, S. and Rosenberg, S.A. (Eds.), Philadelphia, J.B. Lippincott, 1991; pp. 354-377.
46. Tentori, L., Fuggetta, M.P., D'Atri, S., Aquino, A., Nunziata, C., Roselli, M., Ballatore, P., Bonmassar, E. and DeVecchis, L. Influence of low-dose beta-interferon on natural killer cell activity in breast cancer patients subjected to chemotherapy. Cancer Immunol. Immunother. 1987; 24:86-91.
47. Herberman, R.B., Hiserodt, J.C., Vujanovic, N., Balch, C., Lotzova, E., Bolhuis, R., Golub, S., Lanier, L.L., Phillips, J.H., Riccardi, C., Ritz, J., Santoni, A., Schmidt, R.E. and Uchida, A. Lymphokine-activated killer cell activity. Characteristics of effector cells and their progenitors in blood and spleen. Immunol. Today, 1987; 8:178-181.
48. Melder, R.J., Whiteside, T.L. Vujanovic, N.L., Hiserodt, J.C. and Herberman, R.B. A new approach to generating antitumor effectors for adoptive immunotherapy using human adherent lymphokine-activated killer cell. Cancer Res. 1988; 48:3461-3469.
49. Greenberg, P.D. Adoptive T cell therapy of tumors: Mechanisms operative in the recognition and elimination of tumor cells. Adv. Immunol. 1991; 49:281-355.
50. Stevenson, H.C., Lacerna, L.V. and Sugarbaker, P.H. Ex vivo activation of killer monocytes (AKM) and their application to the treatment of human cancer. J. Clin. Apheresis, 1988; 4:118-121.
51. Andreesen, R. and Hennemann, B. Adoptive immunotherapy with autologous macrophages: Current status and future perspectives. Pathobiology, 1991; 59:259-263.
52. Herberman, R.B. Design of clinical trials with biological response modifiers. Cancer Treatment Rept. 1985; 69:1161-1164.
53. DeVita, Jr., V.T. The problem of resistance: Keynote address. In: Drug Resistance: Mechanisms and Reversal. Mihich, E. (Ed.), Rome, John Libbey CIC, 1990; p. 7-31.
54. Rodriguez, G.C., Berchuck, A. and Bast, Jr., R.C. Antibodies in cancer therapy: Monoclonal antibodies for immunodiagnosis and immunotherapy of epithelial ovarian cancer. In: Biologic Therapy of Cancer, DeVita, Jr., V.T., Hellman, S. and Rosenberg, S.A. (Eds.), Philadelphia, J.B. Lippincott, 1991; pp. 550-562.

55. Tong, A.W., Lee, J.C., Wang, R-M., Ordonez, G. and Stone, M.J. Augmentation of lymphokine-activated killer cell cytotoxicity by monoclonal antibodies against human small cell lung carcinoma. Cancer Res. 1989; 49:4103-4108.
56. Berinsten, N. and Levy, R. Treatment of a murine B cell lymphoma with monoclonal antibodies and IL-2. J. Immunol. 1987; 139:971-976.
57. Fefer, A., Truitt, R.L., Sullivan, K.M. Adoptive cellular therapy: Graft-versus-tumor responses after bone marrow transplantation. In: Biologic Therapy of Cancer, DeVita, Jr., V.T., Hellman, S. and Rosenberg, S.A. (Eds.), Philadelphia, J.B. Lippincott, 1991; pp. 237-246.
58. Hemler, M.E. Adhesive protein receptors of hematopoietic cells. Immunol. Today. 1988; 9:109-113.
59. Dustin, M.L. and Springer, T.A. Role of lymphocyte adhesion receptors in transient interactions and cell locomotion. Ann. Rev. Immunol. 1991; 9:27-66.
60. Antoni, G., Presentini, R., Perin, F., Tagliabue, A., Ghiara, P., Censini, S., Volpini, G., Villa, L. and Boraschi, D. A short synthetic peptide fragment of human interleukin 1 with immunostimulatory but not inflammatory activity. J. Immunol. 1986; 137:3201-3204.
61. Corti, A., Fassina, G., Marcucci, F. and Cassani, G. Antigenic regions of tumor necrosis factor alpha and their topographic relationships with structural/functional domains. Molecular Immunol. 1992; 29:471-479.
62. Kobayashi, H. Modification of tumor antigenicity in therapeutics: Increase in immunologic foreignness of tumor cells in experimental model systems. In: Immunological Approaches to Cancer Therapeutics, Mihich, E. (Ed.), New York, John Wiley and Sons, 1982; pp. 505-540.
63. D'Atri, S. Tricarico, M., Margison, G.P., Allegrucci, M., Fuschiotte, P. Grohman, U., Giglietti, S., Bonmassar, E. Antigenic changes of cancer cells following interactions with antitumor agents. In: Drug Resistance: Mechanisms and Reversal, Mihich, E. (ed.), Rome, John Libbey CIC, 1990; pp. 271-294.
64. Stutman, O. Suppressor cells in tumor-host interactions. In: Biological Responses in Cancer: Progress Toward Potential Applications. Mihich, E. (Ed.), New York, Plenum Press, 1982; pp. 23-88.
65. Ozer, H. Tumor immunity and escape mechanisms in humans. In: Immunological Approaches to Cancer Therapeutics. Mihich, E. (Ed.), New York, John Wiley and Sons, 1982; pp. 39-74.
66. North, R.J. Down-regulation of the antitumor immune response. Adv. Cancer Res. 1985; 45:1-43.
67. Tzung, S-P., Gaines, K.C., Lance, P., Ehrke, M.J. and Cohen, S.A. Suppression of hepatic lymphokine-activated killer cell induction by murine Kupffer cells and hepatocytes. Hepatol. 1990; 12:644-652.
68. Riccardi, C., Puccetti, P., Santoni, A. and Herberman, R.B. Rapid in vivo assay of mouse NK cell activity. J. Natl. Cancer Inst. 1979; 63:1041-1045.

69. Barlozarri, T., Lenhardt, J., Wiltrout, R., Herberman, R.B. and Reynolds, C.W. Direct evidence for the role of LGL in the inhibition of experimental tumor metastases. J. Immunol. 1985; 134:2783-2789.

70. Basse, P., Herberman, R.B., Nannmark, U., Johansson, B.R., Hokland, M., Wasserman, K. and Goldfarb, R.H. Accumulation of adoptively transferred adherent, lymphokine-activated killer cells in murine metastases. J. Exp. Med. 1991; 174:479-488.

71. Basse, P.H., Nannmark, U., Johansson, B.R., Herberman, R.B. and Goldfarb, R.G. Establishment of cell-to-cell contact by adoptively transferred adherent lymphokine-activated killer cells with metastatic murine melanoma cells. J. Natl. Cancer Inst. 1991; 83:944-950.

Index

About the Editors

RONALD H. GOLDFARB is Deputy Director for Research and Director of the Program in Cancer Metastasis and Cell Biology at the Pittsburgh Cancer Institute, Associate Professor in the Department of Pathology at the University of Pittsburgh School of Medicine, and Senior Lecturer in Medicine in the Biomedical Engineering Program at Carnegie Mellon University, Pittsburgh, Pennsylvania. The author or coauthor of nearly 80 professional papers and over 60 abstracts, he is a member of the American Association for Cancer Research, the American Society of Immunologists, the American Society for Cell Biology, and the Society for Comparative Oncology, among other organizations. Dr. Goldfarb received the B.A. degree (1970) in biological sciences from Herbert H. Lehman College, City University of New York, Bronx, New York, and the Ph.D. degree (1978) in microbiology and immunology from the State University of New York Downstate Medical Center, Brooklyn, New York.

THERESA L. WHITESIDE is Director of the Immunologic Monitoring and Diagnostic Laboratory at the Pittsburgh Cancer Institute and Professor of Pathology at the University of Pittsburgh School of Medicine, Pittsburgh, Pennsylvania. A Fellow of the American Academy of Microbiology and a member of the Clinical Immunology Society, the American Association for Cancer Research, and the American Association of Pathologists, among other organizations, she is the author or coauthor of over 230 professional papers and book chapters, an Editor of the journal *Clinical Diagnostic Laboratory Immunology*, and an editorial board member of several other journals. Dr. Whiteside received the B.S. degree (1962) in botany, the M.A. degree (1964) in biology, and the Ph.D. degree (1969) in microbiology from Columbia University, New York, New York.

Made in the USA
Las Vegas, NV
16 September 2026

54776935R00190